Rivers Lost
Rivers Regained

History of the Urban Environment

Martin V. Melosi and Joel A. Tarr, Editors

Rivers Lost Rivers Regained

Rethinking City-River Relations

Edited by
Martin Knoll, Uwe Lübken, and Dieter Schott

University of Pittsburgh Press

Published by the University of Pittsburgh Press, Pittsburgh, Pa., 15260

Manufactured in the United States of America
Printed on acid-free paper
10 9 8 7 6 5 4 3 2 1

Cataloging-in-Publication data is available from the Library of Congress

ISBN 13: 978-0-8229-4459-1
ISBN 10: 0-8229-4459-6

Cover photo by Ralf Tietz
Cover design by Melissa Dias-Mandoly

CONTENTS

ACKNOWLEDGMENTS

This book has grown out of two events. The first was a major session on "Cities and Rivers—Long-Term Development of Socio-Natural Sites" at the European Conference on Urban History in Prague, August 2012, organized by Martin Knoll (then TU Darmstadt), Martin Schmid (Alpen-Adria University Klagenfurt/Vienna, IFF), and Dieter Schott (TU Darmstadt), which made the start with eight papers covering a wide range of European and non-European cities and their rivers from the Middle Ages to the twentieth century. Due to other commitments, Martin Schmid could not continue with this project, but we would like to thank him for his valuable contributions and comments in getting this project started. Then, the initiative of the conference session was taken up and enlarged thematically as well as in terms of regional coverage by a conference hosted by the Rachel Carson Center for Environment and Society (RCC) at Ludwig Maximilian University of Munich in February 2013, for which the editors of this volume were the organizers. The editors would like to thank the RCC and its two directors, Christof Mauch and Helmuth Trischler, for their generous support of the conference. The logistical challenge of the event was greatly alleviated by the invaluable help of the staff of the RCC, most notably Sonja Weinbuch.

Sandra Crooms of the University of Pittsburgh Press has guided us patiently and energetically on the long path from conference proceedings to a book. The editorial teams at the RCC and at the University of Pittsburgh Press, to whom we are truly grateful, made this collection a much better book. Martin Mainka has done a marvelous job in producing the index. Photo artist Ralf Tietz was so kind as to grant us the right to use his beautiful photography of the Rhine at Cologne for the cover of this book. We would also like to express our appreciation to Ellen Arnold (Ohio Wesleyan University), Sabine Dabringhaus (University of Freiburg), and Eva Jakobsson (University of Stavanger) for their invaluable comments and thoughtful criticism at the RCC conference. Last but not least, we would like to thank all the participants of the main session in Prague and the RCC conference in Munich for lively discussions in a stimulating atmosphere.

Rivers Lost
Rivers Regained

INTRODUCTION

Martin Knoll, Uwe Lübken, and Dieter Schott

On a bitterly cold February afternoon in 2013 a group of international scholars gazed down from the clock tower of Deutsches Museum Munich on the Isar and attempted to spot traces of renaturalization in the concrete straitjacket in which the river runs through the city. This excursion turned out to be a perfect kickoff for a conference that united scholars from a wide range of countries who all had stories to tell about rivers in relation to cities and the interaction between those two so different actors. While the stories the participants told each other over the coming two days were quite different in time frame, location, and drama, they all agreed on one thing: cities and rivers were (and are) intimately linked to each other; they coevolved and shaped each other in a multitude of ways and aspects. Until recently, this interaction has been framed mostly as a story of loss and spoliation: environmental historians have depicted rivers as victims and objects of comprehensive human attempts to colonize and domesticate and to extract as much material value from them as possible in terms of transport routes, energy resources, and "ultimate sinks" for waste material.[1] More recently, however, the focus of historical river research has shifted toward other, less gloomy narratives, and so it seemed to make sense to open such a conference with an excursion that highlighted progress and improvement rather than decline and disaster. This book does analyze the decline and the spoliation, which have clearly been dominant over the last 150 years, but also explores the resilience of rivers, the amazing recovery of aquatic life after water quality has been improved and, last but not least, the cultural reappropriation of rivers by urban societies that rediscover the multiple pleasures and amenities linked with urban water bodies.

Why bother about cities and rivers? Since rivers are such an essential resource for almost all larger (and older) cities, the river-city relationship provides an angle from which to advance comparative urban history with a focus on the relationship of cities to their natural environments.[2] Despite all attempts to domesticate and regulate urban rivers, cities have never comprehensively controlled them, as the long history of recurring floods and other river-linked disasters clearly shows.[3] Thus, while rivers are an essential conduit of the urban metabolism, they nevertheless constantly (or rather recurringly) remind urbanites that their attempts to control nature and keep it in check could only ever be fragmentary and never entirely successful in the long run. As a consequence, the attempts at stabilizing nature-society interactions and protecting urban citizens from the deleterious effects of riverine dynamics produced a wide range of conflicts and interventions, which yield highly interesting insights into both urban and environmental history.

City-river relations, thus, are a uniquely productive topic and invite a variety of historiographical approaches. The contributions to this volume reflect the heterogeneity and complexity of city-river relations over time. Inviting scholars from different disciplinary backgrounds such as history, architecture, urban planning, literature studies, and ecology, and analyzing case studies from Europe, the Americas, and Asia, this volume offers a broad and innovative exploration of processes of cities losing and regaining their rivers.

Highways and Hinterlands: Directions and Perspectives in River-City Research

Environmental historians ignored urban topics for a long time. Cities were often regarded as the exact opposite of what this field of historical inquiry seemed to be about; namely, a place where nature was to be found only in derivative and ephemeral forms—a "second" nature at best. This changed, however, in the 1990s, when more and more scholars acknowledged that cities can be analyzed from an environmental perspective too. Today, topics such as urban metabolism, animals in the city, and the creation of urban parks are an integral part of environmental history.[4]

The study of the urban environment also benefited from recent attempts to reconceptualize the material foundations of society. While these approaches represent "an eclectic range of non-, post-, and (less commonly) neo-Marxist materialisms," as Chris Otter has pointed out, their common goal is to counter the relative negligence of material/environmental factors that the dominance of cultural studies and cultural theory in the humanities has brought about.[5] As far as the "rematerialization" of urban studies is concerned, rivers play an important role.[6] They supply energy and water and function as a means of transportation and as sinks for urban effluents. As a result of their inherent dynamics, they have enabled, shaped, and transformed urban development.[7]

This does not mean that questions of perception and representation have lost their importance. Rivers are not only energy suppliers and sinks, they are symbols as well, and they can be read. The "river's edge" in particular, as Tricia Cusack has pointed out, "has long been a prime site for religious, monarchical and government buildings as well as a favorite place of residence for elite society."[8] Capital cities such as London or Prague used the urban riverfront to demonstrate the power of nation-states, while promenades along urban rivers such as the Danube River islands of late eighteenth-century Regensburg offered a space to middle- and upper-class urbanites for leisure and representation.[9] In Paris, the Seine has been systematically used since the sixteenth century to foreground architectural grandeur on the river shore—the famous Pont Neuf is a first case in point. Starting with the Collège des Quatre-Nations (1661–1667), a tradition was established of erecting representative buildings that used the river as a "distancing device" to better set in scene the beauty of this architecture.[10] In fictional literature, as Agnes Kneitz aptly demonstrates in her analysis of Charles Dickens's Thames, rivers often act as anthropomorphized players. Urban riverscapes can be and are being modeled in ways similar to how an artist produces a sculpture. A comprehensive perspective on the subject therefore has to integrate a great variety of foci, a task that is taken up within this volume.

One of the earliest conceptualizations of river-city relations can be found in Johann Heinrich von Thünen's theory of the "isolated state." First published in 1826, it provides a concentric model of land-use patterns around urban centers.[11] Framed by the pre-fossil-fuel regime of production and transportation, Thünen's theory acknowledges the role of rivers but underestimates their fundamental influence on the spatial distribution of different types of agricultural and forestal production near urban centers. By facilitating and economizing the transport of bulky goods and accelerating the delivery of easily perishable products, the presence of a river alters and considerably enlarges urban hinterlands in terms of provision.[12] This circumstance is important for the approach that this volume takes toward conceptualizing river-city relations, because it addresses the ways in which many cities are structurally bound to rivers, and it illustrates the functional role of rivers connecting cities with their respective hinterlands—a relationship that is of utmost importance not just for the city but for the hinterland itself. Defining and describing the spatial extent of the legal and economic quality of urban hinterlands as well as the city's socioecological impact, however, are traditional research topics in geography, economic history, and social ecology—starting with the city-hinterland models sketched by Thünen or Walter Christaller, to the "feeding the city" project on medieval London, all the way to more recent calculations of urban metabolisms and footprints.[13]

Rivers open hinterlands. In the case of port cities, rivers have connected cities to hinterlands of global scale for centuries.[14] Late-medieval Flanders, for example, as a boom region where population growth threatened to overstretch the carrying capacity of its immediate hinterlands in terms of food and wood production, relied on water- (and river-)based transport of grain and wood from hinterlands far away in the Baltic to sustain its growth.[15] In this volume we acknowledge that river cities are connected to their respective hinterlands in various ways and we do not look at the history of individual cities as isolated from their natural, social, and political surroundings. In fact, it is one of the main assets of this volume that it highlights these entanglements—most prominently in the first section, which assembles chapters that explicitly focus on these interrelationships.

Organicist approaches, too, have played an important role in explaining river-city relations. One especially influential example is the concept of urban metabolism, which is based on the Marxist metaphor of social metabolism. From this perspective, cities are conceptualized in analogy to biological organisms as maintaining a metabolism with their environments, thereby exchanging material resources, energy, products, and information, and building material stocks, as well as discharging waste. Measuring or calculating material and energy flows across city limits provides a telling perspective on the material dimension of urban development, on city-hinterland relations, and on the sustainability of the urban economy. The concept of "urban metabolism" was first applied in a study on Hong Kong, at that time still a British colony and as such clearly demarcated from the Chinese mainland by political borders—providing the advantage of clearly defined city limits and statistical demarcation of in- and outflows. Recent studies on Vienna show the potential of this approach particularly for tracing the major transformation of European energy regimes over the course of the nineteenth century.[16]

Tracing the material exchange of cities through the use of a quantifying approach will quickly direct attention to the important role of rivers as mediators between cities and their hinterlands. A focus on the metabolism of cities also triggers inquiries into the political control that cities exert in regions far beyond their city limits. The chapters in the first section of this book deal with issues such as how cities can succeed in enforcing their interests via rivers to secure their resource supply beyond their legal boundaries, and where and how extraterritorial interests and control over water bodies have competed with the needs of other—smaller—cities.

An additional challenge for the historical analysis of city-river relations is to come to terms with the entanglements between the peculiar rhythm of a river (as expressed by the irregular sequence of droughts and flooding, for example), the unique spaces and places that rivers create (by morphological changes, sedimentation, and erosion), and the multitudinous human interventions into these natural dynamics. From a *longue durée* perspective, the interplay of these factors triggers questions of periodization and the search for caesuras. Which factors usher in new periods? What is the role of technology, infrastructure, and changing methods of industrial production, like the development of railway systems or the increasing importance of hydropower ("white coal")?[17]

One eye-opening way to address these issues is to look at "life cycles" of rivers—a concept that bridges long-term and short-term perspectives. Can we identify long-term cycles in the utilization and transformation of rivers, such as different socioecological regimes and their respective impacts, or the use and depletion of resources?[18] Furthermore, it is crucial to acknowledge the importance of shorter cycles of seasonality—or, more precisely, different seasonalities: the physical consistency of a water body can make a huge difference in the ways in which a society appropriates a river: frozen and fluid rivers, high and low water levels, or different chemical properties like varying grades of oxygen saturation. These biophysical states correspond with different uses over the course of the year, as several of the chapters impressively demonstrate, most explicitly the analysis of the frozen Neva River in Saint Petersburg by Alexei Kraikovski and Julia Lajus.

Some authors, most prominently Mark Cioc, have argued that rivers have their own "ecobiographies." Cioc has applied this approach very successfully to the Rhine and has shown how such a concept can be used to highlight the multifaceted interrelations between fluvial dynamics in their long-term development and human agency while avoiding dichotomic notions of "natural" forces on the one side and societal impacts on the other.[19]

Another promising approach to tackling the coevolution of cities and rivers analytically has recently been suggested by an interdisciplinary research team investigating the environmental history of the Viennese Danube from the sixteenth to the end of the nineteenth century.[20] This approach focuses especially on the importance of interfaces and connections between the natural and the social world. Founded on the social ontology developed by Theodore Schatzki and the social ecology of the "Vienna School," these connections are described by the authors as "socio-natural sites," i.e., as constantly developing constellations of human practices and material/biophysical arrangements.[21] These nexuses are embodied in infrastructural systems such as bridges, weirs, dams, etc. The major advantage of this approach is the opportunity to study the historical transformation of a riverine landscape in a way that avoids merely anthropocentric or naturalistic notions while acknowledging the dialectic nature of all transformation processes.[22]

Part I: Rivers Controlled: Cities and Their Watersheds

The first part of this book deals with the questions of how and with what effects cities have controlled their rivers and—through these—their hinterlands. Given the crucial role of rivers for the provision of elementary resources stated above, this is discussed for the two European metropolises Paris and London within longer time frames. The chapters by Sabine Barles and Vanessa Taylor reflect how these European capitals managed to safeguard the control over resources and their long-term provision deemed necessary for their survival and functioning. Whereas the policy outcomes are structurally similar—the interests of the metropolis are prioritized through the sanctioning of central state authorities—the perspectives of the two authors differ in that Barles's paper looks at the material and infrastructural outcome of these policies while Taylor is more concerned with issues of democracy and political representation. A completely different setting, the catchment area of two tributaries of the Saint Lawrence in a nonmetropolitan region of Quebec, Canada, is highlighted by Stéphane Castonguay. He shows highly divergent development paths in the way that the major city of the catchment could effectively control and direct this development.

In "Rivers, Industrial Cities, and Hinterland Production in Québec in the Nineteenth and Twentieth Centuries," Castonguay tells a story of how and why two watersheds of tributaries of the Saint Lawrence, the Saint Maurice River north of the stream and the Saint François River south of it, evolved quite differently. Whereas the Saint François watershed showed a significant density of agricultural settlement and a diversified network of towns and villages already in the nineteenth century, the watershed of Saint Maurice remained dominated by logging and, from the 1890s, large-scale hydropower and energy-intensive industries. Castonguay demonstrates that the regional hegemony of one major industry such as logging effectively prevented any effective environmental upgrading at the Saint Maurice before this industry wound up in 1995. The Saint François hinterland, in contrast, with its more diversified settlement and industrial structure and concomitantly a broad plurality of stakeholders with their competing interests, saw environmental improvement on its way beginning in the 1960s. In both cases the major city could not exert such an overpowering influence as Paris or London on their watersheds; other players such as the regional government or large power companies were important stakeholders too. The long-term impact of specific development paths is particularly striking in this case study. The Québéquois case studies highlight impressively the "production of hinterlands" through the mediation of rivers. The river is, as Castonguay puts it succinctly, a "fundamental form of communication and integration," and it is through its effects that hinterlands are being shaped and defined.

Barles demonstrates in her study covering more than three centuries how Paris managed to appropriate essential resources for its metabolism and functioning by making use of the Seine. In the second half of the seventeenth century, wood was the essential resource for the city and Paris was given authority by the French state to establish a regime of monopolistic control over the woods and the trade in fuelwood in the catchment area of the Seine and its tributaries. Legal control and physical adaptation of Paris hinterlands to ensure sufficient wood provision went hand in hand. In the nineteenth century water gained prime importance in order to supply the rapidly growing Parisian population with adequate drinking water. Because the water of the Seine transversing Paris was not considered clean and plentiful enough, Paris turned to springwater from the upper catchment area of the Seine and its tributaries. Despite massive conflicts over this extraction and diversion of springwaters, the capital on the whole was able to buy the necessary resources and land as a quasi-private actor and then to secure the support from the state to divert the waters for the needs of Paris. In the twentieth century the focus on water shifted to a comprehensive attempt to regulate and stabilize the flow of the Seine in order to prevent disastrous floods, as in 1910, as well as severe droughts, during which the state of the river at Paris was frequently reduced to a sewer.[23] Paris and its surrounding *départements* embarked on a long-term strategy to erect large reservoirs at the headwaters of the Seine catchment area that would balance the water flow over the year and nearly abolish seasonality. By the late twentieth century the entire Seine watershed was turned into a comprehensively regulated and technicized hydraulic system. Valuable beyond the Paris case is Barles's coining of concepts: she talks about "institutional imprints" that were exercised by the city of Paris, usually supported and sanctioned by the state on a quite extensive hinterland. Through these projects urban exterritorialities or "out-infrastructures" developed: installations and technological systems that were located far away from the capital but whose only raison d'être was the provision of urban needs. Thus, as Barles postulates, "the river and its tributaries are (almost everywhere) 'urban' in a functional sense."

Taylor reflects on the ways in which water issues were governed and regulated in the Thames River basin from the mid-nineteenth century until the creation of Regional Water Authorities in the 1970s and 1980s. The title of her contribution, "Watershed Democracy or Ecological Hinterland?," signifies Taylor's concern with the history of political debates on the river and highlights a tension that is constitutive for her narrative: she recounts how the hodgepodge of various local government bodies in charge of water issues was replaced in the early 1970s, still under the auspices of more efficient environmental planning, by large Regional Water Authorities. These integrated functions of environmental monitoring and the provision of public water and sewerage on the basis of a supposedly natural scale, the river basin. The downside of this was the loss of local, democratically elected control, which was exacerbated by the subsequent privatization of these Regional Water Authorities by the Thatcher government. But Taylor's account is not just a story of depoliticization, of loss of local democracy; it is more complicated. Through highly differentiated processes of rescaling, new arenas of political action have arisen in which civil society groups have partly replaced the former role of local government. On the other hand, the European Union and its environmental legislation such as the Water Framework Directive of 2000 have provided space for policy interventions and knowledge sharing to a wider range of civil society groups than was available under the old regime of local democracy. Thus, Taylor detects "new modes of participation and accountability for rivers that exceed what was on offer from representative local bodies up to the 1970s."

Part II: Urban Rivers Transformed and Lost

From the mid-eighteenth to the mid-twentieth century (and often continuing to the present), many rivers witnessed unprecedented transformations that profoundly altered the relationship between cities and rivers. The reasons for these changes are manifold but two developments clearly stand out:

industrialization and urbanization. Both radically changed the functions of rivers. As Stéphane Castonguay and Matthew Evenden have pointed out in the introduction to their collected volume *Urban Rivers*, urban and industrial growth were closely intertwined and their joint development "produced not only many more and larger cities in the nineteenth century but also reorganized their internal structure, produced increased problems of energy, materials, and food supply, and delivered new waste streams in the air, soil, and water."[24]

Rivers played a crucial role in this development. They served as commercial arteries, guaranteeing the steady, uninterrupted, and year-round flow of increased navigation; they had to supply drinking water for the rapidly growing urban population; they were increasingly called upon as an important source of energy; and many of them became "ultimate sinks" for the disposal of human and industrial waste.[25] However, in order to properly fulfill these new tasks, rivers needed to be engineered to become more reliable and much less subject to the vagaries of climate, the weather, and river morphology, to become streams uninhibited by rapids and shallow water, by floods and droughts, by narrow bends and meanders.

As a result, many rivers experienced morphological changes to such a degree that they came to resemble canals more than natural streams. They were straightened, shortened, bordered with levees, and harnessed into narrow channels. Hydraulic engineers sometimes openly enjoyed "pushing rivers around."[26] In the twentieth century, "recasting the world's rivers ranks among the signal environmental changes," as John McNeill has pointed out.[27] Consequently, river islands and sandbars disappeared, oxbows were cut through, branches cut off, riverbeds eroded, wetlands adjacent to the main channel drained, biodiversity decreased, and the size of the floodplains shrunk tremendously. The Rhine, for example, was shortened by 105 kilometers—8 percent of its entire length—as a result of several correction works. Until 1975, 90 percent of its floodplain had been claimed by residential areas, roads, farmland, and other anthropogenic purposes.[28]

Rapid population growth in nineteenth-century cities throughout the world but especially in Europe and North America quickly overburdened the traditional systems of water supply and waste disposal. Cisterns and wells no longer sufficed to meet the growing demand for water, so cities began to build waterworks—such as the Longdendale Reservoir near Manchester in 1851—and tap ever more distant sources to fulfill their needs.[29] In the United States, New York City and Boston had already begun constructing elaborate aqueducts in the 1840s. In 1913, William Mulholland famously announced the opening of a 235-mile aqueduct from Owens Valley to Los Angeles with the words "There it is! Take it!" A couple of years later, in what has become one of the best-known episodes in American environmental history, the Hetch Hetchy Valley in Yosemite National Park was flooded in order to create a reservoir for the city of San Francisco.[30]

However, the growing supply and increased water consumption—triggered not just by population growth but also by technological innovations such as the water closet—created new problems. Most importantly, cesspools and privy vaults were overburdened by the increasing amount of sewage and "cities across [the United States] began to drown in their own filth," as Ted Steinberg has pointed out.[31] Consequently, a lot of human and industrial waste found its way into the rivers, creating pollution to an extent unknown and countless sanitary problems. The famous "Great Stink" of London in summer 1858 was to a large degree due to the fact that human excrement from the newly introduced water closets had been washed via sewers into the Thames, where it came to lie exposed to air in the dry June of 1858.[32] As a result, cholera epidemics killed thousands of urban dwellers from the mid-nineteenth to the early twentieth century.[33]

The effects of population growth on urban rivers were not limited to sanitary problems, though. In many towns, marginalized groups often had little choice but to settle in the marginal and vulnerable parts of town, and the riverbanks were certainly among them. Squatters and homeless people in Toronto's Don River valley, African Americans in many US river cities, displaced persons in postwar Munich and Hamburg, and poor citizens in Bogotá, all found temporary homes in neighborhoods that had been given up by the urban elite, in shantytowns, on houseboats, and other more or less informal settlements right next to or on the river.

The urban floodplain was also increasingly utilized for infrastructural and industrial purposes. This amassment of damage potential in an area that is hydrologically a part of the river led to ever more sophisticated and costly flood control projects. This development was especially noticeable in those cities where rivers were "trapped" between embankments or hidden behind massive flood walls. Often, urban flood control works were accompanied by the displacement of the working poor, homeless people, and other marginalized groups living in these liminal spaces. In the name of modernization (for example, in the form of "slum clearance"), such sites were condemned or simply given up and people had no choice but to move, either by force or as the result of neglect.[34] Consequently, riverfronts and public landings—for a long time the central place of river cities—declined in importance and the local populations turned their back to the river.[35] In a similar vein, cities used their river valleys as a repository not just for urban waste and sewage but also "for prisoners, for the institutionalized poor, and for people who in other ways failed to measure up to nineteenth-century liberal values of rationality, moral rigour, and self-advancement," as Jennifer Bonnell has argued.[36] This certainly holds true for many river cities in the age of industrialization and urbanization, as the chapters in this section clearly show.

One of the most striking examples of how a city has "lost" its river can be found in the city of Nantes in western France. Fifty kilometers from the Atlantic Ocean, the city struggled to maintain a shipping channel to the estuary as oceangoing vessels became bigger and bigger over the course of the nineteenth century. Geneviève Massard-Guilbaud describes in intriguing detail the drastic modifications of the Loire in the twentieth century. Engineers filled the river arms and diverted an urban tributary into a tunnel. As a result, the entire structure of the city changed and the river disappeared. Excessively wide streets took the place of the filled river, while authorities and inhabitants remained surprisingly passive in the face of this dramatic transformation of the urban landscape.

In Cleveland, engineers and industrial and transportation interests aimed at modifying the urban landscape too. Their target was the Cuyahoga River, which drains into Lake Erie in northeast Ohio. David Stradling shows how the Cuyahoga has helped to create one of North America's great industrial valleys. Cleveland's steel mills, oil refineries, and chemical and shipbuilding plants all relied on the river and clustered in the narrow flats. Industrialists and city leaders went to great lengths to straighten the "crooked" river, but failed to do so as a result of the complexity of diverging interests circling around the Cuyahoga River. According to Stradling, the story of the Cuyahoga "serves as an object lesson in the city's conflicting interests and complicated politics."

Vladimir Sánchez-Calderón explores the relationship between the Tunjuelo River and the growth of the Colombian capital Bogotá. Sánchez-Calderón identifies three important functions that the river has fulfilled at different times in the development of the city. Firstly, it served as the main water source of the city from the late 1930s to the end of the 1940s. Secondly, the floodplain of the river provided much-needed land for the fast-growing city and especially for its low-income neighborhoods since the mid-1940s. Finally, the river served as a supplier of aggregates, a key component in the production of reinforced concrete, which could be dredged from the river bottom starting in the 1950s. Together, these developments profoundly altered the river's local hydrology and produced several severe incidents of flooding, mostly of the highly vulnerable informal neighborhoods right next to the riverbank.

Christoph Bernhardt traces a century-long transformation period in the relationship of the city of Strasbourg to the Rhine River. The city had benefited from its location between the Rhine and the Ill, a left-bank tributary, since Roman times; however, it also suffered badly from recurring floods. In the early nineteenth century, new flood protection works minimized this danger but also resulted in a decline of the water table and a relocation of the riverbed to the eastern parts of town, away from the city center. The effects of these interventions on river navigation, together with the rising importance of the railroads, were disastrous. Upstream navigation was temporarily abandoned in 1864 and Strasbourg all but lost its role as an important harbor city. The regional metropolis—a German city during the time between the Franco-German War of 1870–1871 and the end of World War I—responded to these challenges by investing massively in the sanitary infrastructure of the city, huge new harbor facilities, and the construction of a Rhine side channel all the way to Mannheim. While these measures heralded a renaissance of city-river relations, they were also highly contested in a city that was culturally divided between French and German influences, as Bernhardt points out.

Dirk Schubert's chapter looks at how the city of Hamburg has historically modified the Elbe River—the link that connects the city both to its upstream hinterland and, via the North Sea, to the entire world. With the rise of railroad and steamship in the nineteenth century, Hamburg saw its future as a tidal seaport, easily accessible by even the biggest ships. Thus, the city invested heavily in several port expansions (including a "free port" area in which housing was prohibited) and the building of modern quays, new bridges, and more. This by and large successful development, however, came to a temporary halt as a result of World War II, particularly due to the severance of much of the city's riverine hinterland, which now lay behind the iron curtain, and the devastating 1962 flood, which killed more than three hundred citizens and called into question the adequacy of contemporary flood protection works. Since then, city-river relations have been characterized by the interplay of an "aggressive industrialization" of the lower Elbe region and ever deeper dredging of the riverbed on the one hand, and increasing environmental concerns and measures to improve the leisure quality along the riverbanks on the other.

Part III: Cultural Dimensions of Urban Rivers

When Felix Fabri, a late fifteenth-century clergyman and travel writer, described the relationship between the small river Blau and the southern German town of Ulm, he used highly anthropomorphic language.[37] Most of the time, he argued, the river and the city coexisted in an intimate and peaceful way, with the river Blau acting as the city's burgher (*civis*) and friend (*amicus*), permanently feeding the city's wells with freshwater, driving the mills, and washing away the effluents. Sometimes, however, Blau turned from a burgher and housemate (*domesticus*) into the most terrible enemy approaching the city with furious violence, breaking and devastating everything it met. Fabri's colorful language in this narrative is by no means a singular example. At any time in history, humans have attributed symbolic meaning to their material surroundings and have thought and spoken about their relations to "nature" in a metaphorical way. Doing so is an important prerequisite of human culture making man an *animal symbolicum*—to use philosopher Ernst Cassirer's term.[38]

Most obviously, rivers are part of this story. They are material structures that can be read by people who live on their shores and make use of them, but rivers can also be imagined, conceptualized, remembered—or forgotten. Rivers are subjects of fictional literature and fine arts. They provided material for romanticized notions of national identities (like "Father Rhine" or the *Donaumonarchie*, a term equating the former Habsburg empire with the Danube, its most prominent river) and urban imaging alike. Social practices related to a river can find a long-lasting place in collective memory and be present in rituals long after the respective social reality and professional contexts have vanished. In Newcastle upon Tyne, for example, the local magistrate surveyed the tidal portions of the Tyne River once a year on Barge Day (celebrated on Ascension) to claim its "conservatorship" over the river.[39] Furthermore, rivers play a prominent role in religious beliefs.[40]

The third part of this volume discusses the cultural dimension of city-river relations. The individual chapters tell us how urban citizens thought about their rivers, what metaphors they used, and what these metaphors show us about the social reality, moral norms, and political ideologies in river societies. They illuminate the roles rivers play within urban and civic rituals and ceremonies. They shed light on how river locations were perceived and how these perceptions changed over time, how urban rivers have figured as core sites of national, political, social, and religious identities. The chapters in this section discuss how these notions were negotiated, who claimed the river and for what purposes. Can rivers be conceptualized as urban sites of memory? And—with regard to Massard-Guilbaud's chapter—what are the cultural implications when, as a consequence of material change, a city "forgets" about its river? The interplay of proximity and distance between urban society and the river, sensual perceptions such as seeing, smelling the water, crossing the frozen river on foot in winter—how do all these perceptions and practices shape urban knowledge and urban identities? The four contributions to this section provide case studies dealing with sites as different as the amphibian world within Chinese shipmen's songs; the dark hell of London's Thames mirrored as sanitary abyss in Charles Dickens's novel; the site of the "holy" Ganges, where contradicting colonial and indigenous notions of purity and sanity were being negotiated; and finally, the seasonal calendar of the river Neva in the everyday life of Saint Petersburg's urban dwellers.

Understanding rivers as prisms of imaginary urban topographies, Igor Chabrowski offers insights into the representation of Chongqing, Eastern Sichuan, in the songs of boatmen sung as they passed or entered the city on the rivers Jialing and Yangzi. For the boatmen, the relationship to water grew out of their daily experience of both the river and the expanding city through which they navigated. Underlining the role of the historically changing socioeconomic context, Chabrowski states that the workers' songs could tell historians more about work on the river than about the stream itself, as they mainly reflected the selective perceptions, hopes, and worries of the boatmen. Thus, lyrics and melody can become valid sources for research in urban identity studies and environmental history.

Awadhendra Sharan inquires how colonial authorities addressed the question of rivers as sinks for urban waste, by focusing on the river Ganges and its relationship to the towns of northern India from the 1890s to the 1930s. For him, an answer to this question has to be based on an appreciation of India as a colonized space. Europeans and native communities developed their respective decision making about the permissibility of discharging sewage into the "holy river" on contradictory concepts of purity, sanity, and the sacred or secular character of water bodies. For this reason Sharan advocates a holistic approach, combining different strands of investigation: tracing native conceptions of sacred rivers (as exemplified in annual bathing rituals), and analyzing the nineteenth-century "modern" problems of pollution in tropical rivers as well as (Western) scientific debates on these problems, finally leading to a reflection on the "ambiguous nature of colonial modernity, and the limits placed on the fuller realization of the modern sanitary project."

Agnes Kneitz presents a reading of the novel *Our Mutual Friend*, by Charles Dickens, from an environmental perspective. Kneitz shows how Dickens interwove his own perception of the contemporary crisis in urban sanitation with his fictional and highly metaphorical story line. From this perspective, the text reveals the author's environmental consciousness as he described the Thames and London as what we would today call a shared ecosystem. According to Kneitz, by focusing on the risks of river pollution and their impact on human lives and by paralleling practices of river use with moral values, Dickens conceived of the river as an actor in bringing death and disease to the city, concluding with the implicit argument that only by restoring its ecological balance (to use an anachronistic term) could London reemerge as a living space.

Narrating urban life in interaction with Saint Petersburg's Neva River, Alexei Kraikovski and Julia Lajus correlate the river's seasonally determined life cycle with its cultural appropriation by urban society. They demonstrate that this life cycle is mirrored in practices of river use, recreation, and ceremonies. Due to the great seasonal differences in temperature the city's topography itself is almost literally in flow. With certain modes of transport possible only at specific times of the year (shipping, walking, and road transport on the frozen surface) and different degrees of the river's accessibility, the Neva's imprint on the urban culture of Saint Petersburg can hardly be overestimated.

Part IV: Rivers Regained

Over the last couple of decades, a steadily growing number of cities have tried to reappropriate their rivers. These efforts have taken many different forms and they represent anything but a linear success story. Nowhere have urban rivers been restored to their preindustrial status—an impossible task anyway in most cities. Often the restoration attempts have followed ideas that reflect the needs and values of current societies more than historical templates. But what all these endeavors have in common is a remarkable shift in the hegemonic understanding of the role the river is supposed to play for urbanites as contrasted to before the 1970s.

Several developments have come together in fostering the "rejuvenation" of urban streams. First of all, the transformation of rivers into "thoroughly anthropomorphized stream[s]" triggered critique from its very beginnings.[41] While the environmental degradation of rivers was by and large accepted as the price that societies, and especially urban societies, had to pay for the benefits of modernization, it was nevertheless perceived as a loss.[42] This feeling of loss intensified as more and more rivers were literally sacrificed and turned into "problem rivers"—highways not only for river traffic but also for the dissemination of pollution and disease.[43] Following modernist reasoning in all its consequences, it was only logical to actually hide these rivers from sight, to fill them completely, to make them flow underneath the surface, beneath concrete ceilings, or to change their character in such a way that they vanished from the public consciousness.[44]

Secondly, river restoration, especially in cities, benefited from the changing economic function of rivers. The loosening grip of railroads, warehouses, and industrial facilities on the urban waterfront opened up new possibilities for other, less ecologically demanding utilizations of the urban floodplain. Many municipalities chose to fill this void by creating urban parks. In Japan, as a result of stalled economic growth after the energy crises of the 1970s, multiple *shin-sui* ("playing with water") parks were opened on the urban riverbanks. "Although these projects were conducted for recreational rather than for ecological purposes, they helped to turn people's eyes back to nature."[45] The same is true for parks that have been the by-product of flood control measures and attempts to improve water quality. One of the most famous examples is Boston's "Emerald Necklace," designed by Frederick Law Olmsted in the late nineteenth century and regarded by many as a "landmark in American park planning."[46] In general, as Anne Whiston Spirn noted in 1984, the "recent profusion of urban parks that serve multiple purposes of flood control, water quality improvement, and recreation do not reflect a new idea, but rather the rediscovery of old solutions," i.e., of solutions stemming from the late nineteenth and early twentieth centuries.[47]

Furthermore, the rise of the environmental movement in the 1960s and 1970s and the increasing percolation of ecological thought into scientific as well as popular discourses also changed the way in which urban rivers were conceptualized. Rivers came to be seen by the urban public no longer exclusively as economic arteries and open sewers but increasingly as fascinating ecological entities in their own right, providing a rich habitat for a diverse variety of fauna and flora and a popular destination for leisure activities right at the doorstep of many urban dwellers. Paradoxically, environmental catastrophes on urban rivers actually played a significant accelerating role in this development. The Cuyahoga River, for example, which functioned as an open sewer for the many industrial companies in Cleveland, Ohio, caught fire several times in the 1950s and 1960s and thus emerged as a potent symbol of urban blight and decay. The last of these conflagrations in June 1969, however, had a profound impact on the nascent environmental movement in the United

States and, at least indirectly, contributed to the passage of the Clean Water Act of 1972.[48] In a similar way, the 1986 environmental disaster at the Sandoz chemical plant in Basel, Switzerland, when tons of highly toxic pesticides, insecticides, and other agrochemicals were washed into the Rhine after a fire and created widespread ecological havoc downstream, catalyzed attempts to clean up the river and triggered the Rhine Action Plan for Ecological Rehabilitation of 1987.[49]

Since the 1990s, these different attempts to clean up and to reappropriate rivers have dovetailed into a significant movement in many parts of the world. In the United States alone, more than thirty thousand projects to restore rivers and wetlands have been initiated since 1990.[50] However, the term "river restoration" comprises a wide variety of activities, ranging "from replanting riparian trees or fencing livestock out of stream corridors to the removal of dams and full-scale redesign of river channels."[51] Urban river restoration is even more complex and complicated due to the manifold superimposed interests and responsibilities involved in the process and the general ambiguities of "urban nature." Here, rivers often had to be rediscovered and physically uncovered, highlighted or daylighted before they could be "restored."[52] Still, there can be no doubt that a large number of urban rivers (or urban stretches of rivers) have yet again witnessed fundamental changes since the 1990s (and often long before) and that many of these changes have resulted in a heightened awareness of the presence of nature in the city. River restoration in the city is much more visible, dramatic, and accessible than it is in a rural context. It "captures people's imaginations" and the attempts to reappropriate a river have quite often served as an ideal stage for the demonstration of local environmental efforts and civil society at large, as all four chapters in this section emphasize.[53]

The benefits of urban river restoration are not limited to ecological, aesthetic, and recreational improvements, but can also include the reinvigoration of economic activities on the waterfront and an intentional or unintentional revaluation of city centers. Furthermore, restored or "repaired" urban rivers also function as outdoor classrooms to educate the public about the value of nature in the city.[54]

The fate of the Cheonggyecheon River in Seoul is a good example to highlight many of the opportunities and problems of urban river restoration. Due to a multitude of sanitary and social problems in the shantytown areas right next to the river, the Cheonggyecheon has undergone a decades-long modernization process that started in the 1950s when the river was buried under a thick layer of concrete. In the late 1960s, an elevated freeway was built on top of the foundation. From 2002 to 2005, however, the city initiated a massive program to dismantle the decaying infrastructure, uncover the river, and create a pedestrian corridor along its banks. Today, the Cheonggyecheon is anything but a natural river. In fact, this "restored river" is still a highly engineered environment, the creation and maintenance of which requires massive and recurring amounts of capital, labor, and political efforts. For example, as the authorities could not bring themselves to restore the river's former flow direction, no fewer than 120,000 tons of water have to be pumped from the Han River every day. Still, the remodeling of the Cheonggyecheon highlights the increasing importance of "repaired" natural places in the city and their functions for urban dwellers.[55]

The fourth and final section of this book reflects the origins, the diversity, and the results of these attempts to regain urban rivers. Harold Platt traces the changes the Chicago River underwent during the transformation from the "age of environmental engineering" to the "age of ecology"—a forty-year-long conflict over the meaning of the river. Platt shows how Chicago remained the last major holdout in the United States against coming into compliance with the Clean Water Act of 1972. Every day, the city dumped 1.2 billion gallons of partially treated sewage into the Chicago River, arguing that its inhabitants had Lake Michigan for recreational purposes. Things changed for the better only when the Environmental Protection Agency in 2011 ordered Chicago to disinfect its sewage and to complete its "Deep Tunnel" plan of flood control. But while environmentalists and city administrators battled over the river aboveground, the river itself produced a second story underneath the surface—a natural history of aquatic life responding to human-altering conditions, as Platt points out.

Michael Toyka-Seid, too, identifies a turning point in the changing relationship between cities and rivers in the late twentieth century. He also questions the linearity of the transformation of rivers into the preferred playground for urban developers, a living space for the better-offs of urban society and a place for leisure and relaxation for everybody else. His research design is especially suited to this task since he looks at two neighboring cities on opposite sides of the Rhine—Mainz and Wiesbaden, state capitals of Rhineland-Palatinate and Hesse, respectively. Highlighting the similarities and differences of the two cities' dealings with the Rhine from about 1970 to the end of the century, Toyka-Seid notes that the reappropriation of the river, though a constant plea throughout the decades, consisted in fact of rather small and isolated developments such as the creation of upscale residential areas, museums on the riverbanks, and a few recreation areas.

Another German river—Munich's Isar—is often referred to as an international model of successful urban restoration attempts. Nico Döring and Georg Jochum trace the important role the river has played in the history of the city and explain the obstacles that local activists had to overcome in their attempts to revitalize the river. The authors describe the origins of the "IsarPlan," an ambitious project to renaturalize an eight-kilometer-long urban stretch of the river that had been initiated by local environmental groups in the 1980s and later been adopted by the state of Bavaria and the city of Munich. After eleven years of heavy construction work, the last part of the project was finished in 2011.

The international tour of urban rivers ends in Fez, Morocco. Shelley Hornstein there showcases the work of architect Aziza Chaouni and her Bureau of Ecological Architecture and Systems of Tomorrow (Bureau EAST) to resuscitate the Fez River, an urban stream that had been buried under a concrete parking lot. Hornstein contrasts Chaouni's work with the many examples of urban star architecture. She interprets the recent restoration projects in Fez and other parts of the world as "monumental" in their own right and hence "worthy of celebrity status." The fundamental difference, though, is that Chaouni's project "involutes—that is, it hides behind and inside, swells, or reflects rather than builds on, up, and out." In Fez, this means that the buried river, a heavily polluted "river of trash," is now being praised and remembered.

The contributions to this volume open up a broad and multifaceted view on the complex history of how cities in different parts of the world have lost and regained their rivers. It is—as so many stories are—an open-ended one. Tim LeCain's example of the restoration efforts on rivers in Montana, formerly straitjacketed for and poisoned by copper mining, reminds us of this open end. When in 2008 the Milltown Dam not far from the city of Missoula was opened, allowing the Blackfoot Rivers to return to their natural flow regime, politicians considered this to be an important step representing "Montana's shift from an extraction to a restoration economy, creating jobs that protect the environment and use the state's natural resources in a sustainable way rather than plundering them."[56] What they did not mention is that the world's need for these resources and their industrial processing has not vanished but the production has been moved to other world regions, poisoning rivers and shaping the river stretches of highly industrialized cities there. How will cities of the global south further develop the relations to their rivers? As historians we are much more at ease with analyzing the past than with predicting the future. The contributions of this volume show that fundamental changes in city-river relations can take place given fortunate constellations of economic conjuncture and political will. Investigating and telling the stories of cities losing and regaining their rivers are demanding tasks. Scholars in this field have to be aware that the coevolution of cities and rivers is an inextricably interwoven process, a complex assembly of human and nonhuman actors, and an arena where material and symbolic worlds merge. With the geographical scope and the broad variety of case studies from different disciplinary backgrounds, this volume tries to tackle this complex and fascinating story.

Part I

Rivers Controlled

Cities and Their Watersheds

Chapter 1

Rivers, Industrial Cities, and Hinterland Production in Quebec in the Nineteenth and Twentieth Centuries

Stéphane Castonguay

Rivers have long contributed to the articulation of urban spaces beyond the boundaries initially marked out by the fortification of the city. They have acted as navigation routes linking the urban milieu to its hinterland, particularly urban industries to sites for the extraction of raw materials and to markets for the distribution of products manufactured by the city's artisans and workers.[1] Just like the installation of water intakes for an aqueduct or the building of sewers within the city, the channelization of the river over long distances for navigation or the diversion of potable water for urban consumption integrated the river into sociotechnical systems in which rural populations confronted urban issues. The same was true of energy facilities that were moved out of the city. Water-based as they were, hydraulic mills had fostered the installation of factories along the edge of urban rivers, where the resources and products of industry were also conveyed by water transport. However, the invention of hydroelectric power generation at the end of the nineteenth century and, more importantly, of technologies to distribute electricity over long distances ended the dependence of the city's industrial elements on a nearby river and opened up the urban space.

Thus, the river was constitutive of the hinterland-building process, not only through the integration of distant territories and their incorporation into networks but also through the supply of elements (energy, potable water, natural resources, and industrial materials) necessary to the industrialization and urbanization processes. Furthermore, by changing the river's hydrology or compromising its quality, peri-urban facilities built to support industrial activity extended not only the territory of the city but also its environment

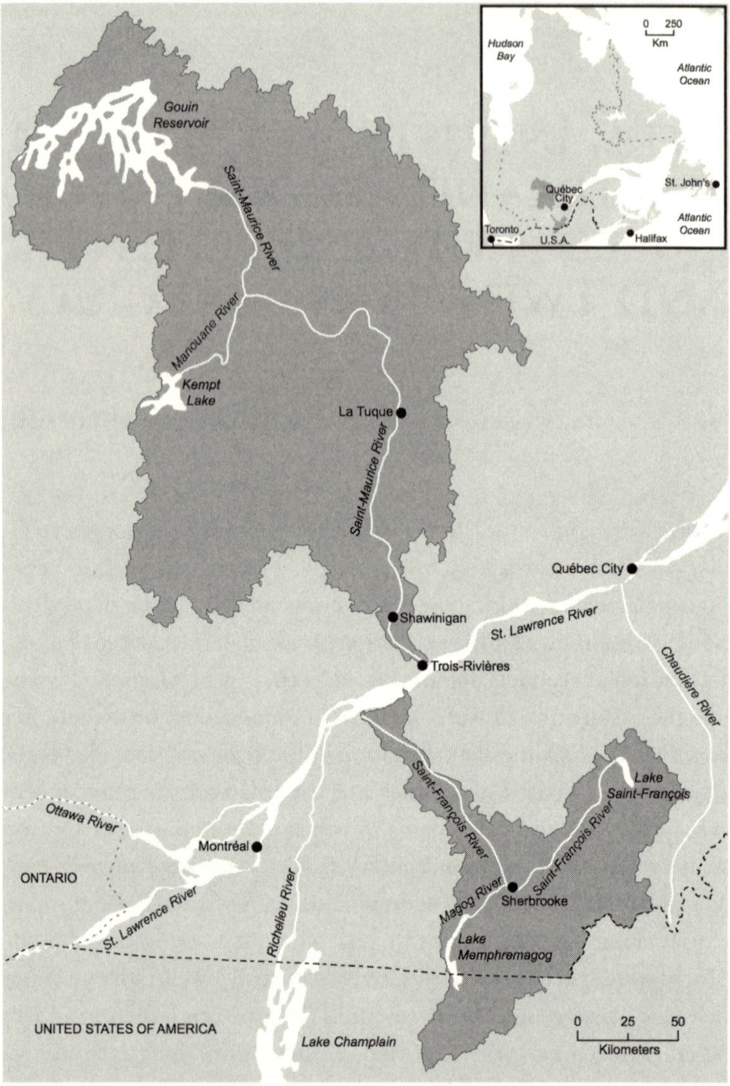

FIGURE 1.1. Watershed of the Saint Maurice and Saint François Rivers. Philippe Desaulniers, Centre interuniversitaire d'études québécoises.

and its political ecology. Flood risks and access to water of sufficient quantity and quality were at the heart of conflicts that arose from ecological changes, and revealed the power relations among the many users located along the river.[2] They also indicated that the transformations in riverine environment that accompanied the processes of urbanization and industrialization were part of the territorial dynamic that shaped urban space. And while the latter ended up loosely corresponding to the contours of the watershed, it nevertheless exhibited a variable geometry, partly caused by ecological phenomena and partly resulting from sociohistorical factors.[3]

Yet studies on the spatial dynamics underlying the extension of the metropolis have only recently included in their analyses of land-occupation schemes the riverine elements and environmental changes arising from the influence of urban and industrial milieus on their emerging hinterlands.[4] While environmental historians have recast the study of city-country relations by focusing their analysis on the environmental transformations that accompany the establishment of metropolises, the exchange networks underlying these relations remain essentially understood in their land forms—roads, railways, highways.[5] The fundamental form of communication and integration that is the river needs to be further examined, along with the spaces commanded by its usage and representations, as well as its materialities.

By contrasting the extension of the urban space of two cities located in different watersheds, this chapter seeks to understand riverine transformations in relation to urbanization and industrialization processes and to identify the role of the hydrology, ecology, and socioeconomy in hinterland formation. Sherbrooke and Trois-Rivières, the two cities under discussion, are located on tributaries running south and north of the Saint Lawrence River: the Saint François and the Saint Maurice Rivers (fig. 1.1). These rivers share many commonalities, both hydrological and historical. They both exhibit an important declivity, and a series of dams punctuate their hydrological regimes (figs. 1.2 and 1.3). Both rivers underwent a series of ecological and structural changes as of the middle of the nineteenth century following the establishment of sawmills and riverine infrastructures for wood floating. In the first decades of the twentieth century, a governmental agency proceeded to transform their respective headwaters into large reservoirs to ensure a regular flow; the construction of dams for energy production and flow regulation, along with other features of the second industrial revolution, further transformed their fluvial landscapes. Dams, mills, and factories were built at the confluence of their many tributaries, and urban centers evolved from these industrial nodes. These human transformations of the riverine environment accentuated the natural phenomena that annual precipitations provoked, such as erosion and floods, and that modified the beds and banks of both rivers. However, the two rivers also differ in length, degree of declivity, and pace of flow. Their watersheds also present specific features that affected the rate and density of their settlement, the Saint Maurice River being located in a rocky, sandy valley of

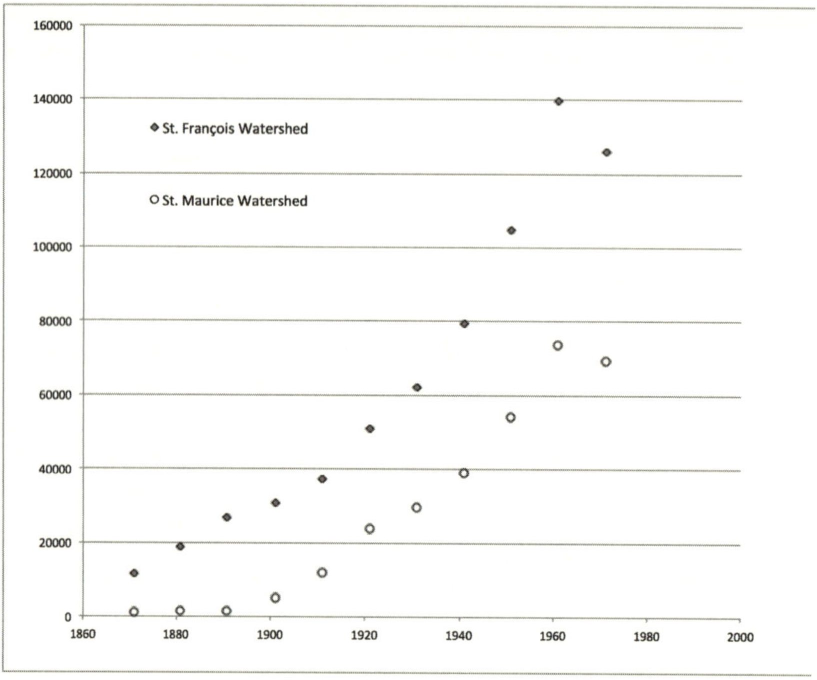

FIGURE 1.2. Population growth along the Saint Maurice and the Saint François Rivers. Source: Canada Census (1871–1981).

poor fertility, while the watershed of the Saint François River offers a more varied topography, partly fit for agriculture and more suited to colonization. Because their riverine environments presented different possibilities for settlement, the watersheds of the Saint François and Saint Maurice Rivers were subjected to diverging patterns of urbanization and industrialization, with a consequent demographic growth and land-occupation scheme that influenced cultural representations of the river. Therefore, while similar driving forces transformed the Saint Maurice and Saint François Rivers into urban, industrial rivers, distinct natural and social factors influenced the rhythm of hinterland formation throughout the two watersheds.

Settlement, Urbanization, and the Early Industrialization of Rivers

Urbanization of the riverine landscape followed different patterns along the Saint Maurice and Saint François Rivers. Topographical features on the north and south shores of the Saint Lawrence River influenced the rate and type of settlement, and the hydrology partly encouraged the selection of specific

FIGURE 1.3. Hydroelectric Improvements on the Saint Maurice River. Source: Shawinigan Water and Power Corporation, "The St. Maurice, a 2,000,000 h.p. river, Montreal, June 29th 1942."

locations for the growth of urban centers. However, the intensity of colonization activities and the extent of settlement on the riverbanks determined the vulnerability of riverine populations to inundations, and the establishment of communication routes along the shores and their consequences on land use patterns. Most of all, these variables framed how, on the north and south shores of the Saint Lawrence River, the Saint Maurice and Saint François Rivers were transformed and experienced different patterns of industrialization. To capture these transformations, I refer to a process described by historian Eva Jakobsson, who coined the phrase "industrialization of rivers."[6] While Jakobsson strictly referred to fluvial transformations executed to inscribe a river into a sociotechnical system for the production of hydroelectric power, my use of the phrase differs slightly since my analysis emphasizes the formation of an economic system based on the integration of different industrial systems—for the production of hydropower, pulp and paper, and aluminum, as well as log driving and touristic and recreational activities.

Located between Montreal and Quebec City on the north shore of the Saint Lawrence River, the city of Trois-Rivières benefited from the resources of the Upper Mauricie area, where the abundance of furs directed its commercial growth since the seventeenth century.[7] The Saint Maurice River acted as a route toward these resources. Yet except for a few parishes established on the lowlands of the Saint Lawrence River and Trois-Rivières—which had become the head of a judicial district and a diocesian siege—the opening of the hinterland gave way to little settlement beyond the immediate banks of the river. By the late eighteenth century, the pioneer front of colonization in New France had stopped at the edge of the Precambrian Shield.

Forest exploitation made continental penetration possible when, following the continental blockade during the Napoleonic Wars, Canada and its provinces started supplying white pine to the British Empire, and then, some decades later, to the construction industry in the United States after the American Civil War. Connected to international markets and served by an important harbor located at the confluence of the Saint Lawrence River, Trois-Rivières benefited from the growth of the forest industry throughout the Saint Maurice valley. Crown lands were quickly transferred to entrepreneurs who oversaw the clear-cutting of vast forest areas until the mid-1870s, when pine trees became scarce and an economic crisis caused world trade to slump.

To be exported, lumber needed to be driven down the watershed, but given the inexistence of adequate means of transportation before the advent of the railway in the Upper Mauricie area, the Saint Maurice River proved to be the fastest, least expensive way of driving wood to the confluence of the Saint Lawrence River. Yet the route from the Upper Mauricie to the harbor of Trois-Rivières was a difficult one, and navigability was interrupted in several locations, partly because of rapids but mostly because of steep reaches where rafts were repeatedly destroyed and forest entrepreneurs lost logs and profits.[8] The construction of slides at these locations required financial and technical assistance from the provincial government. Starting in the middle of the nineteenth century, heavy public investment by the colonial government of the province of Canada enabled the transformation of the river and its insertion within a technical system for timber exportation. In its annual report for 1852, the very same year that the first paper mill was installed on the Saint François River, the Commissioner of Crown Lands insisted on the need to transform the river for the transportation of logs,[9] and thereafter a series of installations enabled the timber of the Upper Mauricie to reach the Saint Lawrence River and the harbor and sawmills of Trois-Rivières. At Shawinigan, where falls and rapids prevented the floating of log cribs any farther down the river, the government proceeded to construct a 335-meter log slide and 4,500 meters of retaining booms, one above the falls and the other below in the lower bay, with pillars to retain the booms despite the current.[10] While it heavily subsidized wood-floating infrastructure, the government proved to be less generous toward the funding of road improvement and railway construction. Investment in these domains might have facilitated the colonization of the Saint Maurice valley, but on the other hand colonization and forest exploitation seemed to compete for public money and for public land.[11] Forest camps moved according to the availability of white pine, and the transformation of the forest cover from fire and intensive exploitation deterred settlers from establishing villages. Instead, they followed the forestry entrepreneurs in search of new logging sites. Settlers not only earned their living by supplying farm products to forest camps, they also sold the logs from their land-clearing activities. Once the soil fertility was depleted and the lumber camp had moved on, the settlers relocated.

While the preindustrial history of the Saint François River was also stimulated by the lumber economy, land occupation in the Eastern Townships followed a different dynamic. Agricultural colonization drove initial settlement of the area, where prosperous agriculture enabled the growth of villages and towns thanks to the establishment of dams and mills. Partly orchestrated by the British American Land Company (BALC), settlement initially involved Loyalists fleeing the American Revolution who squatted or acquired small plots of land to colonize a large area south of the Saint Lawrence valley just north of the American border.[12] There, soils of varying fertility and forest resources offered some prospects for the establishment of a pioneering community.

Numerous sites along the Saint François River offered the possibility of harnessing hydraulic energy, and these gave rise to several towns and villages in the Eastern Townships. One of these was Sherbrooke, where the steep gorges surrounding the Magog River turned out to be well suited to hydropower.[13] In 1834, the BALC erected a dam upstream from the falls, at the head of Petit Lac Magog, which was transformed into a water reservoir.[14] The dam supplied hydraulic energy by mechanical transmission to a sawmill built by the BALC in 1837. Thereafter, four additional dams were erected on the Magog River, and the BALC rented riverine land to manufacturers that set up woolen mills, smelters, and tanning yards. The city of Sherbrooke evolved from these installations.

Because of its declivity and difficult passages that necessitated long portages, the Saint François River was of limited navigability. Nevertheless, it provided the pathway for the construction of a railway link between Montreal and the American seaport of Portland, Maine. The advent of the railway accelerated the peopling of the Eastern Townships, not only along the Saint François River but also across the territory, as the railway tracks crossed the Eastern Townships; smaller villages emerged and accommodated services and small industries. Thereafter, the exchange network linking the Eastern Townships to its suppliers and markets encompassed a more diversified range of products and resources, as factories in towns like Sherbrooke relied on the importation of staples such as cotton from the southern United States, iron from Nova Scotia, wool from Australia, and rubber from South East Asia.

The advent of the railway in 1853 further stimulated the urbanization of the Saint François River that influenced the siting of the villages.[15] New towns sprang up, such as Windsor Mills and East Angus, built at the confluence of the Saint François River and the Watopeka and the Eaton Rivers, respectively. Other sites on the Saint François River benefited directly from sufficient gradients for dams to be built, enabling the growth of small industries that attracted people and resources. At Brompton Falls, a few kilometers downstream from Sherbrooke, C. S. Clark and Company built Canada's largest sawmill to benefit from the railway's proximity to the river.[16] The company acquired the cutting rights of the Upper Saint François area, where during the winter it sent its workers to cut down trees to be carried down the river in springtime. Along with other sawmill owners, it installed slides for floating and driving logs downriver.[17] These installations increased the intensity of flooding events and the resulting damage, if only because ice jams proved more frequent. Furthermore, the destruction of log cribs caused damage—farmers complained about the logs littering their agricultural fields and the destruction of their farm buildings by flooding caused by poor maintenance and management of dams and sluices.[18] By transforming the river and the forests, industrialists operating at the confluence of the Saint François River and its tributaries therefore provoked changes throughout the watershed that affected riverine populations removed from the urban centers.

Specific relationships with the river resulted from patterns of land settlement and riverine transformations. In the Saint Maurice valley, as financial resources were devoted to improving the riverine infrastructure to facilitate log driving, the lack of incentives to settle in the valley discouraged the agricultural colonization of the area and moved natural resources toward Trois-Rivières, where wealth became concentrated, along with the urban population of the Saint Maurice valley. In contrast, in the Eastern Townships,

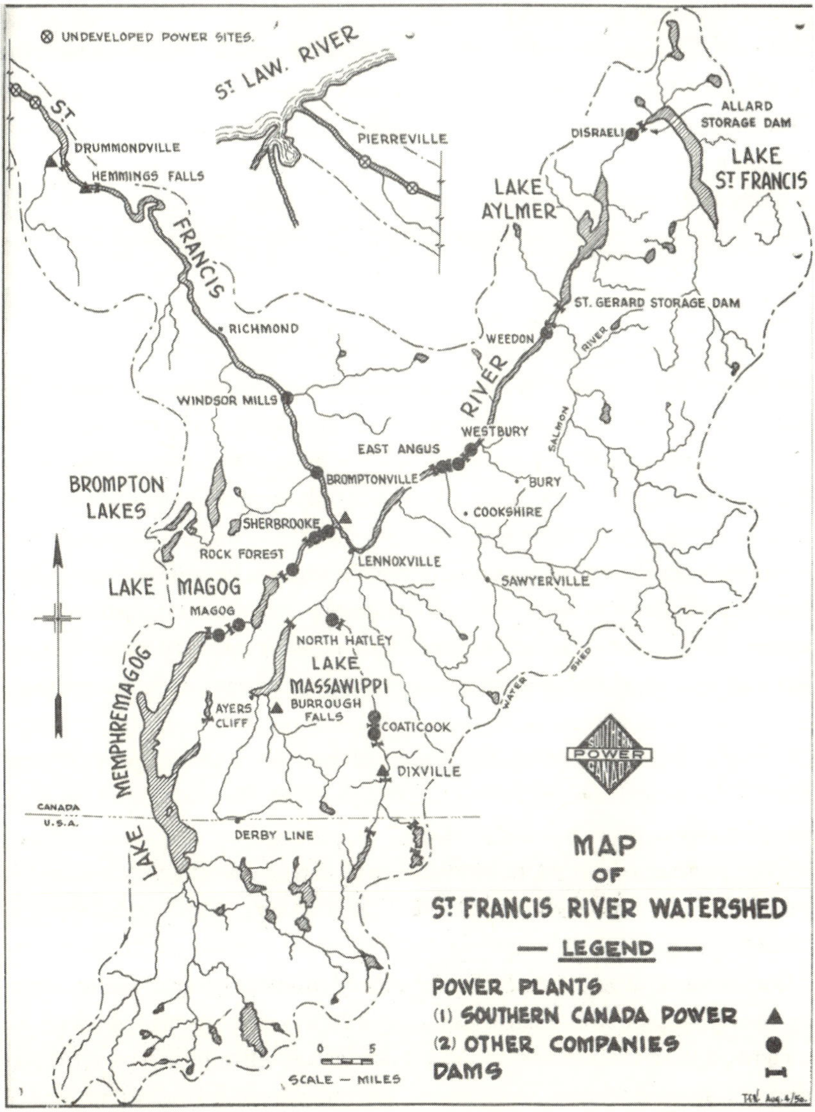

FIGURE 1.4. Hydroelectric Improvements on the Saint François River. Source: Dan Anderson, "St. Francis River Stream Flow Control," vol. 9, *The SCP News*, September 1950, 9.

the much younger city of Sherbrooke welcomed a variety of industries that drew their resources from a diversity of locations, as did other communities similarly located at the confluence of a tributary and the Saint François River. By the beginning of the twentieth century, the urban population of the

Saint François River was slightly higher than that of the Saint Maurice valley (fig. 1.4), although it remained spread out in a series of towns throughout the watershed. Therefore, the spatial dynamics of the urban centers in the Eastern Townships were much more complex with regard to the extension of their hinterland and the social interactions with the riverine environment and its ecological changes.

The Second Industrial Revolution and the Acceleration of Urban and Riverine Transformations

Modification of the Saint Maurice and Saint François Rivers and their watersheds accelerated and reached an unprecedented scale in the first decades of the twentieth century, following the introduction of technologies related to the second industrial revolution. Technological changes in the hydroelectric and pulp and paper industries contributed to the formation of a new riverine environment, not only because of the changes in the aquatic ecology and hydrology but also because of the urban growth spurred by industrial development and manufacturing. They materialized at the headwaters of the Saint Maurice and Saint François Rivers, where a governmental agency supervised the construction of regulation dams and the creation of reservoirs in the late 1910s. They also affected the urbanization of the riverine landscape, the chemical and biological composition of the river, and the river's role in connecting urban societies with the hinterland.

Perspectives on the Saint Maurice River changed at the very end of the nineteenth century, a few years after the construction of a hydroelectric power plant on Niagara Falls in 1893 and, more generally speaking, with the advent of the technical system peculiar to the second industrial revolution. In a span of five years, the site of Shawinigan Falls, a little-known location thirty kilometers north of Trois-Rivières, became one of the launching ramps of the second industrial revolution in Canada.[19] The site was initially acquired in 1887 for the construction of a hotel and the development of a tourist area for the appreciation of its sublime features, but a fire destroyed the venue that same year. In 1899, a Boston banker bought the site of Shawinigan, as it came to be known, to construct a dam and a power plant. To acquire capital to build the hydroelectric power plant, he secured contracts with Pittsburgh Reduction (an aluminum company), Canadian Belgo Pulp and Paper, Shawinigan Carbide, and Montreal Light, Heat and Power Company. Except for the Montreal Light, Heat and Power Company, a utility that supplied an urban market mainly composed of light industries, what these companies had in common was the energy intensiveness of their manufacturing processes: all needed a large, cheap supply of energy for the production of pulp and paper, aluminum, and calcium carbide. Starting in 1899, Shawinigan Water and Power, Pittsburgh Reduction, and Canadian Belgo Pulp and Paper selected a site in the Shawinigan bay to build their plants and tap the hydropower of the Saint Maurice River. The Shawinigan Water and Power Company built intake canals to supply the aluminum and pulp and paper plants with thousands of hydraulic horsepower, while it provided the Montreal Light, Heat and Power Company and Shawinigan Carbide with electricity produced on the site. In the next three decades, and especially during the Great War, new turbines would be installed to supply more energy to these companies, which also enlarged their facilities.

The effects on the Saint Maurice River were manifold, not the least being the outlook on the riverine landscape. Thanks to the production of hydroelectricity, manufacturing plants and urban centers appeared north of Trois-Rivières along the river. The Shawinigan Water and Power Company's plant was not the first one built on the river, but its advent modified the scale of the changes that other companies implemented and fostered the urbanization of the Saint Maurice valley. In Grand-Mère, a small village a few kilometers north of Shawinigan, where the Laurentide Pulp Company had been operating since 1889, a new power plant was constructed in 1915 to support the increased productive capacity of the pulp and paper plant, which had been enlarged over the years. While most of its workers originated from the nearby parish of Sainte Flore, the town of Grand-Mère was created in 1901 to accommodate the growing population and its industry. Some 140 kilometers north of Shawinigan, an American company, the Brown Corporation, formed the Saint Maurice Industrial Company to build a dam and run a pulp and paper mill, eventually leading to the formation of the town of La Tuque in 1908.[20]

Following the creation of these industrial centers, the proportion of urban dwellers to the overall regional population rose above 50 percent in 1921, from 25 percent in 1911; in comparison, urbanization rates for the Eastern Townships were 32 percent in 1891, 46 percent in 1911, and over 50 percent in 1915.[21] A railway finally ran along the Saint Maurice River in 1900 and integrated its population and industry into a larger Canadian network in 1911, further diminishing the hegemony that Trois-Rivières had exercised throughout the watershed. During that period, Trois-Rivières saw the construction of four new pulp and paper plants that benefited from cheap energy and wood driven down the Saint Maurice River, and its harbor recovered from the lumber depression of the 1880s. Yet the Shawinigan Water and Power Company drove most of these changes and dominated—for a time—the urban network of the Saint Maurice valley, especially once it obtained from the provincial government access to the hydroelectric potential of the upper reach of the watershed.[22]

Despite all this power production, the economy of the area still rested on a traditional activity, albeit transformed in line with the technology of a new industrial process: the pulp and paper industry. And for that activity to be successful, immense areas of forest needed to be cut and driven down the river to where pulp and paper mills awaited the gathering of the winter crop. Hence, the first industrial transformation of the river remained embedded in the riverine landscape following the advent of the new industrialism, and slides and gates were incorporated in hydroelectric dams to facilitate the passage of logs. Further negotiations among the different users of the river for hydropower production and log driving led to the construction of dams for the regulation of river flow. Along with the pulp and paper manufacturers, the Shawinigan Water and Power Company formed the Saint Maurice Hydraulic Company to build a series of dams on an important tributary, the Manouane River, to regulate the lower Saint Maurice River. That conglomerate also petitioned the government to regulate the flow of the Saint Maurice River on a larger scale by assigning a governmental agency, the Quebec Water Streams Commission, responsibility for the transformation of the headwaters of the Saint Maurice River, where it built a regulation dam at the Rapid la Loutre, and formed the Gouin Reservoir. Retaining water upstream would facilitate the coordination of the different uses throughout the year, especially during wintertime, when hydraulic energy for power generation was at its lowest level.[23]

Cooperation along the Saint Maurice River was achieved with more success than on the south side of the Saint Lawrence, where some users attempted to assert their primacy over others. True, the Saint François possessed a smaller flow than the Saint Maurice River, but the political power dynamics differed markedly, since the settlement process and its precocious urban materialization had transformed the riverine landscape, and a larger number of actors—from the industrial, agricultural, civic, and municipal sectors—battled to determine priority of use of the Saint François River. Furthermore, the more established urban settings increased their reliance on certain river resources, be it potable water or hydroelectric power. Finally, the second industrial revolution also had an impact on the riverine landscape of the Saint François valley. First, the advent of the pulp and paper industry further stimulated the industrialization of a river already modified by the infrastructure sustaining the lumber economy. Because this industry relied on a steady supply of a high quantity of energy for its paper mills, large dams replaced the small sawmill dams of the first industrial wave, necessitating larger investments and faster economic returns. Second, the privately owned Southern Canada Power Company competed with a flurry of municipally owned public utility companies to supply an area whose urbanization was accelerating with burgeoning manufacturers, principally in the textile sector, that acquired sites on the Saint François River to build hydroelectric dams and power plants. Given the shallow depth of the Saint François and the sometimes dry conditions that prevailed in summer, pulp and paper companies, as well as municipalities and private power companies, sometimes faced droughts that forced their power plants to close and slowed industrial activities.[24] Starting in 1914, the provincial government initiated the building of storage reservoirs at Lake Saint François and Lake Aylmer to ensure a constant flow and a consequently regular supply for hydroelectricity production.[25]

As a large regulation reservoir was built at their headwater, both drainage basins therefore encountered a similar fate. However, in the case of the Saint François River, the provincial government agency responsible for launching these projects, the Quebec Water Streams Commission, insisted specifically on the need for a regular flow to prevent power breaks, as well as to protect the riverine populations and urban infrastructures from flooding events. These goals coalesced in the conservationist rhetoric of the Quebec Water Streams Commission, wherein regulation dams would enable the conservation of water otherwise lost during the inundations. The governmental agency subsequently supervised the construction of the Allard Dam to close Lake Saint François and, insofar as it was possible, provide certain control over water release to regulate flow, to the benefit of industrialists and municipal power companies that bore the costs of drought and low water but also to the benefit of municipal governments and riverine populations that bore the costs of inundations.[26]

The two projects, Allard Dam and Gouin Reservoir, provoked environmental changes throughout both watersheds, and most notably at the headwater, with the inundation of the surrounding forest area to create the reservoir. Farmers around Lake Saint François complained about the flooding of their fields with the building of a reservoir dam. Government authorities argued that the collective benefits outweighed individual interests, and proceeded to build the dam with the strong support of regional political and economic elites. While Lake Saint François simply enlarged its circumference to meet the requirements of a regulating reservoir, the formation of the Gouin Reservoir turned out to be a rather destructive enterprise in ecological perspective, as tributaries and smaller lakes were assembled into one large water mass. The consequences of flooding on the fauna were manifold, as beavers and muskrats drowned and waterfowl and moose lost their feeding areas and nesting habitats. These immediate effects were followed by the mass migration of animals and the depletion of the hunting and trapping grounds of the Attikameks, the First Nations people who had established their reservation at the headwater, and whose fishing and hunting grounds were inundated, along with their trading routes.[27] For the construction of the Allard Dam at the headwaters of the Saint François River, only a few farmers around Lake Saint François were expropriated, with financial compensations.[28]

The regulation dams contributed to the industrialization of the Saint Maurice and Saint François Rivers, integrating them into sociotechnical systems for the production of goods and services and expanding the outreach of their metropolises, albeit to different extents. Quebec Premier Louis-Alexandre Taschereau certainly perceived their role in the provincial economy, as he claimed that both dams enabled the government to augment the power upon which the industrial development of the province rested.[29] The Gouin Reservoir further encouraged the Shawinigan Water and Power Company, which went on to build other dams along the Saint Maurice River, with power plants at La Gabelle in 1924, Rapide Blancs in 1934, La Tuque in 1940, La Trenche in 1950, and Rapide sans Nom in 1958. Along the Saint François River, the private Southern Canada Power Company, which had absorbed all but one of the municipal power companies, and the public utility company of the City of Sherbrooke acquired different sites upstream, at Weedon (1921) and Westbury (1928), where they built dams and power plants. However, other sites obtained downstream through emphyteutic leases between the government and private power corporations, such as Ulverton and Two-Miles Falls, were never developed because of opposition by the local population, given their experience with floods and their disappointment with the ineffectiveness of the Allard Dam in that regard.[30] The situation eventually forced these two power companies to connect their network to that of the Shawinigan Water and Power Company on the North Shore.[31]

The Effluents of Progress: Pollution, Nuisance, and the Rehabilitation of Riverine Environments

The construction of hydroelectric dams and power plants stimulated urban and industrial development and modified the relationships that urban societies had with the river. Urbanization created new energy and water requirements, and rivers were integrated into technical systems for large-scale hydropower production and the disposal of greater quantities of industrial and urban waste. New urban centers emerged in the Saint Maurice valley, and the population welcomed the industrialization of their river, which, according to the regional elite, transformed their region into the equivalent of the Merrimack valley, the cradle of American industrialization, or the Ruhr area because of the intensive industrial and economic activities burgeoning along the water course, whatever its ecological impacts.[32] In the Eastern Townships, older urban centers and smaller villages along the Saint François River were enlarged, requiring the construction of reservoirs and hydroelectric dams upstream to meet the increasing domestic and industrial energy demand. Urban populations expected local benefits from riverine transformations afar, yet they proved insensible to the polluted effluent, especially when it drifted toward other municipalities. Riverine transformation provoked an estrangement from a river that ceased to be the receptacle of nonproductive activities.[33]

Because of their respective industrialization and urbanization processes, populations along the Saint Maurice and Saint François Rivers had different experiences of riverine transformations arising from the fouling of rivers by industrial activities and urban growth. Perceptions were uneven across the two watersheds because of the extent and intensity of ecological damage that differing densities of urban population provoked. Furthermore, the prehistory of leisure and nautical activities and the position of rivers in regional imaginaries and imageries also affected the perception of nonproductive uses of the river. For example, the pulp and paper industry contributed solidly to tainting the water of both rivers with chemical compounds, decaying wood, and polluting effluent. Yet the Saint Maurice was one of the last rivers in Quebec to be decontaminated, more than thirty years after controversies surrounding the quality of water had erupted around the Saint François River, leading to a series of institutional innovations at the provincial level to tackle the problem.

In the years following the end of World War II, the Saint François River came under public scrutiny, for its riverine populations sought to enjoy the benefits of nautical and fishing activities, but the recreational use of the river was prevented by the fouled condition of the water stream from domestic and industrial effluents. Controversy erupted over limiting the amount of waste discharged into the Saint François River and ensuring the provision of clean water to municipal aqueducts. As water pollution became a prominent social issue along the Saint François River from the pressure of local constituents and citizens, municipalities were pitted against each other over effluent decontamination, while riverine associations and public bodies pushed for private corporations and manufacturers to implement or improve their water treatment facilities. Fish and game clubs and civic associations in the Eastern Townships urged municipal and provincial governments to clean and rehabilitate the Saint François River, and pushed for decreasing the centrality of productive uses in defining the urban relationships to the riverine environment.[34] Spurred by organizations promoting outdoor recreation and local tourism as an economic activity to overcome the industrial decline that had set in during the 1950s, industries and local governments encouraged this transformation of the riverine environment by publicizing recreation and tourism activities based on river amenities and infrastructure development toward that goal.[35]

The Saint François River, then, was the site of numerous mobilizations that eventually broadened the debate beyond the issue of water pollution, some tackling general goals such as the reforestation of the watershed, others leading to the ecological restoration of specific areas of the watershed.[36] The Saint François became the first river in the province to be thoroughly studied by government scientists for its ecological conditions when the Biological Office of Quebec—a provincial agency run by fish scientists that had previously been dedicated to pisciculture—started investigating water pollution problems and the state of fish populations from the late 1940s on. This also drove local populations to demand greater involvement from public bodies to decontaminate the river and build water treatment facilities. This process turned out to be long and tortuous, lasting more than three decades before encouraging signs were seen on the river and fishing and swimming were once again practiced throughout the watershed.

Downstream municipalities were specifically challenged by upstream municipalities that sat at the confluence of the Saint François River and one of its many tributaries and discharged their domestic effluent into the main stream.[37] Urban centers like Sherbrooke turned to a tributary, the Magog River, for a sanitized waterfront and an enjoyable urban park. Consequently, the Saint François River was left on its own, and urban populations downstream lived with the accumulating foul matter and odors.[38] For their part, manufacturers claimed that in the absence of decisive action to prevent water pollution from domestic effluent, they lacked any incentives to process their own discharge. By the early 1980s, many municipalities had rehabilitated a tributary of the Saint François where their urban population could enjoy green spaces and beaches and pursue leisure activities. Only a decade later, with the financial and technical support of the provincial environmental department and the mobilization of local and regional environmental groups and civic associations, did municipalities acquire their own water treatment facilities, and the Eastern Townships develop a recreation and tourism industry around the amenities of the Saint François River.

In a context of quasi-monoindustrialism, the Saint Maurice River could not garner enough supporters to address pollution problems until the late 1980s. Given its location at the bottom end of the stream, the city of Trois-Rivières was especially hard hit, receiving all the waste discharged along the course, on top of its very own domestic effluent. Moreover, the Saint Maurice River still reached out toward an extended supply of timber that floated throughout the watershed and contributed to saturating the river with suspended solids. Nonproductive activities took place, but were confined to certain times and places. For example, from 1934 on, an international canoeing competition was held between La Tuque and Trois-Rivières and, along with motorboating on certain summer weekends, enjoyed the support of the Saint Maurice River Boom and Drive Company, which ceased its activities for brief periods of time. Fishing in certain areas was tolerated, but swimming was prohibited, especially in front of Trois-Rivières, where vacationers flocked to two islands. Municipalities on the Saint Maurice River also encountered sanitary problems related to drinking water, but their smaller numbers, lesser populations, and access to tributaries and aquifers made the issue less dramatic; while municipal governments on the Saint François River accused each other of causing water pollution, the smaller number of riverine communities in the Saint Maurice valley made it less pressing to call upon a municipal neighbor to cease the disposal of untreated domestic waste.

Scattered mobilizations around the Saint Maurice River delayed any real interventions for its decontamination. Ecological surveys started systematically in the 1980s—some thirty years later than on the south shore—and soon revealed the impoverishment of fish populations in terms of diversity and quality—not to say toxicity.[39] The high volume of water flowing down the river did not wash away the compounds that pulp and paper mills released, and organic matter from decaying wood logs deprived aquatic life of oxygen. Despite the introduction of clean technologies following provincewide debates related to this polluting industry in the 1980s, smaller volumes and lower toxicity of industrial effluent did not prevent the occasional accident, such as the accidental spill of PCBs.[40] Furthermore, the Saint Maurice River Boom and Drive Company still drove more than one million cubic meters of logs on the river annually during this decade, ceasing its activities only in 1995. While log driving challenged nautical activities because of the risk of a free-floating log hitting a boat, it was mainly the chemical processes of the pulp and paper industry and, to a lesser extent, those of the aluminum industry that prevented recreational use of the river, as they inhibited aquatic life from thriving in the Saint Maurice River.[41]

What sheltered the Saint Maurice River from public scrutiny and environmental concerns? Why was the population of the Saint Maurice River valley prepared to keep its distance from the river that gave the region its name and identity? A large pool of lakes and tributaries provided an alternative for the population to engage in leisure activities. Moreover, the river had played an important role in the formation of the regional identity, and because of the economic activities that it supported, it was appreciated and praised from an industrial perspective—quite like the "chimney of progress."[42] As the pulp and paper industry declined because of international competition and other structural factors, questioning its role and that of log floating in the destruction of aquatic habitats and recreational life would have dealt a final blow to a dwindling economy.[43] The regional population likely drew their sense of belonging more from the region's forests and its river, as witnessed today by the plethora of local museums dedicated to raftsmen, lumbermen, and the pulp and paper industry. Furthermore, the region's name—*La Mauricie*—itself was of recent use (1930s) and derived from the name of the river; it encompassed the economic region whose industrial progress resulted from the transformation of the river. To contest its polluted state was to compromise the economic basis of the people and their identity. Only once the decline of the pulp and paper industry appeared irrevocable were protests against environmental abuses possible. The local population was then in a position to afford another view of the river that, for the past century, had been cast as a provider of industrial growth and prosperity thanks to the hydropower and the wood that it supplied to the pulp and paper mills downstream. Furthermore, not only was the well-being of the riverine population less conditional on the viability of disappearing industries but it appeared that its economic survival rested on the cleaning up of the river and the cessation of polluting activities. Yet late into the 1960s the recreational potential of the Saint Maurice valley was barely acknowledged in that regard, except for private fishing clubs that benefited from emphyteutic leases on Crown lands and a few modest summer cottages along the banks of the river.[44] Provincial and federal governments were in the process of creating national parks, but La Mauricie remained "inadequately equipped to receive large numbers of tourists," partly because of the large timber concessions granted to forest companies. According to one geographical study of that period, what was basically lacking was improved road infrastructure, "the opening to the public of large attractive lakes," and greater involvement by governmental agencies in promoting local tourism. This was what the Eastern Townships had been improving and benefiting from for decades. Equidistant locations from Montreal—the largest urban center in the province of Quebec—did not render the Eastern Townships and La Mauricie equally attractive for tourism. Riverine populations had nurtured specific attachments to local environmental amenities, and that affected their willingness to gain control over the river and the forms of rehabilitation that took place.

While the promotion of recreational activities for tourist industry development in the Saint Maurice valley was still absent in the 1970s, a few individuals and organizations pursued boating and fishing activities, and there were yacht clubs in several locations, such as Cap-de-la-Madeleine, Grand-Mère, Saint Jean des Piles, and La Tuque. It was around that time that these users, along with environmentalists, moved to the forefront of various legal and political battles that eventually brought an end to the degradation of water quality and the riverine environment. In 1991, the Saint Maurice River Basin Management and Development Corporation—a local body set up by the civil society—lobbied for an end to log floating and the cleaning up of the water.[45] Corrective efforts gained pace as industrial decline accelerated. Pulp and paper mills were being modernized, and while industrial effluents were cleaner, the number of employees decreased as labor-saving technologies were introduced. In certain industrial sectors, such as the chemical industry, factories had been closing down or moving away since the 1960s. By the early 1990s, unemployment reached 25 percent in some areas. Industrial development did not enjoy the preeminence it had once had in the Saint Maurice valley to legitimate the industrial state of the river. In fact, the general economic devastation just staved off opposition to wood floating on the basis of the necessary industrial development. The Saint Maurice was the last river in Quebec in which log floating was practiced, and it ended in 1995. Thereafter, the Saint Maurice Boom and Driving Company cleaned the riverside and riverbed of sunken logs whose decay tainted the water with phenolic acids and suspended solids; the cleaning of the riverside between Rivière aux Rats and Grand-Mère and in the La Tuque area was completed over the two years that followed.

On the two rivers, then, urban populations dealt differently with the infrastructural and ecological changes that had obliterated certain potential uses over the years. The area's industrial past framed some of the postindustrial uses, while at the same time the riverine society projected certain experiences that it had selected to define the heritage activities that it wished to promote. The rehabilitation and type of recreational uses of each river differed according to the type of population who accessed the riverine environment and sustained specific ecologies and ecological requirements—motorboating being of importance on the Saint Maurice River, while the Saint François River enjoyed a long tradition of sport fishing.

Riverine transformations were part of an urbanization process that extended the sphere of influence and the space of power of regional poles. The production of a hinterland by a fluvial route resulted both from the metabolic needs of urban industries and populations that consumed energy and materials and discharged waste, and from the physiography of the watershed and the resources it possessed. This comparative case study reveals how decisive the process of settlement and colonization that framed the urban and industrial history of these two rivers was in defining the extension potential of the urban space.

Under specific economic and technological contexts, at different periods, the formation of the urban space of Trois-Rivières and Sherbrooke framed the industrial transformations and the rehabilitation of the Saint Maurice and Saint François Rivers. Several factors influenced the relationships between society and river throughout the two watersheds at different stages: agricultural settlement in the Saint Maurice valley was not as intensive as it was in the Eastern Townships; frontier resources, while less readily available, were in large supply over an extended area throughout the watershed of the Saint Maurice River; and the urban and industrial transformations of the Saint Maurice River were largely influenced by the second industrial revolution, and were much less structured than in the Eastern Townships by the civic requirements for public services of an extended urbanization process in terms of energy requirements and water quality.

This chapter underscores the difficulties surrounding the use of concepts such as hinterland and metropolis, although the latter remains useful to capture the dynamics surrounding the formation of a region. Among the difficulties encountered when studying the relationships between a river and a city throughout a watershed is that it appears that at different times, more than one city can claim the title of the metropolis expanding its hinterland or integrating other industrial or resource centers into its urban space. For instance, in the first half of the twentieth century, Shawinigan occupied a much more strategic place in the formation of the Mauricie hinterland than did Trois-Rivières, which had structured the whole area for centuries, and which regained its central role in the development of the region in the last decades of the twentieth century. At stake here is not so much determining the regional capital of a watershed as identifying where the transformations of the river originate, and who is being served and disserved by a transformation such as constructing a regulation reservoir at the headwaters or polluting the water with domestic effluents. In that regard, the situation on the south shore was more complex, if only because a more extensive settlement process gave rise to a multiplicity of urban centers. Furthermore, the Saint Maurice valley became part of the hinterland of different towns on the south shore, as industrial and urban development along the Saint François River relied on the importation of electricity and wood from the north shore in the second half of the twentieth century. Throughout the period under study, Sherbrooke emerged as the principal urban center along the Saint François River, not so much because of its local use of the river where it crossed the city—except for discharging its effluent—as because of the energy it produced on different sites along the water course. Its tributary—the Magog River—turned out to be more useful for providing freshwater and wood, as well as hydraulic energy that was initially tapped by dams and mills and enabled the growth of its industrial basis. If a local political and economic elite from Sherbrooke managed to expand the space of the city's power throughout the watershed, it was not strictly from the hydrologic benefits of the proximity of the Saint François River but from a series of factors such as the branching of tributaries, a diversified urban network, and a concomitant railway. The Saint François River made these factors, as well as the expansion of the urban space of Sherbrooke, possible.

CHAPTER 2

THE SEINE AS A PARISIAN RIVER

Its Imprint, Its Ascendancy, and Its Mutual Dependencies in the Eighteenth through the Twentieth Century

Sabine Barles

The histories of Paris and the Seine, the river that cuts across it, are intimately linked. Historiography provides much evidence of their relationship, but it mainly stresses what Paris owes to its river, the modifications in the Seine's course through the city, and the significance of river transport.[1] Research into what the Seine owes to Paris, outside the limits of the capital, is much scarcer, apart from certain historical works on technology.[2] But the needs of Paris have been the cause of many transformations of the river and its tributaries, leading to actual forms of urban dependence. These also include institutional domination, in terms of the control and authority exercised by the Parisian administration over a considerable territory. They complement the powers of central government (the French State), and are manifestations not just of an environmental imprint but also of external influences of Paris. These imprints, this influence, the dependencies and related issues, are analyzed here, by examining the way in which Paris has been supplied with wood and water, as well as its more recent protection against flooding.[3] Lastly, it will be shown how Paris has appropriated not only mobile resources but also land in varying ways from one century to another, yet always leaving a strong, large, and persistent imprint on the river basin, so that the river and its tributaries are (almost everywhere) "urban" in a functional sense. The conclusion will summarize this analysis through key notions that can help describe city-river relationships.

The dependence of Paris on the Seine in getting its supplies expanded throughout the ancien régime. Following the transport of cereals, wood floating developed as of the sixteenth century. Marie-Hélène Bourquin-Simonin has shown that floating wood already occurred in Normandy on the Andelle in 1490.[4] It was only in the 1540s, however, that it really emerged as a means of supplementing "wood in Paris's depleted supply areas with wood from distant forests."[5] These processes subsequently spread, so that Paris was supplied with wood from the Yonne, Marne, and Oise, as it was "much recognized how such shipment by rafting was beneficial to the supplies of Paris and to trade in distant provinces."[6] The Yonne saw its importance rise, providing two-thirds of the wood consumed in Paris.[7] Accordingly, Vauban[8] could write in 1698, "It is still one of the 'nursing mothers' of Paris, as it brings all the wine from Burgundy, floated wood from Morvan, and much wheat and oats."[9]

Timber rafting operations, as well as the preceding and succeeding operations, are very complex, lasting two to three years. They require many infrastructural modifications very high upstream, as log drifting requires a sufficiently strong current that small rivers cannot provide. By adapting ponds to release water at specified points of time, it is possible to create enough of a current to get logs moving: "thanks to the help of these ponds . . . to the help from such artificial sources . . . it is possible to bring wood down from distant mountains a hundred leagues [one league is approximately four kilometers] from the capital, and . . . produce 200,000 *voyes* [one *voye* is approximately two cubic meters] in Paris."[10] Canals, also called "artificial streams," were also created to shape water courses and facilitate the movement of logs.[11] Paul Benoit and his team, who studied upstream developments in particular, thus identified ninety-nine ponds on the map of the Abbot Delagrive published between 1732 and 1737, and noted that their number increased further over the eighteenth century.[12] A four-foot-wide path runs along the river to allow workers to follow the logs.[13] Further downstream, landing stages (made of piles and poles) blocked logs at the confluence of streams and rivers in order to "prevent wood from going down the large river."[14] Sluices and dams were constructed to regulate water levels and facilitate the circulation of wood (or boats). In Paris itself, infrastructure included special construction sites and wood-carrying boats in the ports. Timber rafting thus led to landscaping with some permanent, temporary, and seasonal features, whose emblem was the

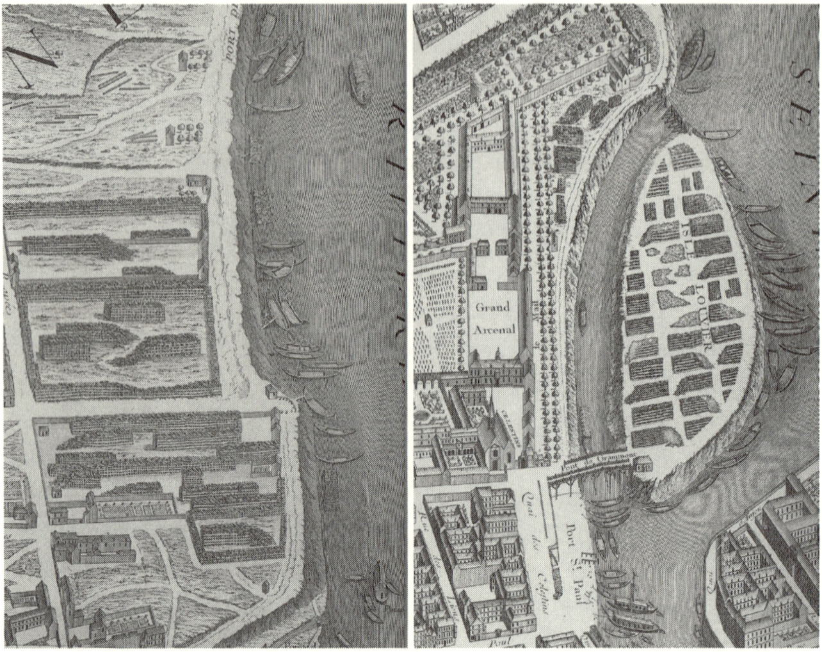

FIGURE 2.1. Piles of logs upstream from Paris (left) and on Isle Louvier (right). Turgot's map, 1734–1736.

segment and line (logs and their dark color). This must be imagined as an animated, noisy landscape that characterized much of the basin to serve Paris. There is therefore something clearly "urban" about the basin. But it was made even more urban due to power relationships and underlying dependency.

Historiography has focused on the role of the Royal Order of Water and Forests of 1669 (l'Ordonnance royale des Eaux et Forêts), which gave the State wide-ranging powers not only over forests but also over navigable and rafting waterways. "Rivers and streams capable of use by boats, [whose] beds are not worked" also became Crown lands, "except concerning fishing, mills, water-tanks, and other uses private persons may make of them through titles and valid possessions."[15] For navigable and rafting waterways, the flow of water was given the absolute priority, so that no one could "build mills, dams, locks, stake nets, sluices, walls, trees, piles of stones, earth and fascines, nor other edifices or hindrances to the flow of water."[16] Nor was anyone allowed to "divert water from navigable and rafting waterways."[17] Also included were tow easements—twenty-four feet wide for roads and thirty feet with no planting or fencing on one bank of the river, and ten feet wide on the other.[18]

This Order of 1669 was both unifying and foundational. It reflected the control the State assumed over forests and waterways, through a specific government body, the Administration des Eaux et Forêts. However, local conditions could differ from the overall model, as the Order itself envisaged when it specified that the jurisdiction over waters and forests related to all navigable and rafting waterways, though "without prejudice to the jurisdiction of provosts of merchants in towns in which they may have knowledge of all or some of these matters."[19] This was especially the case in Paris. Daniel Jousse added that the Order for the City of Paris of December 1672 (Ordonnance concernant la juridiction de l'Hôtel de Ville)[20] attributed to the Provost of Merchants the responsibility for contracts undertaken to supply the city with wood and that responsibility of the City of Paris was restated by the Parliamentary Ruling of December 31, 1760.[21]

This interpretation gave Paris an extraterritorial role, but it was quite controversial.[22] Nevertheless, it was confirmed by the reading of the 1672 Order mentioned by Jousse. The issue of wood supplies and more generally the provision of Paris was dealt with at length, and two key factors stand out: 1) the control of the city over the wood trade, from the very first transaction (the purchase of cut wood from wood merchants) to the very last (the sale of logs and fagots to Parisians); and 2) control over the whole of the Seine and its tributaries in as far as they contributed to supplying Paris. Thus, the 1672 Order not only complemented the 1669 Order but actually substituted or superimposed itself on the latter in certain areas, so that for certain matters Paris gained precedence over the State, especially concerning the river.

Relating to navigation and rafting, the 1672 text and especially its first chapter about "navigation to supply Paris," was not very different in its demands from the 1669 text.[23] Yet it was more precise, explicitly putting the river under the jurisdiction of City Hall: "And for the full execution of the above, the maintenance of commercial freedom and ease of navigation, provosts of merchants and deputy-mayors should visit the Marne, Yonne, Oise, Loing, and Seine Rivers, as well as other navigable and rafting waterways . . . while the orders of these provosts of merchants and deputy-mayors and the rulings they make should be executed by provision, as is done with the police."[24] The execution of "all aspects relating to the supply of the city of Paris by water" was placed under the supervision of a general commissioner of navigation and his inspectors, "who are employed by the city of Paris and nominated by the Provost of Merchants."[25] Chapter 17 of the Order relates to "merchandise of new wood, driftwood, and wood for construction."[26] Among other things, it was stipulated that "all merchants [were forbidden] to short-circuit goods destined to supply Paris, and to buy them *en route*."[27] As a result, the assignment of any logs to Paris was irrevocable, clearly demonstrating the intimate link between Paris and its feeder forests.

It is easy to see here the control Paris could exercise over its rivers: it extended to all waterways of the Seine basin likely to supply the city either for navigation or rafting. This meant that the control or *institutional imprint* could vary over time, according to needs and developments. In other words, opening up a river for rafting meant handing it, and adjacent forests, over to the care of Paris. This question was debated on the eve of the Revolution concerning the river Aube, whose "two banks were lined with considerable forests all the way upstream to its source, and which at present contribute little to the supply of Paris,"[28] but whose development was strongly opposed by local ironmasters. They were aware that they would immediately lose precious fuel, which would enter the *arrondissement* of Paris, the expression widely used in the texts studied here.

The term *arrondissement* usually refers to one of Paris's twenty districts. But it originally comes from the word *arrondir*, meaning to round off or to round out. According to Antoine Furetière, "it was also common to talk of a Prince rounding out his crown when taking control of small state enclaves in his lands."[29] The same was true for the word *arrondissement*, as shown in the following usage: "To round out (*faire l'arrondissement*) your Lordship's estate, this inherited enclave must be bought."[30] Thereafter, all dictionaries quoted these two usages. "To round out" thus became a process for enlarging, enriching, and homogenizing property or land. Legal dictionaries used the term little, apart from the dictionary published by Prost de Royer between 1781 and 1788, according to which "the word is given to all tracts of land made up of villages, hamlets, gaps, etc. surrounding a main town with a judge, notary, trustee, controller, tax collector, etc., whose functions extend to all parts of this circular space."[31] The Paris *arrondissement* would thus include its feeder lands and lands over which it had power. Indeed, it was the wood crisis that gave birth to a precise definition in the ruling by the Court of Parliaments of December 30, 1785, confirming an Order by the Provost of Merchants dated December 1. The Court "order[ed] the Officers of specific authority of the Waters and Forests administration in the *arrondissement to record woods within six leagues of the Seine and its tributaries* in their registers of all wood intended to supply Paris; reserving a share they deem necessary to provide heating for inhabitants in each canton. Specific measures will be taken for such reserves, which are not expressly intended for use by Paris."[32]

Water Supplies and Water Sources

The profound transformations that affected supplies in the nineteenth century—the energy transition toward fossil fuels and of course political changes—somewhat distanced Paris from its river. Or, more specifically, it limited the actions of Paris's administration outside its borders. The closure of the Office of the City (Bureau de la Ville) in 1789 and the following institutional reorganizations led to the transfer of the management of supplies to Paris by water to the Ministry of Interior. This paid for staff, who were then paid by the City as of *l'an XI* (September 23, 1802, to September 23, 1803, of the Republican calendar). This situation was the source of a long-running conflict between the State and the City concerning the external administration of navigation (in contrast to the internal administration, i.e., within the Seine *département*, which covers Paris and its immediate suburbs). From the beginning, the City wanted "management of the administration to be carried out by the Paris municipality."[33] But this was not the case, and the standoff concerning the payment of inspectors' salaries by the City of Paris lasted at least until 1833.

However, as the century wore on, Paris progressively relinquished control over river transport to the State, increasingly looking to the river as a source of water and means of sanitation instead—that is, its water content, rather than the riverbed and banks, became most important. Even during the ancien régime, supporters of using the Seine's waters opposed those arguing for the use of springwater to supply the capital. Various diversion projects were proposed, though none was carried out. As Frédéric Graber has shown, these projects led to protests by people living on the banks of the various rivers in the basin (especially the Bièvre), which ultimately led to the projects being scrapped after the Revolution in the name of private property.[34] Under the First Empire (1804–1814), when "washing the streets was, in the eyes of all, the main question, eclipsing all others," in terms of water supply, all hopes were directed at the Ourcq canal.[35] This met with local opposition, but according to Graber, archives suggest that this protest received little support, and did not lead to the project being abandoned.[36] Paris thus appropriated for itself, in all senses of the term, a water resource and waterway network, which it shaped for its own needs. In the words of the decree of August 4, 1864: "the Prefect of Paris has the right to order works, aimed at ensuring the supply of the city of

Paris, on the bed and banks of the Ourcq canal while the channeled section of the river Ourcq should be seen as part of an integrated whole."[37]

This appropriation culminated during the Second Empire (1852–1870), when "it was felt that considerable improvements needed to be undertaken, and the idea emerged of distributing water to homes on a large scale, as in London."[38] Quite rapidly, Baron Haussmann (the prefect of Paris) and Eugène Belgrand (the engineer in charge of Paris's water supply) decided in favor of springwater, leading to a controversy over the relevance of this choice.[39] From a technical point of view, supporters of using water from the Seine argued that a modern solution lay in powerful machines to raise water, instead of relying on somewhat archaic aqueducts, which were also associated with diversions from springs: "This is a primitive way of moving water from one point to another: don't we have steam power now? Don't we have machines?"[40] Concerning quality, they argued that river water could be filtered. Lastly, they argued that it would cost less to draw water from the Seine. In contrast, those opposed to using water from the Seine primarily focused on water quality, so that even Belgrand stated that "water from diversion flows had to be taken from springs," where it is "pure, clear and fresh."[41]

Apart from such technical arguments, water diversion projects led to several legal battles relating to the appropriation of water, which are studied here in detail. In 1854 Belgrand carried out an inventory of "springwater sources capable of supplying Paris at 70 meters altitude." Among the numerous possibilities provided by the water basin, he identified those of the Somme-Soude, a tributary of the Marne in Champagne Pouilleuse (with a diversion of 214 kilometers).[42] However, "in the *département* of the Marne which covers the Somme-Soude valley, a veritable coalition has grown up opposing any kind of diversion. Mr. Belgrand showed without success, on several occasions, the safety of drainage ditches, which were to receive only overabundant water. Such water was lost at that point in subsoil strata which supplied the sources of the Somme and the Soude."[43] Auguste Mathieu led the opposition. He was a deputy of Épernay and member of the parliamentary majority, a well-known lawyer, who was alarmed about the consequences of a project that would affect fifty-four communes, nearly twenty thousand inhabitants, nearly forty thousand hectares of grassland, and sixty-five industrial establishments. He envisaged a desolated landscape as the result of the diversion project:

> The day water is diverted from Somme-Soude for the benefit of the city of Paris, solitude and desolation will spread by the force of circumstances. Deprived of

> water, grasslands will disappear, as will cattle, and with cattle the manure it provides. Day by day the earth will return to its primitive aridity, and the discouraged population will flee from this barren land which labor and sweat will henceforth be unable to render fertile. . . . And our whole territory will have been expropriated from its inhabitants without any possible compensation! Are such consequences, which in reality are so fatal, possible under the law?[44]

Not surprisingly, his answer to this rhetorical question was negative, and based on two major arguments. The first relates to the status of running water, "one of the things which in the negative community of humankind remains *res communes*." Given this status, water cannot be expropriated because it belongs to no one. Expropriation by the city of Paris would mean that the status of running water would be changed from common to private: "This would mean exclusively attributing to the needs of the inhabitants of Paris, something which in its nature belongs to nobody and whose use is common to all." The other argument related to the extraterritorial nature of the expropriation, since "the water courses in question [did] not flow in the lands of the City of Paris, nor those of the Seine *département*." All these reasons "prohibit the City of Paris from taking control of water courses in Champagne."[45]

Mathieu's case got a hearing both with various ministers and with the emperor. His lobbying against the project was so extensive that Haussmann reported, "His Majesty found himself subject, during his visits to the Camp of Châlons, [to] incessant demands about the matter and ended up saying to me, 'Why do you insist so much on taking the water from these people when, according to studies whose results you have submitted to me, there are other equally good sources?'"[46]

Once the Somme-Soude project had been abandoned, Paris turned to Dhuis (or Dhuys), a small but impetuous tributary of the Surmelin, which is itself a tributary of the Marne. Haussmann got the City of Paris to buy the headwaters of the Dhuis in July 1859 for sixty thousand francs.[47] In 1862 (March 4) the public utility of building an aqueduct was decreed,[48] which led to a new debate that mainly took place in the Senate and was reported in the daily newspaper *Le Temps*. A petition signed by local inhabitants, people living in the valley, and local mayors challenged the constitutionality of the decree. The case was also defended in the Senate by two lawyers: Denis Victor Tourangin and Adolphe Le Roy de Saint-Arnaud. They more or less repeated the arguments of Mathieu, pointing out to the representatives of the State: "Do you find precedents elsewhere? But you have only undertaken diversions in the open and without ever leaving the locality from which you partially divert water."[49] Again the argument of extraterritoriality was put forward as it related to common goods: "This is a kind of common good, which cannot be taken away from those it belongs to," who numbered hundreds or even thousands and included not just those living on the banks of the river but people living in villages—"but what caused the creation of these villages? The river of course."[50] From this argument it was concluded, "if you take something away which belongs to everybody, then everybody has to be compensated."[51] Two factors were put forward. First was the fact that "the City does not have different rights from those of the seller."[52] Being a private agent, the seller "could not divert water flowing from its source," so the City, buying from a private agent, was in the same situation.[53] Second was the fact that "a town cannot expropriate something which it already has in analogous form": i.e., Paris had water from the Seine.[54]

Apart from Le Roy de Saint-Arnaud, the commission nominated to examine the issue, ruled in favor of the constitutionality of the decree, arguing that it did not relate to a diversion of the Dhuis, as the decree only concerned the acquisition of land to build the 131-kilometer aqueduct. The commission added that if local inhabitants had rights to assert, they could do so in the courts, to which Le Roy de Saint-Arnaud replied, "Once the City of Paris has committed its 18 million, once work has started, what will happen if the City runs up against an obstacle in the courts? I can clearly see what the government will do: it will divert the Dhuys."[55] Senator Noël Lefebvre-Duruflé, a supporter of the emperor, described the Dhuis as a "water course at the bottom of a wild ravine, which transforms itself into a torrent at certain moments," and which local inhabitants did not use.[56] Far from "sacrificing the countryside in favor of the luxury of the towns," to use Tourangin's words, it was a case of "satisfying hygiene requirements and public health," according to the well-known chemist Jean-Baptiste Dumas, who was also a member of Paris's municipal council.[57] The utility of the project was undeniable, especially given that "all the formalities had been met,"[58] as Pierre Magne (minister without portfolio) recalled, while the inquiry had been conducted according to the rules. But, as some recalled during the discussion, the inquiry had concerned persons to be expropriated, who were owners of land affected by the construction of the aqueduct, and not people living in the Dhuis valley. With weariness, and following a last declaration by Baron Haussmann—for whom a journalist in *Le Temps* proposed for epitaph "he diverted, he diverted, he diverted"—the decree was maintained.[59]

FIGURE 2.2. The Dhuis aqueduct, twenty kilometers upstream from Paris, present situation. Photo by the author, 2012.

Protests picked up again after several years with the "fatal diversion" of the Vanne (173 km),[60] when the city of Paris had actually shown more foresight by buying up not only thirteen springs but also the land, mills, and factories of persons affected, for the sum of 3.2 million francs.[61] This conflict essentially pitted the town of Sens against the city of Paris, right through 1880.[62] However, it did not prevent the declaration of public utility, nor work going ahead. After a period of calm at the start of the Third Republic, new projects started in Normandy this time, with the diversion of the Avre and the Vigne (103 km), covering four springs. The preliminary project was presented in 1885, but work did not start until 1891—"we know how long the parliamentary process can be, what obstacles local oppositions can raise," Vernier wrote in *Le Temps*[63]—following the law of July 5, 1890, that limited the diversion to 1,280 liters per second, and provided for compensation for users.[64] Other sources were also diverted at the start of the twentieth century: the Loing and Lunain (the law of July 21, 1897)[65] and the Voulzie (1926).[66]

FIGURE 2.3. The Vanne aqueduct, south of Paris (Arcueil and Cachan), present situation. Photo by the author, 2012.

This new form of urban extraterritoriality led to the creation of new types of control and influence, both scattered and linear as well as infrastructural. In the latter case, they may be defined as technical urban annexes or "out-infrastructures"—infrastructures located outside the city they serve. The aquatic landscape was profoundly modified (especially as these developments came on top of transformations linked to navigation, which were led by central government), as it covers much more than riverbeds and walls along ever straighter lines: the Ourcq canal and the Dhuis aqueduct still meander to some extent (notably the Dhuis). Subsequent aqueducts were essentially straight. Bends in the water courses were replaced by an increasing number of engineering works, as seen with the Avre.

The Level, Flow, and Dams

Despite these works, the water supply and wastewater treatment system experienced a profound crisis, which began at the end of the nineteenth century and was precipitated by flooding in 1910 and the drought of 1911. Growth of the population of Paris—and of the suburbs, whose projects the capital ignored for a long time—combined with the rise in the unit consumption of water (which was even more dramatic when industry is taken into account) led to increased pressure on the Seine. In the 1920s, 28 m^3/s of water were taken by the conurbation from the river (on top of which came water used by power stations), though the flow of the river at low water was barely 35 m^3/s.[67] The French State refused to allow Paris, which at the time was consuming 15 m^3/s on average, to increase its use.[68] The capital was thus no longer able to ignore the problem by relegating it to being merely a "question for the suburbs." Furthermore, the quality of water continued to deteriorate: greater consumption automatically led to great urban and industrial waste. Given such greater waste, the low flow of the river (due to higher consumption), was no longer able to dilute waste. At the same time, the power stations led to a worrying increase in temperature.[69]

Thus, several projects were relaunched: the diversion of water from the Vals de Loire, Lake Neuchâtel, and Lake Geneva, along with the construction of reservoir-dams upstream from Paris to support low summer water flows of the Seine and to ensure sufficient depth (for navigation) and flow (for supplies). One of the objectives was to dilute the increasing water pollution. This latter problem was explored by Henri Chabal, an engineer from École Centrale: in 1920, he put forward a plan based on creating twenty-three reservoirs in the Seine basin, ranked by three levels of decreasing emergency, with the final level seeking only to protect Paris from floods, as in 1910.[70] This project was reactivated after flooding in 1924, and in 1925 a new commission was entrusted to Inspector Dusuzeau, which confirmed the principle of reservoirs: "These reservoirs are to accumulate winter waters and release them in the summer. They therefore have a twofold function of preventing flooding and improving flows at low water, while incidentally generating electricity."[71] To their supporters, they provided all the necessary advantages: they would bring benefits in moments of high water or low water, allowing the regulation of problems that were both quantitative (excessively high or low water levels and flows) and qualitative (diluting pollution). In addition, their costs were reasonable and they would improve navigation, raise agricultural output, produce energy, and later help develop regional tourism.[72]

A new project, based on the Chabal project, was adopted by the Ministry of Public Works in 1926, having been presented to the General Council of the Seine *département*.[73] In 1928 the Dams and Reservoirs Unit (Section des barrages-réservoirs) was created, being subordinated to the Technical Bureaus of the Paris Port Authority of the Seine Prefecture (Service technique du Port de Paris de la préfecture de la Seine).[74] First, emergency dams were granted public utility status in 1929: Crescent and Chaumençon in Morvan, Champaubert-aux-Bois in Marne, Pannesière-Chaumard (later rechristened "Pannecière") in Yonne. These projects were directed by the Technical Bureaus of the Seine Prefecture, whose prefect rapidly sent requests for subsidies to the *départements* concerned, stressing the interest they had in the project. An amount of 263,000 francs was requested from Marne, for example—equal to 1.5 percent of estimated total spending.[75] This was presented as a modest contribution that "expresses above all the solidarity which it is desirable to create among the beneficiaries of the operation."[76] The reaction of the *départements*, however, was fairly unanimous. Initially, they all expressed considerable reservations and they subsequently turned down requests for subsidies when they were resolicited in 1931. So the dams were finally financed only by the Seine *département* and central government. The reservoirs at Crescent, Chaumençon, and Champaubert-aux-Bois began operating in the 1930s, while the construction of a fourth reservoir at Pannesière-Chaumard started only in 1938. This followed difficult negotiations with local land- and houseowners (seventy-four houses were flooded, and 540 hectares had to be acquired).[77] The project was completed only in 1950, after a long interruption due to the war.

These four dams made it possible to provide twelve cubic meters per second of water, spread out during the four months of low water. It was stressed by the Economic Council in session on December 20, 1955, that while the "flow of the Seine was thus notably reinforced . . . this contribution of 1 Million m^3 per day was still insufficient to ensure the good dilution of water, which became more polluted as the population grew and industries developed."[78] The Economic Council had chosen to examine this issue, adding that the capacity of the reservoirs remained insufficient in terms of protecting against floods and was incapable of restoring to the Seine "a quality which the water course lost during the summer months when it became a sewer at great cost to surrounding inhabitants and to . . . fish."[79] Indeed, the reservoirs stocked only 120 to 140 million cubic meters, whereas Chabal's project had envisaged ten times as much. Shortly after the war, new floods occurred in 1944, 1945, and 1955, the last being followed by a severe summer drought. This put the issue back on the agenda with an overall aim of increasing storage capacity to one billion cubic meters, thanks to the construction of a second generation of works, whose capacity was much greater than their predecessors. For the Technical Bureaus of Paris, the main aim was to "reinforce flows during low water," while combating floods had become a "secondary objective."[80]

Four sites were studied on the Aube, Marne, Seine, and the Serein (a tributary of the Yonne in Morvan). The latter was envisaged as of 1945 by the Technical Bureaus of Paris, but provoked "violent protests" throughout the *département* of Yonne, causing a real "outcry"[81] and the creation of a Defense Committee of the lower Serein valley, so that study of the work was suspended. The reservoir of Marne was approved by the General Council of the Seine *département* in 1951, a project that would lead to the flooding of three villages, which did not prevent the Ministry of Public Works from accepting it in principle, in 1953. However, it did lead to "strong agitation" in the *département* of the Marne and the study of various alternatives.[82] The Seine reservoir was approved by the General Council of the *département* of the Seine in 1954, and was accorded public utility status in 1959.[83]

Renamed the lac d'Orient, situated on the Seine upstream from Troyes, the reservoir was the first to be completed, possibly because it was the least contested.[84] There were only a small number of buildings that had to be expropriated and families that had to be rehoused, though this did not prevent some transactions from being very protracted.[85] The Marne reservoir, which became Der-Chantecoq, absorbed the reservoir at Champaubert-aux-Bois, and raised far more difficulties, drawing the attention of the Economic Council in 1955. The council concluded after examining the issues that "while it is not possible to eliminate the social consequences of developing the reservoir" it is important "to reduce the painful consequences to the population as much as possible." The Economic Council thus proposed reestablishing operating farms by purchasing other farms for sale, or creating and reinstalling farms in "available holdings throughout France, and especially in the South West."[86]

FIGURE 2.4. Orient reservoir, present situation. Photo by the author, 2010.

In any case, at the end of the 1970s the average potential summer water flow had reached sixty-five cubic meters per second spread over four months, and eighty cubic meters per second in 1989.

These new urban dependencies led to both new hybrid landscapes (infrastructures and ecosystems) and new influences by Paris. After the institutional reforms of the 1960s that led to the disappearance of the Seine *département* and to the creation, around Paris, of three new *départements* (Hauts-de-Seine, Seine-Saint-Denis, and Val-de-Marne), they brought about the creation in 1969 of the Interdepartmental Institution of Dams and Rivers in the Seine Basin (Institution interdépartementale des barrages réservoirs du bassin de la Seine, IIBRBS).[87] This authority is managed by four *départements* from the Greater Paris conurbation: Paris and its three surrounding *départements*. Such extraterritoriality is a new proof of the institutional and environmental imprint of Paris, which with later extensions covers more than ten thousand hectares. It includes a number of constructions that are integrated into the extensive hydrographic network that can hardly be called natural.

After having played a major role in shaping, managing, and controlling the Seine and its tributaries in the whole basin up to the end of the eighteenth century, Paris's interest in the river Seine has been concentrated within its administrative boundaries. After a long period of dereliction from the interwar period to the 1970s,[88] the city launched a program dedicated to the recovery of the embankments, and in 1991 they were added to the UNESCO World Heritage list. But this represents only a very small aspect of Paris-river interactions and the main result of this chapter is that these interactions can be traced much more outside than inside the city.

What we observe during the last three centuries is the changing focus of the city from the river, its tributaries, and its surrounding areas as a means of

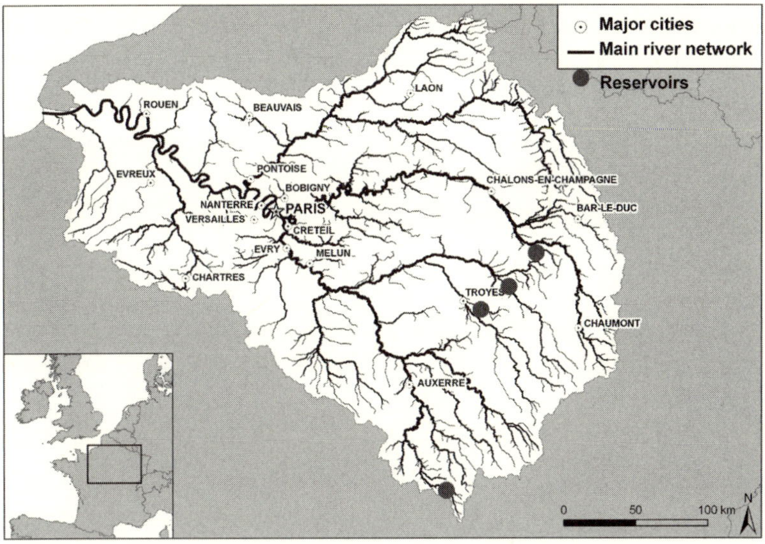

FIGURE 2.5. The four main reservoirs. From north to south: Reservoir Marne (Champaubert et Der-Chantecoq), Aube (Temple and Amance), Seine (Orient), Pannecière. Adapted from the map by Sylvain Théry, PIREN-Seine.

transport to its content—that is to say, the water. This interest grew from the beginning of the nineteenth century onward, and had in turn huge impacts on the waterscape through river derivations, canals, dams, and aqueducts all over the Seine basin. Taking into account the discharge side would confirm, and even reinforce, this picture of the city colonizing water and, as a result, rivers.[89] These remote urban impacts are visible in the waterscape that is partly shaped by the city and its needs, resulting in the urbanization of the waterscape not only within but also outside the city, and in a complex urban environmental imprint. These infrastructures that are not located within the city but whose only raison d'être was "urban" can be considered as technical urban annexes, or out-infrastructures. These out-infrastructures have been designed as such, but played (and still play today) also an ecological role that has been recently emphasized, especially thanks to the debates about biodiversity. That makes them hybrid, ecotechnical objects.

The transformations of the waterscape and of the river system are also connected to the question of the decision-making power. In the case of Paris, we observe a huge and changing institutional imprint together with the environmental one. During the ancien régime, this institutional imprint concerned the control of the river, its uses, and the works it needed. After the French Revolution, with the shift from river as means of transport to water, Paris bought several springs, rivers, and pieces of land.[90] This resulted in the multiplication of urban extraterritorialities: areas located outside, but governed or even owned by the city.

CHAPTER 3

WATERSHED DEMOCRACY OR ECOLOGICAL HINTERLAND?

London and the Thames River Basin, 1857–1989

Vanessa Taylor

This chapter examines the changing relationship between London, the Thames, and upstream communities during a time of fundamental change. The 1973 Water Act replaced local water undertakings in England and Wales with large regional authorities organized around river basins. In London, powerful bodies responsible for water supply, sewerage, and pollution control ceded these functions to a new Thames Water Authority.[1] The move to river basin management sparked widespread protest about the loss of local democracy here and elsewhere. Debates around these key urban functions had been central to London's political life since the early nineteenth century. A strong body of urban and public health history addresses earlier debates, but little attention has been paid to the historical implications of this new regional development even within discussions of a late twentieth-century "decline of urban governance."[2] This radical shift deserves examination, both for its implications for democratic governance and for the ways in which rivers and communities in the Thames region were reconceived during this time.

Two main issues are addressed here. The first concerns the relationship between London and the rural and urban communities of the upper Thames—part of the capital's "ecological hinterland."[3] Did the move toward integrated river basin management balance resources more evenly across the five-thousand-square-mile Thames catchment? Or did it allow for London's greater dominance, in the context of long-standing tensions between London and upstream communities? The second concerns the impact of regional, technocratic river management on localized democratic governance in London and across the region. Was a cohesive Thames region achievable or desirable? Is there any scope for seeing this Thames catchment or "watershed" as "the natural home of democracy" of Donald's Worster's vision? Or did this development "hollow out" political processes in the Thames region, resulting in what Olivier Graefe has called a "postpolitical environmental region"?[4]

A "spatial mismatch" has been noted between local government boundaries and the complex interdependencies of large urban spaces.[5] But the boundary-crossing nature of rivers presents particular challenges for governance. Rivers provide a vivid example of the way cities are connected to their economic and ecological hinterlands and "involuntary neighbours" by flows of water, sewage, pollution, and refuse, as well as trade. They force you to look beyond what Barry Doyle has referred to as the "sealed" boundaries of local governments.[6] This study of the shifting relationship between a city and its river basin aims to expand the terrain of urban history.

In critical analyses of the UK water industry, the rise of river basin management from the 1970s has been associated with the erosion of local democracy—a development seen as completed by water and sewerage privatization in 1989. This period saw the death of a long-established connection between water ratepayers and rights to representation in relation to local authority undertakings. (In Britain, domestic water rates were linked to property values, like other local taxes, rather than to metered consumption.) Castro, Kaïka, and Swyngedouw express an influential view (referring to developments up to 1983): "These measures significantly reduced the public accountability of water management . . . which increasingly became the preserve of a techno-scientific and managerial elite."[7] Understanding the impacts of regional managerialism on democratic processes requires further investigation of what public accountability meant before these developments.

The mercurial quality of "democracy"—its diverse and contested meanings across water sectors in different contexts—complicates this discussion. In the United States, for example, researchers have looked at the promise and failures of popular participation in river basin management, from the visionary nineteenth-century watershed movement to the "grassroots democracy" claims of the Tennessee Valley Authority (established 1933), alongside assessments of current arrangements. There has been no comparable historical exploration of river basins in the United Kingdom, though public participation has become a focus within social science literature on natural resources management.[8] Since the expansion of piped domestic water in nineteenth-century Britain, conflicts over democracy in water supply have involved broadly two issues. Firstly, debates have concerned rights to water use, including access to clean, sufficient water at an affordable and fair price, just allocation of water resources, and a say in water conservation decisions—for example, over the siting of reservoirs. Secondly, debates have centered on modes of governance and the best means of securing these rights—frequently identified with forms of public ownership of water, and with local government representation of water consumers' interests.[9] Debates around the river Thames, too, were infused throughout this period with notions of the river as a public resource. But democratic claims around water and river management have long been contested: supporters of public ownership and representation have battled with those advocating private ownership and expert-led management. The "public" itself has included a multitude of conflicting interests.

A narrative of lost democracy in discussions of UK river basin management may have relied on idealized versions of earlier structures of representative local democracy and public accountability. There have always been significant problems with public access to decision making for rivers and water resources. The Thames catchment in the twentieth century was a tangle of statutory and nonstatutory agencies, "blurred accountabilities," and "backroom . . . consultations"[10] involving national government, semirepresentative public bodies, industry associations, local authorities, and interest, amenity, and community groups. This chapter traces an evolution from nineteenth-century calls for catchment boards to river basin management in the 1970s, and the severance of local links to water and river services that followed. It ends with a brief discussion of water privatization in 1989, which further complicated the regional picture. To what extent did the reorganization of water and river management from local to regional units transform political space?

London and the Nineteenth-Century Thames

Table 3.1 indicates the changing governance structures for three stretches of the Thames from 1850 to 2000: the freshwater river, the tidal river in London, and the estuary downstream. A range of public and commercial bodies was responsible for river-related functions, including navigation, the port, conservancy, pollution control, water supply, sewerage, land drainage, flood protection, fisheries, recreation, and riverside development. The Thames Conservancy, established in 1857, was the first of its kind. A semirepresentative body comprising sectional interests and key local authorities, it was responsible for navigation and conservancy on the Thames from its Gloucestershire source to the sea (after 1909, up to the tidal limit at Teddington). The first citywide authority for London appeared in 1855: the Metropolitan Board of Works was a

Table 3.1. Main Statutory Organizations for the Thames River Basin, 1850–2000

	Freshwater River	Tidal River (London)	Tidal Estuary (below London)	International
1850	Local authorities (LD, FP). Commissioners of Sewers (LD, FP). Internal drainage bodies (1861–1930) (LD, FP). Local authority undertakings & private companies (WS).	City of London Corporation (to 1857) (N, P). Private dock companies (P). Vestries/local authorities (LD/FP). **Metropolitan Board of Works** (1855–89) (S, LD, FP). 8 Private companies (WS).	Local authorities (LD, FP). Commissioners of Sewers (LD, FP). Internal drainage bodies (1861–1930) (LD, FP). Local authority undertakings & private companies (WS).	
	Thames Navigation Commissioners (1771–1866) (N, RC) Cricklade—Staines (tidal limit). **Thames Conservancy** (1857/66–1973) (N, RC, F) Staines to sea (Cricklade to sea from 1866). See also **Lee Conservancy Board** (1868–1947) (N, RC).			
		London County Council/LCC (1889–1965) (S/LD/FP).		
1900		**London Boroughs** (1900–) (including 14 riparian: LD, FP). **Metropolitan Water Board/MWB** (1904–73) (WS).		
	Thames Conservancy continued (to tidal limit from 1909) (N C some catchment).	**Port of London Authority/PLA** (1909–) (P, N, RC, PC) Teddington (tidal limit) to sea		
	Local authorities, including County Councils (LD, FP). **Catchment Boards** (x 47: 1930–48) (LD, FP); **Internal Drainage Boards/IDBs** (1930–) (ld, fp). Local authority undertakings & private companies (WS).		Local authorities, including County Councils (LD, FP). **Catchment Boards** (x 47: 1930–48) (LD, FP); **Internal Drainage Boards/IDBs** (1930–) (ld, fp). Local authority undertakings & private companies (WS).	
	River Boards/RBs (x 32: 1950–64) (LD, FP, RC, F).		**River Boards/RBs** (x 32: 1948–64) (LD, FP, RC, F).	

<table>
<tr><th></th><th>Freshwater River</th><th>Tidal River (London)</th><th>Tidal Estuary (below London)</th><th>International</th></tr>
<tr><td rowspan="11">1950</td><td>Local authorities (LD, FP, P:L).
Local authority undertakings & private companies (WS).
River Authorities (x27: 1965–74) replace RBs (LD, FP, WR, RC, PC, F); IDBs continued (ld, fp).</td><td>Greater London Council/GLC (1965–86) replaces LCC (S, LD, FP, P:S/RD).
London Boroughs (LD, FP, P:L).
MWB continued (WS).</td><td>Local authorities (LD, FP, P:L).
Local authority undertakings & private companies (WS).
River Authorities (x27: 1964–74) replace RBs (LD, FP, WR, RC, PC, F); IDBs continued (ld, fp).</td><td></td></tr>
<tr><td colspan="3">1973 Water Act establishes 10 Regional Water Authorities (WR, WS, S, PC, LD, FP, RC, N, F, R/A).
31 private companies continued (WS) / IDBs continued (ld/fp).</td><td rowspan="10">Ramsar Wetlands Convention (UK 1976). EU Directives: Fish (1978); Birds, Shellfish, Drinking Water Abstraction (1979); Drinking Water, Groundwater (1980); Environmental Impact Assessment (1985); Urban Waste Water Treatment (1991); Habitats (1992); Water Framework/ WFD (2000); New Groundwater (2006); Floods (2007)</td></tr>
<tr><td colspan="2">Thames Water Authority/TWA (1974–89) (WR, WS, S, PC, LD, FP, C, N, F, R/A) (absorbs Thames Conservancy / Lee Conservancy Catchment Board, est. 1930).</td><td>TWA; Anglian WA; Southern WA</td></tr>
<tr><td rowspan="2"></td><td colspan="2">PLA continued (N, P, c: refuse/oil).</td></tr>
<tr><td colspan="2">London Docklands Development Corporation (1981–98)</td></tr>
<tr><td colspan="3">1983 Water Act ends statutory local authority representation / public access rights at RWA meetings.</td></tr>
<tr><td></td><td>GLC abolished 1986 (LD, FP, R/A to TWA).</td><td></td></tr>
<tr><td colspan="3">1989 Privatization of Water and Sewerage in England & Wales
New national regulators: OFWAT; National Rivers Authority (WR, RC, PC, LD, FP, F) (Environment Agency from 1996); Drinking Water Inspectorate.</td></tr>
<tr><td colspan="2">Thames Water Utilities Ltd (TW) replaces TWA (WS, S).
Private "water only" companies continued (WS).</td><td>TW; Anglian Water Ltd; Southern Water Ltd (WS, S).
Private "water only" companies (WS); IDBs continued (ld, fp).</td></tr>
<tr><td rowspan="2"></td><td>Greater London Authority (2000–) (P:S/RD, ld, fp).
London Boroughs continued (P:L, ld, fp).</td><td></td></tr>
<tr><td colspan="2">PLA continued (N, P, rc: refuse/oil).</td></tr>
</table>

C Conservancy, **F** Fisheries, **FP** Flood Protection, **LD** Land Drainage, **N** Navigation, **P:L** Planning: Local Riverside, **P:S/RD** Planning: Strategic/Riverside Development, **P** Port, **PC** Pollution Control, **R/A** Recreation/Amenity, **RC** River Conservancy, **S** Sewerage, **WR** Water Resources, **WS** Water Supply. Note: Lower case indicates some specific powers new tier above local authorities with responsibilities for sewerage, land drainage, and flood protection. This was replaced in 1889 by the more powerful London County Council (LCC). "Water London," an area extending beyond LCC borders, was supplied by eight private water companies. Water supply, sewerage, and pollution were part of a wider "London government problem." How was London's growing population to be governed and supplied? Who should own and control its utilities? How should power be balanced between metropolitan and local government, and between London and outlying areas?[11] Water featured prominently in "Progressive" Liberal campaigns of the 1890s, calling for public ownership of supply and control by the LCC as a democratic body. When water was made public in 1904, however, it was transferred to a "public corporation"—the Metropolitan Water Board (MWB)—with the LCC and other local authorities allocated board representatives. The LCC was bypassed again in the public takeover of London's private dock companies through another public corporation: the Port of London Authority (PLA), established in 1909. The PLA managed navigation, conservancy, and pollution control on the tidal river. For Progressives in the early 1900s, the opportunity for democratic control of the river's port, conservancy, and water supply had already been lost.

From the early years of the twentieth century, therefore, the river and water supplies in and around London were managed by bodies that were closely associated with the capital's political elites but whose public accountability was highly circumscribed. These organizations formed part of a wider, fluctuating pattern of "delegated governance" in Britain. Contemporary criticisms focused on their nonaccountability and "particularism," being structured around representation of particular interests (such as shipowners on the PLA board).[12] Historians of London have also viewed the MWB, PLA, and Thames Conservancy in this light. Up to 1973, these organizations sat awkwardly alongside the universalist aspirations of metropolitan government, itself engaged in struggles with local authorities and the public inside and outside its own borders.[13] The river generated conflicts of interest and competing communities that could not be addressed adequately by the forms of public accountability and participation offered by local government or public corporations.

Catchments and Communities of Interest

The mid-nineteenth century saw the emergence of demands for river catchment conservancy boards in England with a series of parliamentary reports and bills appearing from the 1860s, prompted by concerns over agricultural land drainage and flooding. Drainage problems pitted upstream against downstream, uplands against lowlands, and agricultural landowners against towns. Conservancy board advocates argued that land drainage should be a collective responsibility funded by local taxes on the grounds of a "community of interest." At an investigation in 1877, many called for "representative bodies" with control over the "whole . . . water-shed . . . of each river." Upland landowners objected that there was no such community; individuals should manage their own land.[14] The pressure for conservancy boards also represented a growing recognition of the interconnectedness of river and groundwater systems. Those interested in the capital's water supply talked increasingly of the water lying in the chalk and limestone of the upper Thames region. London's future water needs were a persistent focus for public debate: the freshwater Thames itself could not meet the needs of a city using over 150 million gallons a day by the 1880s. But interconnectedness created conflicts of interests between communities. As the *Times* put it following a failed scheme to pump water to London from a gravel bed at Dorney, Buckinghamshire, in 1888: "those who are interested in the land forming the river basins from which our supplies . . . are drawn are becoming indignant . . . under a system which grew up . . . when scientific knowledge concerning the sources of the water which fills our rivers was in its infancy."[15]

Conservancy boards did not become widely established until 1930, when land drainage and flood protection boards were created for English and Welsh catchments. Their functions were expanded with the creation of river boards in 1948, then again with river authorities in 1963 (with the Thames and the Lea Catchment Conservancy Boards surviving this process).[16] Central government was at the same time moving gradually and unevenly toward the amalgamation of water undertakings, greater regional planning, and larger local government units, but these were distinct developments—there was no consistent strategic link in the early years between the management of river catchments and land use development.[17] Well into the postwar period there were divergent strategies for providing water for urban growth, as demonstrated by two separate groundwater schemes in the Thames area.

Firstly, in the 1940s the MWB developed an ambitious plan to expand the "Greater London Water Area" and increase the available pool of water (figs. 3.1 and 3.2). The area selected (inside the dotted line, fig. 3.1) roughly coincided with the regional "Greater London" boundaries of Patrick Abercrombie's postwar reconstruction plan (1944). The area formed, the MWB emphasized in 1947, "a single hydro-geological unit." Importantly, this was not the Thames

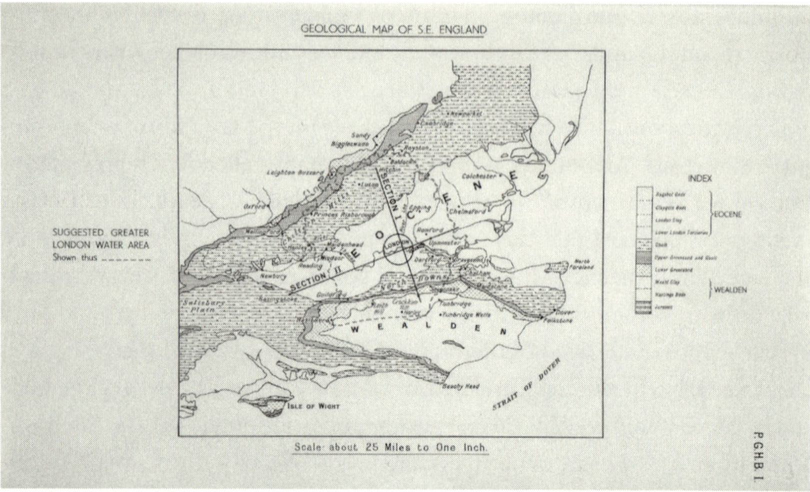

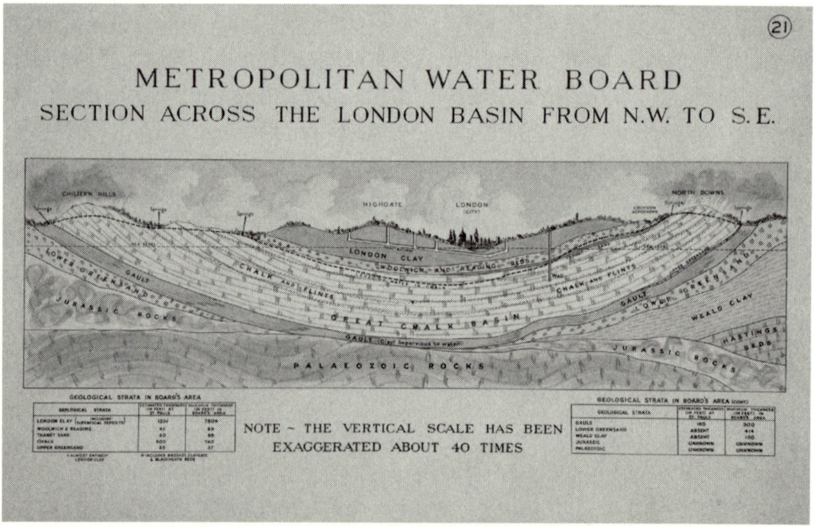

FIGURES 3.1. AND 3.2. Water Area and Section: "London Basin" scheme for Greater London Water Supplies (1947–1949). Source: TNA: HLG 50/2264: P. G. H. Boswell, *A Review of the Resources and Consumption of Water in the Greater London Area* (London: MWB, 1949). Reproduced with permission.

catchment but part of the London Basin, a geological structure covering much of southeastern England, an area said to be "unique" for its abundant groundwater and for being "so favoured economically by having . . . supplies naturally and continuously . . . delivered . . . from a peripheral, sparsely-populated catchment zone into a central, densely-populated district." There was, the

MWB suggested, "a considerable . . . community of interest and interdependence throughout the area within 30 to 40 miles from the centre of London." Economies of scale and equalization of water rates were just two of the promised benefits. At a 1947 government committee considering the plan, Sir Cyril Radcliffe spoke on behalf of water undertakers around London. Radcliffe is better known for his role in a more high-profile act of community engineering that year: the partition of India. Here he challenged the MWB's claims about a "natural area" with "one community." Despite regional planning developments, he said, there were still "many communities . . . run as separate communities." It was clear to see that the area was centered on London.[18]

It was the scheme's governance that split the committee, however. The majority favored separate divisions led by boards representing local authorities; the minority, a single authority with an expert committee. London boroughs sought representative management on the basis that "water supply is not so much a public utility . . . as a public health service." Under Labour leader Lord Latham, the LCC backed the small, appointed committee: water was "a service largely of a technical and 'commercial' character akin to those of electricity and passenger transport," at that time being nationalized. The Ministry shelved the matter pending the Labour Government's (unrealized) water nationalization plans.[19] The London Basin scheme was divided, then, both by competing claims over resources and by conflicting views of the nature of water—divisions that were to resurface in the 1970s. The LCC's position indicates the distance it had traveled from its early campaigns for public control of water.

This was followed by a less grandiose plan that represents a significant shift toward a catchment approach. The Thames Conservancy had been surveying groundwater in the upper Thames since the 1950s, after a reservoir proposal had been blocked by local opposition. Their borehole scheme, cosponsored by the MWB, was officially announced following the 1963 Water Resources Act. Through a system of "stream flow augmentation," groundwater was to be pumped into tributaries such as the Kennet and Lambourn to supplement the flow of the Thames toward London. The scheme aimed to offer an additional two hundred million gallons a day, and was promoted as cheaper and more economical in land use terms than surface reservoirs, but it created numerous conflicts around community entitlement to groundwater.[20]

Significantly, these were more piecemeal conflicts than the opposition to the earlier MWB plan. When the Lambourn Pilot Scheme was announced in 1965, the local water board and farmers around Newbury complained that local rivers would dry out. Lord Hurd, a leading Berkshire landowner, objected

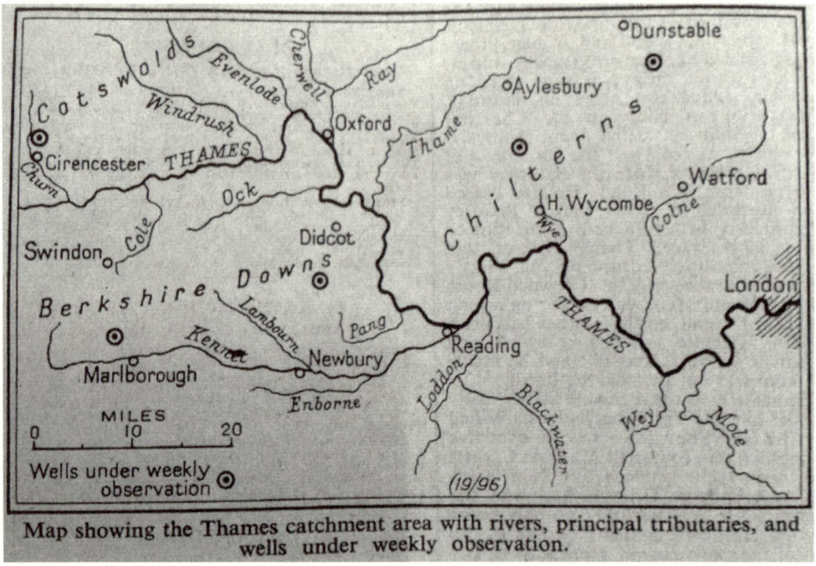

Map showing the Thames catchment area with rivers, principal tributaries, and wells under weekly observation.

FIGURE 3.3. Harnessing the upper catchment for London Thames Conservancy's "stream flow augmentation" scheme. Source: *Times*, June 24, 1963, 6. Reproduced with permission.

to the "intrusion," calling for "a competent body of local people sitting with the committee considering the pilot." The Thames Conservancy pointed out publicly that Berkshire already had a member on their board, but there were opportunities for "backroom consultations" with powerful vested interests: local National Farmers Union members gained financial compensation and the Lambourn Scheme went ahead in 1965. The full-scale Thames Ground Water Scheme (Stage I) began after a public inquiry in 1972 and was rolled out by the newly formed Thames Water Authority (TWA). Further public inquiries followed the announcement of the scheme's expansion in the Cotswolds (1980) and at Gatehampton, Berkshire (1987), at the time "the largest ground water abstraction project in Europe."[21]

The Thames Ground Water Scheme indicates both the momentum that could be gained with official sanction and the growing importance of river basins for regional plans. Subcatchments increasingly lost out to river basins; urban hotspots identified in regional plans shaped the priorities and outcomes of public inquiries. The mid-1960s also brought a shift in the balance of power between riparian owners (with customary abstraction rights) and water agencies. Under the 1963 Water Resources Act, abstractions for private supplies were now to be licensed by river authorities with new powers to develop water

FIGURE 3.4. Watercress beds in the Kennet Valley (undated). Source: Barbara Croucher, *The Village in the Valley: A History of Ramsbury* (Marlborough: Croucher, 1986). Reproduced with kind permission of Duncan Croucher.

resources.[22] A plan to augment supplies for Swindon, Wiltshire, in the 1950s and 1960s illustrates these changes.

When in 1957 the Borough of Swindon sought to pipe groundwater from the rural parish of Axford in the Kennet valley (around sixteen miles to the south), it met with strong opposition from the local landowners, fisheries interests, and residents of the Kennet Valley Protection Association (KVPA).[23] Swindon's application was tacitly supported by the Thames Conservancy, which had a long-standing policy of opposing exports beyond the catchment while allowing transfers across subcatchments.[24] The KVPA's consultant geologist challenged the validity of this policy, pointing out that water piped to Swindon would return not to the Kennet but to the Thames (via the river Ray): "The Kennet is entirely a Chalk-stream, and the removal of its water (or that of its catchment area) to a part of the Thames drainage-system that comes from . . . the Jurassic Rocks, involves an unreplaced loss of water from the Chalk area."[25]

The Axford community was up against the regional plans of central government—Swindon was a major growth town—and a public inquiry went against the Association. When Lady Burdett-Fisher (a KVPA member with watercress beds) refused permission for a borehole to be sunk on her land, Swindon used powers acquired under the 1948 Water Act to proceed. An abstraction of 13.1 megaliters a day was licensed in 1965 (fig. 3.4).[26]

The development of multifunctional river basin agencies was not inevitable even at this stage, however. In preparing the Water Resources Bill in the early 1960s, government officials had struggled to address the anomalous historic position of the PLA as pollution authority for the tidal Thames and considered transferring its powers to the Thames Conservancy. While this would have created a single authority for the river (bringing the Thames into line with other rivers), it would also "have the effect of uniting two areas with little or no common interest and . . . producing a large majority representing urban London, an area which would have little concern with some of the authority's principal interests—land drainage and fisheries—but would be the chief source of river pollution." It was considered that "the difficulties of establishing a satisfactory river authority for the whole Thames basin would be quite without parallel."[27] Within ten years, however, the multifunctional single authority approach had been established nationwide.

Democracy, Local Government, and River Basins

Under the 1973 Water Act, river, water, and sewerage agencies in England and Wales were absorbed into ten multifunctional Regional Water Authorities (RWAs). Thames Water Authority (TWA) took over ten water undertakings and 163 sewerage operators.[28] The act represented both a long-term trend toward larger units in water and sewerage governance and a radical policy shift. Its key purpose, as Jordan et al. have shown, was the integration of water quality and supply management at a time when the industry was turning to water reuse as the principal method of expanding water resources to meet escalating demand. Sewerage agencies were expected to prioritize effluent treatment more effectively if they were also responsible for water quality and supply.[29] But the shift to river basin scale had as much to do with a wider move toward regionalism—with precedents in the 1940s energy nationalizations—as with hydrological imperatives. River catchments were not inevitably the optimal scale for water services. As his department prepared to report on water reform in 1968, Labour's local government minister Lord Kennet noted that "too much stress should not be laid on the importance of integrated river basin management; there were many river basins in England and Wales, but only three represented distinctive large areas, and therefore it was not feasible . . . to think of taking river catchments as the sole determinant."[30]

Once the principle of regional authorities had been established, their boundaries were decided on distinctly "hydrosocial" grounds in relation to land use planning. As an official noted in 1971: "Given a minimum level of financial strength the significant criteria for determining the boundaries . . . seemed to be population and, to a lesser extent, area." London was part of the "'central deficiency zone' of southeast England." The decision to make the Thames stand alone, however, rather than merge with the river authority to the north—though desirable in unifying the "South Eastern Planning Region"—was to avoid "creating an unwieldy and too powerful South Eastern regional body." TWA was the only single-catchment RWA.[31]

Alongside questions of scale and efficiency, ministry officials also had to consider public reactions to terminating local government's traditional environmental functions. The 1969 Kennet Report implied that for local democracy the horse had already bolted: "bearing in mind that water operations are subordinated to planning . . . there is not much scope for local democracy in what are primarily technological matters, and . . . it cannot be seriously maintained that local democracy nowadays functions in this field significantly in practice. It is on conservation . . . that local authorities' and the public's interest principally centres. . . . But geography prevents local authorities being responsible for conservation." A related working paper stated that "in large matters of this kind decisions will always effectively be taken by central government and not by river authorities."[32]

The reformers' reliance on assumptions of local authority inefficiency and on management models provided by electricity and gas has been noted by Jordan et al. Political questions about water resources were silenced by reference to "technological matters" and "geography."[33] Official policy initially settled on small, managerial RWA boards but this underestimated the effect of a transformation on this scale, which generated widespread opposition. The pre-1973 water industry was a mixed economy of public and private undertakings. Discussing in 1971 whether to replace water and sewage undertakings with multi- or single-purpose authorities, civil servant Jack Beddoe considered it "doubtful whether either solution will create much general public interest." "Only last year," he noted, "five local authority water undertakings transferred to a statutory company in Essex without a vestige of public interest."[34] But the "gigantic" scale of the authorities and the rupture of localized provider/consumer relations were the problem for many, alongside questions of ownership. The Conservative Government's decision to retain private statutory water companies while replacing public undertakings fueled cross-party clashes.[35]

Among the most vocal of local government bodies opposing the plans was the Greater London Council (GLC), created in 1965 with an area five times as large as the LCC, who insisted on the "special circumstances of London." GLC leader Sir Desmond Plummer wrote to the environment minister that "London's river" was "an absolutely fundamental part" of the council's environmental planning. It was "inconceivable" that the "river and the water services which affect every Londoner personally should be managed by some other unelected authority," when the city provided "70% of the population and financial base." As "the natural choice of organisation," the GLC had two proposals. Their suggestion of a GLC-run river basin was rejected by the minister on the grounds that for the GLC to manage "water services in . . . [an] extensive rural area and in towns like Banbury and Cirencester" would "from the point of view of . . . overall democratic control . . . be much inferior to the original proposal." Their alternative, a "London Regional Water Authority" for the catchment of the tidal river below Teddington, did not address the need to expand London's supplies.[36] The TWA was established as a separate organization, though the GLC gained river amenity functions, justified by the government on the grounds of "the unique position of the River Thames as a national[,] rather than a purely local, asset."[37] The major frustration for the reformers nationally, in failing to transfer land drainage to RWAs, also affected London. The GLC retained land drainage and flood defense for the Thames and other "main metropolitan water courses."[38]

In promoting the river basin as the "natural" unit for water management, the 1973 Act de-naturalized the relationship between local authorities and the environmental functions that had become integral to their role. Local government links with water and sewerage proved politically impossible to sever altogether at this point, however. A compromise allowed for a majority of local authority representatives on the RWAs, though as Keating and Rhodes have said, this created an ambiguity: representatives were expected to ignore local interests that were also their "distinctive contribution." Labour Member of Parliament Dennis Howell later pointed out that "the great difficulty created by the Water Act . . . was that it was not based on any known unit of democratic accountability." Peter Black, TWA chairman and GLC Conservative, tried to sound optimistic in 1978, referring to central government edicts: "Despite the problems which will arise, the authority believes that the unique constitution of water authorities—part nationalized industry/part local government—ensures that they can respond wisely."[39]

The exclusion of local authorities from water, sewerage, and river management was finally accomplished by the 1983 Water Act, which ended statutory local authority representation on RWAs and implicitly reframed water citizens as customers through the establishment of Consultative Consumer Committees. In 1984 TWA Chairman Roy Watts, who steered the authority to privatization, introduced the new appointed board as a seamless evolution. With fifteen members (replacing sixty-two), it was "more independent and . . . intended to add to the *basic and unchanging requirements of continuously improving service to customers*, an even greater pursuit of efficiency as well as some of the disciplines of the market place. It seeks to deal only with policy, strategy and performance monitoring and to do so without the proliferation of Committees which marked the previous arrangements."[40] The GLC itself was abolished in 1986 and its land drainage, flood protection, and river amenity functions transferred to the TWA.[41] At this point London's formal political relationship to the Thames environment was lost.

The rise of river basins, then, contributed to the decline of long-standing local and metropolitan government functions. RWAs reflected a new managerial imperative that also drove the removal of the GLC and other bodies from the PLA board during these years. They facilitated increasing central government control—historian Asok Mukhopadhyay called them the "regional arms of Whitehall."[42] The democratic structures of local and metropolitan government, however, had never been sufficient either for the boundary-crossing nature of rivers or for competing interests among river users, riparian owners, water consumers, and amenity and environmental groups. Conflicts around water and river rights have been a breeding ground for interest groups emerging in the gaps between formal routes for public participation and accountability.

Civil Society, Governance, and the Thames

Some civil society organizations sprang from the closure of formal representative channels. After the abolition of the GLC some of its leading members and officials sought to influence policy for the river, docks, and riverside through the London Rivers Association (1986–2007). Numerous local organizations were mobilized by the closure and commercial redevelopment of London's upriver docks in the 1970s and 1980s, as central government transferred the planning powers of riparian boroughs to the London Docklands Development Corporation (LDDC, 1981–1998).[43] But interest groups have a longer history in the Thames region. Sometimes they were a response to crises. Severe flooding in East Molesey, Surrey, in 1968 prompted action from the Molesey Residents' Association (1965–) and a Flood Victims Campaign.[44] The River Thames Society (RTS, 1962–) and Thames Planning and Amenities Forum (TPAF, 1965–1996) represent more long-term engagement with planning and recreational issues and pressure for statutory protection. The River Thames Society supported GLC bids for continuing membership of the Port of London Authority board in the 1960s and early 1970s, to promote neglected amenity interests on the tideway. TPAF, representing both local authorities and amenity groups, for a time sought planning authority status to provide a "wide vision for the whole river."[45]

After the formation of RWAs in 1974, civil society groups continued to produce collective responses to problems of river and water management. TPAF and the Cotswolds River Action Group opposed TWA's plan to increase abstractions from the Cotswolds limestone at a public inquiry in 1980. Most vocal was TPAF member Frederick Freeth, Wiltshire County Council's "rebel" on the TWA. But the river basin has proved a problematic scale for popular protest. The persistence of localized pressure groups and, especially in recent years, the proliferation of tributary or subcatchment organizations suggest that social activism is still more readily mobilized at the local level. But the context for such organizations is crucial: for all this vibrant local activity, the ability to affect decisions has been highly variable between different groups and over time.[46]

Recent literature on multilevel governance provides a framework for considering changing forms of political engagement along the Thames and its tributaries in the context of a shifting relationship between civil society and the state. The years following the Conservative election victory in 1979 have been widely associated with the death of consensus over the role of the state in welfare and other services in the UK, the decline of local government, and the transfer of key state functions to the marketplace and civil sphere. Geographical and political science literature has explored a shift "from government to governance," implying a move away from fixed representational modes of government toward a range of new institutions, public-private partnerships, and networks at a local level. "Rescaling" is central to these discussions and several, sometimes contradictory, trends have been identified: "scaling up," suggesting both the loss of local powers to central government and the erosion of national sovereignty by Europe; "scaling out," entailing the transfer of public services to the marketplace but also greater involvement of "non-state actors" (commercial bodies, stakeholders, interest groups) in decision making; and "scaling down," including an expanded role for local groups.[47]

Water and sewerage privatization in England and Wales in 1989 scaled out key regional functions to the market. It also led, however, to unprecedented levels of national regulation for water resources and rivers, with an economic regulator (OFWAT) and two environmental regulators: the National Rivers Authority (Environment Agency since 1996) and Drinking Water Inspectorate. Environmental concerns have been tied to market mechanisms within this framework. Increasingly, too, environmental expertise was internalized within water agencies, with important consequences for the management of pollution, abstraction, land drainage, flood risk, and ecologically sensitive sites.[48] River basin management has coincided since the late 1970s with the scaling up of regulation for rivers and water: European Union environmental directives have been powerful drivers in expanding national regulation, especially since privatization, also providing potential supranational pathways for local activists. Tower Hamlets Federation of Tenants in east London cited the Drinking Water Directive when in 1986 they successfully complained to Europe over nitrate levels in their TWA supply. The Water Framework Directive (WFD, 2000) has required the production of river basin management plans, introduced targets for the "ecological status" of rivers, and obliged member states to "encourage the active involvement of all interested parties," implying a degree of unspecified scaling down.[49]

The ongoing controversy over Axford illuminates some important changes in the role of civil society organizations over time. Since the 1950s the river Kennet has been a site of conflict between the perceived interests of water consumers and those interested in the Kennet as an environment, amenity, and local resource.[50] The private water users and amenity interests of the Kennet valley lost out to Swindon's consumers in the 1960s. Despite influential local support, they lacked the resources of Swindon Borough backed by the Thames Conservancy and central government's regional policies. When leading consultant engineers declined to advise them, they resorted in 1959 to a water diviner to try to find alternative water sources for Swindon. Officials pronounced confidently on the likely impacts of the abstraction: "There should be little or no effect in Kennet river as there appeared to be no direct communication between the water in the underground formation and the water in the river."[51]

Over fifty years later, the Axford pumping station continues to be controversial, however. The impacts of groundwater abstraction on rivers are now acknowledged to be imperfectly understood. In the 1990s the Environment

Agency (EA) and Thames Water (by then a private company) clashed over the renewal of Axford's abstraction license. Following the grant of a new temporary license in 2008, a joint study by Thames Water, EA, and Natural England (another government agency) is said to have concluded that "abstraction was contributing to low flows but, due to the complex nature of the systems, it was difficult to demonstrate specific evidence of direct adverse impact on the environment."[52] A local organization, Action for the River Kennet (ARK), has campaigned since 1990 against deteriorating river conditions, calling for an end to the abstraction. They have worked in recent years with influential environmental NGOs such as the Royal Society for the Protection of Birds and the World Wildlife Fund (WWF), as well as the statutory EA and Natural England. A current policy of reducing Swindon's reliance on Axford is tied to pressure for the Kennet to reach "good ecological status" under the EU's WFD.

Today's more porous borders between the civil and statutory spheres in river and water management highlight the brittle separations of earlier decades. Oppositional groups today have levels of access to information on water policy that would have been unthinkable in the 1960s. The WFD's ecological benchmarks for rivers underpin a reappraisal of subcatchments in the Thames basin and elsewhere.[53] This situation currently hinges on a tense balance between the EU, national government, and the regulatory agencies it funds. The rise of influential environmental NGOs—with expertise and access to national and international policy circles—has played a crucial role in shaping the changing regulatory landscape.[54]

River basins in England and Wales are neither watershed democracies nor postpolitical. They are the product both of their regulatory contexts and of the range of official and unofficial participatory avenues open to different groups at any given time. The impacts of river basin management on political space in the Thames region between the 1960s and 1980s were not straightforward. London's political institutions undeniably lost both decision-making powers and the symbolic power conferred by environmental management of the river. The opening of the GLC's Thames Flood Barrier in 1984 just before the GLC was eliminated symbolizes the abrupt end. In terms of the allocation of water resources, however, London and other key towns were the winners in the Thames catchment during this period, with expanded ecological hinterlands. The localized priorities of less-populated rural districts in the catchment were downgraded. It remains to be seen what effect Defra's recent "[sub]catchment-based" initiative has on this local/regional balance.[55] While localized communities of interest were subsumed within new regional divisions, new communities of interest have continued to emerge in the civil sphere.

During the heyday of "representative" management up to the 1960s, there was frequent criticism of the river's fragmented governance and the failure to represent a full range of interests and communities. The potential for competing interests along the river is as strong as ever, and the uneven battle between the priorities and resources of local interest groups and those of central government remains a key problem for environmental citizenship. The power of different groups is still tied to their consonance with official policies. The government can do much to change the balance of central and local power at any time.[56] But this is not a story of rivers lost to their citizens. The emergence of greater consensus over environmental issues combined with an expanded regulatory structure and more powerful activist organizations has shaped new modes of participation and accountability for rivers that exceed what was on offer from representative local bodies up to the 1970s. However technocratic much of the current discussion and literature on river basins undoubtedly is,[57] public pressure continually bubbles up through civil society organizations and the tactical use of political and legal avenues that are now more widely available. River basins are not (yet) the cohesive engines of governance that the water reformers of the 1970s may have been hoping for. Fragmentation of decision-making processes for the Thames and its catchment remains an obstacle to effective action but—especially in a multilevel context—it also provides valuable space for political contestation.

Part II.

Urban Rivers Transformed and Lost

Chapter 4

The City Whose Rivers Disappeared

Nantes, 1850–1950

Geneviève Massard-Guilbaud

This chapter shows how decisions regarding the management of a river, taken single-handedly by "experts" probably well intentioned but using their legal powers without taking their "incompetent" fellow citizens' opinion into account, can lead to an environmental catastrophe and drastically and definitely upset an urban landscape. The city of Nantes was for a long time both a vivid Atlantic harbor and an important river port on the Loire, the main French actor in the appalling "triangular trade" that made the city very wealthy in the eighteenth century. By the turn of the century, when this trade came to be seen as morally wrong and was finally declared illegal (1814), the city reacted by progressively developing a new industry strongly bound to the harbor. It became famous for its food-processing industry (sugar, oil, biscuits, canned fish and vegetables) related both to tropical imports and to the needs of the merchant navy. Beyond the making of biscuits and the canning of peas and sardines, its industry included shipbuilding, chemicals, and metallurgy. But Nantes is sixty kilometers away from the ocean. While in the nineteenth century the drafts of ships kept increasing, the estuary silted up, in spite of all efforts by engineers to "clean" it and prevent the return of sand and sediments by building sand traps, channelization, and rip-raps of the banks. Something more radical had to be done (or so the traders said), or the ships would stop sailing up the river and the harbor would die.

The period 1830 to 1880 saw a protracted technological controversy over the best way to save Nantes Harbor. Finally, the decision was made to dig a canal parallel to the section of the river that was the most difficult to manage. It took ten more years to complete it. The Canal de la Martinière opened in 1892 and at first it revived Nantes Harbor. But the improvement was short-lived and very soon the big ships stopped taking it both because the draft was no longer deep enough (six meters, whereas the boats now needed seven or eight meters) and because a new technology now made it possible to dig the riverbed as deep as necessary, apparently proving the case of those who had fought the canal and advised the channelization of the river. The canal closed in 1913, only eleven years after its opening, and became a cemetery for old boats. It had been an enormous waste of money.[1]

So far, the story is quite similar to that of many other estuary cities. But what followed might be less so. To reach the depth requested by the biggest ships, drastic digging of the riverbed resumed at the beginning of the twentieth century.[2] These works made it possible for all the boats to reach Nantes, but the harbor itself was also silting up. Another round of works began, to "improve" and deepen the harbor.[3] All these works dramatically modified the river rate of flow, allowing the tide to enter the riverbed much farther inland than before, and significantly lowering the water level at low tide, as fig. 4.1 shows. The natural balance between the sea and the river had been broken, and it did not take more than ten years before this turned into a disaster for the city of Nantes.

This chapter will first present the consequences of these works, a strictly economically minded operation that ignored the actual functioning of the river and the results its disruption would have. It will show how the different authorities reacted to the catastrophe they created and describe the balance of powers between the State, its engineering body the Ponts-et-chaussées,[4] and the elected city council and mayor who defended the city interests. It will ask why and how such gigantic works, which slashed the cityscape, could be imposed upon its inhabitants, and show whose interests this policy served. Finally, it will analyze which rationale led the city to implement works that one century later came to be considered such a mistake that a program of river restoration is currently being implemented.[5]

Collapses

The "improvement works" carried out in the estuary and the harbor at the beginning of the twentieth century quickly proved disastrous for the city infrastructures. They increased the speed of water passing under the central arches of bridges, thus undermining their stability. This had been expected by the Ponts-et-chaussées, and the replacement of the oldest bridges, deemed too weak to resist the river's new rate of flow, had been planned. But as World War

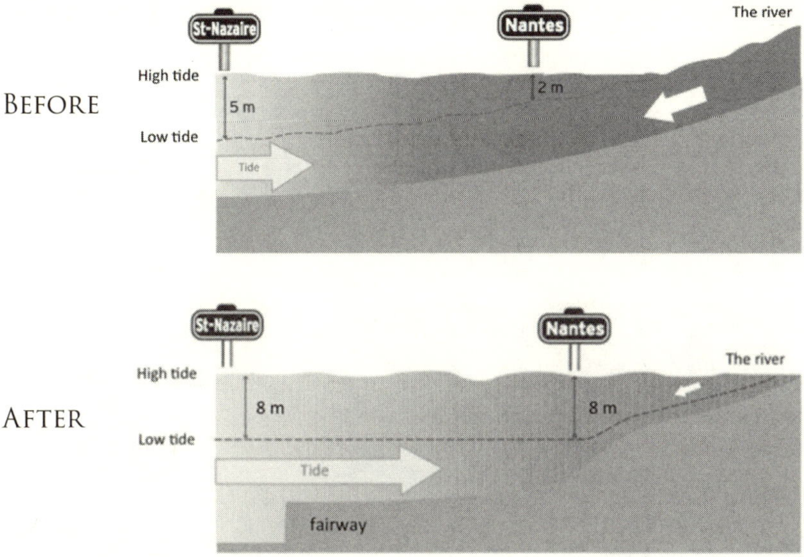

Figure 4.1. Simplistic profile of the estuary channel before and after the works. Diagram by the author, after a drawing featured in "Programme interrégional 'Loire grandeur nature,' Programme Amont, Restauration de la ligne d'étiage en Basse-Loire, Voies navigables de France," 2003.

I broke out, these expansive works were postponed. Furthermore, as would be acknowledged later, the distribution of responsibilities among the different administrative bodies was unclear.[6] Even after the war was over, nobody felt really responsible for the monitoring of these bridges.

As another consequence of the works, at low tide the water level in Nantes dropped between 2.5 and 3 meters lower than it did before the works. Again, this had been expected. The engineers thought that this would dry up the shallower branches of the river, the two north branches. This could have sanitary drawbacks, to which they planned to attend "in due course"—or so they said. What they had forgotten was that, due to the alluvial and unsteady nature of the ground, a good part of the city was built on wooden stilts. These stilts could not tolerate such a drop in the water level. Exposed to air twice a day, at low tide, they began to rot. This the engineers had not foreseen. Finally, the new river regime also intensely eroded the retaining walls of the quays, which in several places began to be seriously eaten away. This again had not been anticipated. All this resulted in the collapse of several quays and two bridges between 1911 and 1925 (see locations and chronology of the collapses

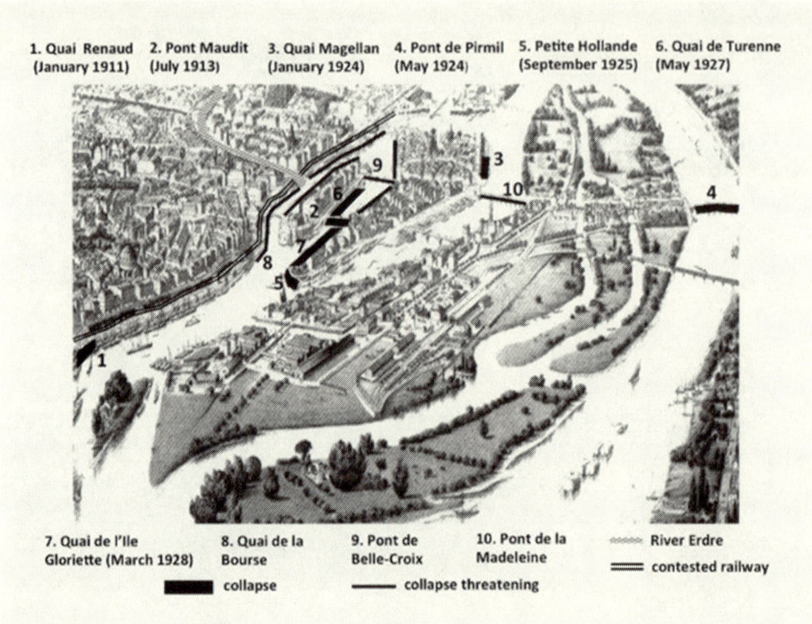

FIGURE 4.2. Localization of the collapses and their dates. Bird's-eye view of the center of Nantes, from the west. This engraving was chosen as a background because it gives a good idea of the place, but it dates from the middle of the nineteenth century. In the 1920s, the islands at the bottom of the picture were built up. Initial engraving: Hugo Alesi, *Vue sur Nantes vers 1868*, Musée du Château, Nantes.

in fig. 4.2). Other buildings and quays also sank or caved in and threatened to collapse in turn, in spite of the emergency rip-raps that were repeatedly implemented at great cost.[7]

The first accident due to the works done in the estuary was the collapse of one quay, the Quai Renaud, which dragged down several harbor-lifting devices into the river, in January 1911. In July 1911, one of the last houses remaining on the Pont de Pirmil, the old bridge built in the sixteenth century, collapsed. Then came the Pont Maudit's turn.[8] This bridge, the busiest of the city, used by the tramway for the crossing of the river, collapsed into the so-called Hospital branch of the Loire in July 1913. In January 1924, the Quai Magellan fell down. The most dramatic was probably the collapse of two central arches of the Pont de Pirmil, which was also the longest bridge of the city, in May 1924 into the southern branch of the river—the widest one.[9] In September 1925, part of the Place de la Petite Hollande, where a busy market was held regularly, also slumped into the river, pulling down a pile driver that had been stored there to dig in new consolidation stilts. Part of another quay, next to the Pont Maudit, broke down in May 1927 and the Quai de l'Île Gloriette caved in in March 1928.[10] Other places, such as the Quai de la Bourse, the Pont de Belle-Croix, and the Pont de la Madeleine, showed faults and movement announcing that they would soon follow the same way. If the collapses caused much trouble to the traffic, by chance there were never any casualties.[11]

Already in 1921, the well-off inhabitants of the Quai de la Bourse (8 on fig. 4.2) became really anxious: their quay showed visible signs of subsidence, while the load it had to support was particularly heavy.[12] Following a questionable decision made in the 1850s, this quay (like all the other quays of the same bank) supported a railway that crossed the city like an east–west bisector, following the river Loire (fig. 4.2). The tracks had even been doubled in 1917 to speed up the transportation of American troops that had landed in Saint-Nazaire. After the war, the railway company had retained this second track, despite the agreement passed during the war that it should be removed after the hostilities. The mayor kept asking the railway company to respect their commitment, to no avail. Thus, between the buildings lining the quay and the river passed a street, two different railway tracks, and a tramway line. There was also a passenger railway station, the Gare de la Bourse. Twenty-three level crossings accompanied the passing of the train through the city center, causing many accidents and casualties. In 1920, the wise decision to reroute the railway to the north of the city had been made, but it had not been implemented yet. In spite of this decision, the State was not in a hurry to pay for the diversion. Unquestionably, had the quay collapsed, a tram or a train and its passengers or those waiting for it would have been thrown into the river.

The Pont Maudit had been rebuilt and the Quai Renaud restored, but something obviously had to be done to prevent the even more significant catastrophe that was pending with the busy Quai de la Bourse. Inhabitants of the quay and the local press put pressure on the municipal council and the mayor. The latter was aware of the danger and keen to prevent it, but he had no authority on the river, which was governed by the Ponts-et-chaussées administration. Repeatedly, the mayor wrote to their chief engineer, pressing him to act.[13]

While the first collapses had clearly been attributed to the deepening of the estuary and harbor, in 1921 the Pont-et-chaussées did not acknowledge that the city's subsoil instability and collapses could have the same cause.[14] They accused the press and city residents of exaggerated pessimism and emphasized that the rip-raps they were putting in place were strong enough to prevent any accident. Clearly, they were playing for time and they insisted that anyway, a definitive solution would be found only with the filling of the northern branch of the Loire, the Bourse branch.

This idea of filling the northern branch of the Loire (or the two northern branches, depending on the proposed project) already had a long history. It had been proposed for the first time in 1859 and had reappeared every now and then, but each time with a different rationale—sanitary improvement, gain of space, improvement of street traffic—which shows that the filling of a river was seen as a panacea, a fashionable way of resolving any trouble you could meet. This filling had always been rejected by the city officials, however, not without excellent reasons: aesthetics, cost, problems for the river boats. The downstream section of the river Erdre was the outlet of the canal linking Nantes to Brest, widely used at that time for goods transportation. If the Bourse branch of the Loire, at which the canal terminated, was to be filled, another outlet for the river Erdre and its boats would have to be found. In 1902 a detailed study had even been presented, but no decision was made because in spite of the engineers' statements, there was no convincing solution.[15]

In March 1919, an act, the Loi Cornudet, made it compulsory for French cities of more than ten thousand inhabitants to adopt an "enlargement and embellishment plan." The first draft designed for Nantes in 1920 had seen the *n*th coming back of this filling project,[16] the space reclaimed being supposedly transformed into a large open green space.[17] It was just a first draft, nothing had been formally decided yet, but the city had not rejected it, this time agreeing that a large green public space, cleared out from the railway, might compensate for the loss of this river branch, that the first "improvement works" in the estuary river had tended to dry up already. This branch being the shallowest, it was also the first that fell dry at low tide or during low water periods. The Hospital branch and the Erdre would keep the presence of water in the center of the city. From then on, the Ponts-et-chaussées seized any opportunity to reassert that this filling was a necessity, "a simple matter of good sense," and that it should be implemented as soon as possible. The sinking of the Quai de la Bourse gave them another chance to reaffirm this, although no good solution had been found yet in respect to the Erdre-Loire confluence. Different plans were imagined for that; however, none was really convincing. The passing of the river through an underground pipe would make boat traffic impossible and necessitate a transshipment between the Erdre and the harbor. The city gasworks, sitting on the right bank of the Erdre, could no longer be supplied with coal by barges coming from the harbor. Between the harbor and

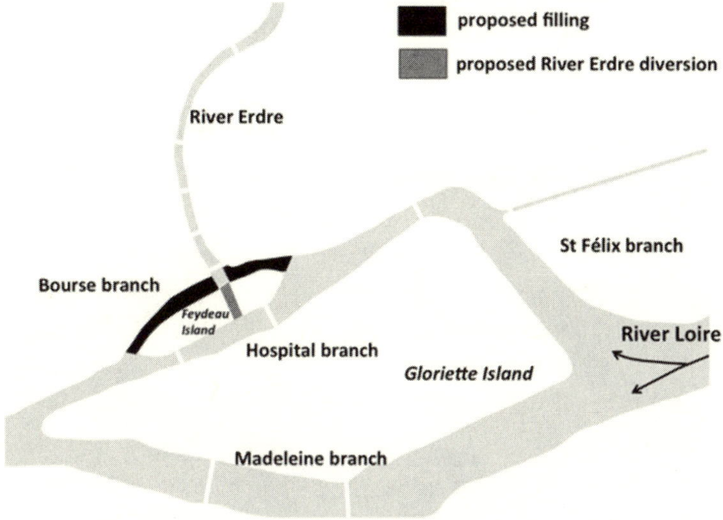

FIGURE 4.3. The Ponts-et-chaussées proposal (c. 1921). This figure and the following one show only three branches of the river Loire, which had four. For reasons of scale, the south branch—the widest—has not been represented here. Drawing by the author.

the plant, tons of coal would have to be carried on carts through the city center! The mayor protested that this would increase the price of gas used by city dwellers for heating. The last proposal regarding the Erdre confluence consisted of breaking through the Feydeau Island eighteenth-century housing estate to bring the tributary to the hospital branch (fig. 4.3). This suggestion could in no way get a blessing from the city, which rightly considered the beautiful buildings of the Île Feydeau as a precious heritage deserving protection (see this island in fig. 4.7). Actually, the Ponts-et-chaussées records attest that as early as that date, the plan they had in mind was far more "ambitious": filling the two branches of the river that surrounded the island. I'll come back to it later; this plan was not made public at that time.

The city chief engineer, for his part, did agree that the flimsy Quai de la Bourse was to be immediately secured, of course, but most important in his view was to proceed as fast as possible to the diversion of the railway (already decided upon, but not yet implemented), which would relieve the quay and the city as a whole. In agreement with the city architects, the city chief engineer

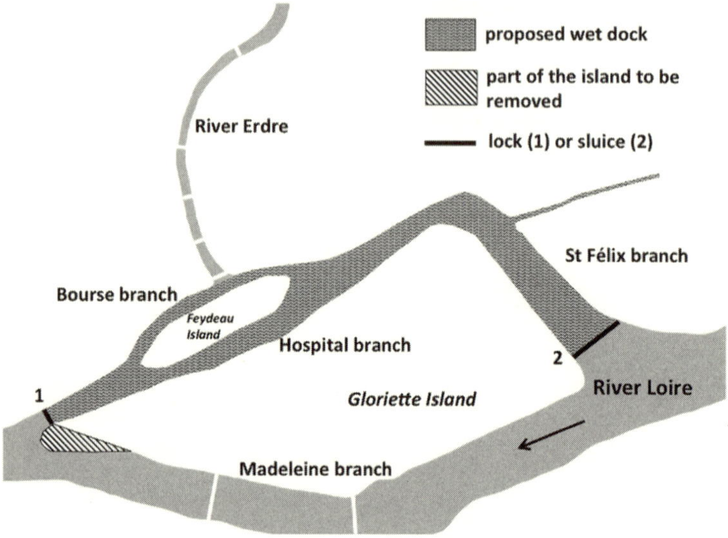

FIGURE 4.4. The city chief engineer's proposal (c. 1921).

opposed the filling, which, he said, would in no way solve the problems created by the Loire deepening: the Quai de la Bourse subsidence would stop, indeed, but the erosion problems would go on elsewhere. To stop it definitively, he suggested a totally different plan: transform the two northern branches of the Loire into a wet dock terminated by locks (or a lock and a sluice, to reduce the costs) at its two ends (fig. 4.4). This would protect the quays from erosion and the building foundations from uncovering at low tide and rotting. The traffic of the boats from the Erdre to the Loire (and the reverse) would remain as it was.[18] This project also protected the aesthetics of the city, as the rivers not only would remain where they were but would enjoy a high level of water whatever the season or the tide.[19] The various activities that took place on it (angling, bathing, swimming, washing) would be allowed to continue. The mayor and the municipal council approved of this plan. No official criticism of this alternative plan coming from the Ponts-et-chaussées can be found in the records at that time, although they obviously knew it. The truth is that this plan was both simple and convincing.

According to law, any decision regarding the river fell within the capacity of the Parisian Conseil général des Ponts-et-chaussées (Higher Board for Bridges and Roads), the national authority leading the corps. In spite of the obvious emergency, no decision was made yet. To answer the mayor's concerns, the Ponts-et-chaussées built further finger obstacles (*épis*) to break the river flows and protect the Quai de la Bourse. Maybe because the plan drafted by the city engineer was so much more convincing and difficult to contest, the Ponts-et-chaussées chief engineer was probably waiting for a better opportunity to impose his plan upon the municipality.

FILLINGS

In 1924, trains were still passing on the Quai de la Bourse. Three thousand meters of quays and all the bridges crossing the north branches of the river were now deemed liable to collapse shortly. Rip-raps and other obstacles helped, but they did not wear well and could not be a lasting solution. The sinking of the Quai de la Bourse and the movement of its retaining wall had become "most serious."[20] Suddenly, during the first days of June, the Ponts-et-chaussées made unilaterally and "in emergency" the decision to fill that part of the river facing the most threatening subsidence, downstream from the Erdre-Loire confluence (fig. 4.3).

Although it seems difficult to believe, the mayor was not consulted nor even officially informed. He just heard incidentally about the decision on June 14, in a letter he received from the Ponts-et-chaussées that let him know that the city would have to pay for the new sewers to be built to compensate for the disappearance of the waterway, used so far as a drain for wastewaters.[21] As the works were to begin immediately, the letter went on, this expense should be budgeted very soon. The mayor's answer was clear: he expressed his anger about the method—such an important decision made without the city being even consulted!—and his total disapproval of the filling—a catastrophe for the city, he said, from all points of view, including aesthetics and heritage protection. He added that the city would in no way take part in the funding of works (the building of sewers) that resulted from decisions taken singlehandedly by the Ponts-et-chaussées. Understandably, the mayor did not want to be held responsible for a catastrophe, so he added that he could not nor would not in any way oppose the filling if the Ponts-et-chaussées deemed this was the only solution to guarantee the security of his fellow citizens. However, he went on, he did not understand why a decision that had been pending for years was now a matter of days. He insisted that other plans existed, including the one drafted by the city engineers. He demanded that a committee be created to evaluate these different possibilities—if security could provisionally be guaranteed for a few more days. The municipal council unanimously supported this stance.

FIGURE 4.5. The Pont de Pirmil after its collapse, with the tramway rails hanging in the gap. Reproduced with the kind permission of the Archives Municipales de Nantes.

Indeed, it was not by chance that the Ponts-et-chaussées chief engineer deemed that the decision he had postponed for years must now be made in emergency. A few days earlier, the Pont de Pirmil had collapsed. A police officer walking by the old bridge had noticed that a gas pipe running on it was broken, due to a movement of the structure. The traffic was immediately interrupted. The Ponts-et-chaussées chief engineer, who had been urgently summoned, was taken to the bridge in a boat by a fisherman to check the size of the new fault. Just half an hour after the traffic had been interrupted, the bridge collapsed. The two men in the boat barely escaped the catastrophe.[22]

Nantes had now seen two of its bridges and two of its quays fall down into the river. All the infrastructures lining the north branches, that is, the city center, threatened to meet the same fate shortly. Nonetheless, although further consolidation works were rapidly implemented, any definitive decision regarding the quays was postponed once again: to meet the mayor's demand, the committee that had been set up to investigate and decide upon the actual causes of the Pont de Pirmil collapse would also compare the different plans proposed to secure the north branches infrastructures and decide which was the best.

This committee was headed by an inspector general of the Ponts-et-chaussées. The analysis and recommendations they made, as well as the definite and technocratic tone in which they were done, are surprising given the liability of this body in this issue. All the collapses and the threats weighing on the quays were formally put down to the "improvement works" done earlier in the estuary and the harbor (that, at least, was true, and new). Not only were these harms bound to happen, the committee went on, but they were "the very proof of the 'improvement works'' success"![23] The only mistake the committee found in the engineers' work was that the rebuilding of the oldest bridges, which had been planned as soon as 1913, had not been implemented. Postponed because of the war, it had been neglected afterward, while it was known that these bridges could not withstand the new strength of the flow.[24]

Anyway, the committee declared in no uncertain terms, the situation would only get worse, as the deepening of the river was now to enter a new stage: as early as 1913, it had been decided that the Loire would be deepened well beyond the city of Nantes, allowing the tide to come into the riverbed as far as eighty kilometers inland. According to an engineering theory called the "theory of the funnel," the bed would also be given the shape of this instrument. This was supposed to reduce the hydraulic friction and therefore speed up the river current.[25] Twenty kilometers still had to be deepened and reshaped to reach this goal. New works would soon begin, which would "improve" the flow even more and tear the city down. What was happening to the city was just a normal side effect of a plan that was not questioned in the least. That it could be disturbing, or at least questionable, for the population to see their city falling apart or their rivers filled was not even mentioned.

To face the forthcoming "improvement works," and according to a plan recently designed by the Ponts-et-chaussées chief engineer, the *two* northern branches of the Loire (not only the most northern, as in the previous public proposals) were now to be filled, as the urban infrastructures could in no way resist the forthcoming rate of flow. According to this new plan, the river Erdre could no longer reach the Hospital branch after crossing Feydeau Island, as this branch would now also be filled. The Erdre would have to be canalized into a tunnel passing under the hill on which the old town had been built in the Middle Ages, and come out in a section of the former Saint Félix branch that, for this reason, would escape the filling. In this section, the Erdre would therefore run *in the former bed of the Loire*, but *in the opposite direction*! The Ponts-et-chaussées had a wild imagination when it came to upsetting nature.[26]

The alternative project proposed by the city chief engineer (fig. 4.4), which would have put an end to erosion and the collapse of urban infrastructures without filling or diverting any waterway, was not really taken into consideration. Two terse arguments seemed enough to disregard it: the lock closing the proposed wet dock would never resist the weight of such a volume of water, and anyway, the project would be too expensive. Neither of these arguments is likely to convince the historian—this sort of dock is a rather common infrastructure, and it was not demonstrated in any way that the lock could not resist the water weight. As for the expense, the argument is even less convincing: the city project included the building of two locks. But while these two locks would have been the only works to be paid for in the city's plan, there was much more in the Ponts-et-chaussées': one lock, but also the digging of a long tunnel, the remaking of the city sewer network, the filling and redevelopment of the former waterways, quays, etc. There is no question, the city project could only be the cheapest, and the reason why the Ponts-et-chaussées plan was chosen is to be found elsewhere, as we will see in the last part of this chapter. Regarding the cost, let us note that the estimates provided were superficial, as if done on the back of an envelope on the basis of a gross evaluation (no detailed process has been found in the records, nor any quote by a builder). Not surprisingly, these estimates would not be respected.

On March 13, 1926, the Ports Secretary of State adopted the plan proposed by the Ponts-et-chaussées. However, as the cost estimate went over fifty million francs, only its first part was accepted; that is, works strictly necessary to secure the city infrastructures, which reduced the bill to twenty million.[27] Once again, the mayor of Nantes was not informed of the decision until he received a letter from the Ponts-et-chaussées letting him know that "joint meetings" (*des conférences mixtes*) were now to be held to decide how the bill would be shared between the different "partners": the State, the City of Nantes, and the local Chamber of Commerce.[28] But more trouble was still to come for the poor mayor: the Ponts-et-chaussées soon explained that according to law, the City of Nantes should be represented in these meetings by . . . the Ponts-et-chaussées engineers! The legal department of the city proved that this was not true, and the mayor could finally attend and represent the city. But clearly he was repeatedly treated as an insignificant, negligible partner.

It is not possible to tell what happened next in this short chapter. However, before I shift to an analysis of what this already tells us, I would like to summarize briefly what was done: the fillings, begun in 1926, went on until 1946 (fig. 4.6), and the development of the freed spaces took a few more years. These works were not simply "major works," but an incredible upheaval of the whole cityscape, which a simple drawing cannot really show.

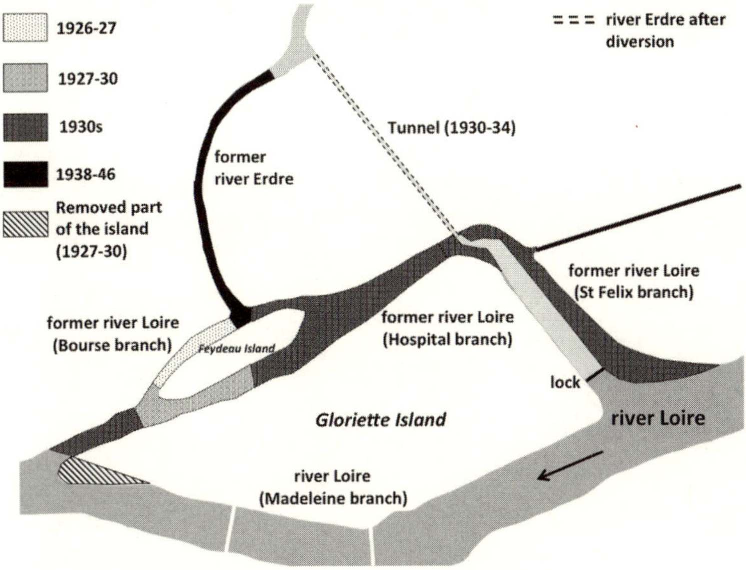

FIGURE 4.6. Nantes fillings chronology, 1926–1946.

A few pictures showing the places before the fillings and after them will also give the reader an idea of the magnitude of the changes. Figure 4.7 gives a sense of the space occupied in the city by the rivers that were filled.[29] Figure 4.8 shows the Loire as the Nantais could see it in the city center. This section as well as others near the city center would disappear between 1926 and 1939. Figure 4.9 shows the other river, the Erdre, in its crossing of Nantes. This river too would disappear in a tunnel and its former bed be filled. Finally, figures 4.10 to 4.12 show the fillings at different stages of progress and the landscape as it looked like after them.

WHO GOVERNED THE ENVIRONMENT AND HOW?

If this paper has focused on what happened between the two world wars, it must not be overlooked that the pressure to deepen the river initially came, in the nineteenth century, from the local chamber of commerce, which represented shipowners, shipbuilders, and industrialists. The latter progressively increased their weight to the detriment of the former two, but all shared an interest in the river's navigability.[30] It is understandable that having managed, after the slave traffic ended, to redirect the trade toward new destinations and products—sparing the city the economic decline that loomed—they worried

FIGURE 4.7. The center of Nantes and the Île Feydeau, photographed in 1911. Reproduced with the kind permission of the Archives Municipales de Nantes.

FIGURE 4.8. The Loire in front of the Dukes of Brittany castle (Saint Félix branch). Reproduced with the kind permission of the Archives Municipales de Nantes.

FIGURE 4.9. The lock on the river Erdre, just before the confluence with the Loire. Reproduced with the kind permission of the Archives Municipales de Nantes.

FIGURE 4.10 The filling of the Hospital branch, just under the brand-new Pont Maudit, rebuilt at great expense after the collapse of the old one. Reproduced with the kind permission of the Archives Municipales de Nantes.

FIGURE 4.11. The filling of the Erdre (1940). Reproduced with the kind permission of the Archives Municipales de Nantes.

about the silting up of the estuary. It is also logical that they received, in their quest for solutions, the support of the city council. But the long-term economic history of the estuary shows that alternative solutions to this endless struggle against the river existed—those that would be adopted later: an association with the city and harbor of Saint-Nazaire and a shift of Nantes's activities toward the mouth of the river. Today, the Port Atlantique Nantes-Saint-Nazaire is the premier French Atlantic harbor, but it has been relocated downstream from Nantes. In the city itself, any harbor activity has disappeared, apart from leisure boating. But in the nineteenth century, instead of choosing an association with its Saint-Nazaire counterpart, the Chamber of Commerce of Nantes opened a war against it and persevered in the belief that trade should go on as it had in the past.

This hostility was due to the refusal by Nantes's economic elites to accept the economic development of Saint-Nazaire at the very time when things had become so difficult for them. As early as the beginning of the century, the village of Saint-Nazaire had applied for the right to build a wet dock (it had to be voted on by Parliament). After long negotiations it finally won its case and the dock opened in 1856. The building of this dock had been supported by Nantes, whose chamber of commerce thought that it would become its primary port and improve its own trade. At that time, Saint-Nazaire had only 5,743 inhabi-

FIGURE 4.12. Dried up, carved up in its middle by a street as wide as a motorway and deprived of the nice old fish market that ornamented its eastern end (bottom of the picture), the beautiful Île Feydeau has now become a convenient parking lot! Reproduced with the kind permission of the Archives Municipales de Nantes.

tants while Nantes reached almost 110,000.[31] Although Saint-Nazaire had been an independent town since 1793, Nantes still considered it as an annex. But of course, the municipality of Saint-Nazaire had other views and did its best to develop its own trade, competing with the old historic city. This seemed an unacceptable treachery to Nantes entrepreneurs. This is why, when the silting of the river began to seriously threaten their trade, they proved ready to try any solution the engineers would propose to keep up the navigability of the river instead of associating themselves with their Saint-Nazaire counterparts who had, in the meantime, created their own chamber of commerce.[32]

The pressure put by Nantes on the Ponts-des-chaussées engineers was strong, and the problem they had to solve a difficult one. They certainly did their best to take up the challenge they were confronted with. They traveled to see what was done elsewhere, comparing the case they had to manage with other cases.[33] Divergences between engineers were important, as the history of the Canal de la Martinière shows. They attest to the complexity of the problem.[34]

What is amazing when you analyze the attitude of the Ponts-et-chaussées is *not* that they tried every method they could think of—this was, after all, what was expected from them. What *is* amazing is their disregard for social issues or for any issue that was not strictly the one they were in charge of. The catastrophe that happened to Nantes was not an engineering mistake. Engineers made no miscalculation. As soon as they developed the project they called a "tide basin," that is, to open the river widely to the tide, they knew that the rate of flow would be unbearable for the bridges and that in Nantes the lowest water level would be considerably lowered. They knew it and spelled it out in writing. But they paid no attention to the consequences this would have, focused as they were on their unique objective: preventing the sand from blocking off the river. That was an engineering problem, the only one that deserved interest. As for the trouble it could create for the city, they would see in due course, and find a solution. Any problem would eventually be addressed by technological solutions. The catastrophe that happened to Nantes is a result of the kind of training provided in the French State engineering schools (Polytechnique, Ponts-et-chaussées, Mines): excellent in mathematics and physics, overconfident in the power of technology, but absolutely ignorant of any other matters and especially of social and human issues and scornful of the views of those who had no scientific training. No course was ever provided in these fields. The prestige of these schools gave those who studied there an enormous power and unlimited confidence in the rightfulness of their views.

Repeatedly, Nantes's mayor, architects, and engineers argued that the filling of the river(s) would be an untold loss for the Nantais, reminding the Ponts-et-chaussées that the branches they wanted to fill were not just a secondary branch of the river, passing in the middle of nowhere, but urban rivers along which the old city had been built for centuries, equipped with bridges, quays, a lock, fluvial harbors, and a swimming pool. It was essential for the washerwomen, attended by anglers and walkers, constitutive of the shape of the city and of the landscape, and a source of beauty, pleasure, and entertainment as well as necessary to many workers, part of the local identity and culture. The Ponts-et-chaussées did not pay the slightest attention to these pleas, nor did they bother to respond to them. Several professional societies and voluntary associations protested, and the most widely read local newspaper, *Ouest-Éclair*, organized referendums that clearly showed that a large majority of the population disagreed with the works they witnessed.[35] One of the most influential local industrialists of the time, Louis Lefèvre-Utile (of LU biscuits), struggled bravely against the filling project, paying for the creation of an alternative plan.[36] Other entrepreneurs also protested,[37] as did several civil and professional societies. The Ponts-et-chaussées behaved as if the issues these citizens raised did not exist at all. They had to make navigation possible in the estuary, they would do so, the rest did not matter.

This attitude shows an overconfidence close to arrogance, an unlimited faith in technology, a way to think about the environment as something man could always submit to his will, and the world as to be governed by economy (the river "improvement" aimed at maintaining the harbor activity) and science, not by values such as aesthetics or culture. They lived and acted in total ignorance of a large part of social and human issues. This behavior was not particular to one leading engineer; in the period under scrutiny (about one century), several Ponts-et-chaussées chief engineers succeeded each other—Lechalas, Kauffmann, Marcheix. All kept to the same course of action. It was the behavior of the body as a whole, the most prestigious Ponts-et-chaussées, created in 1669 by Colbert, under the reign of Louis XIV, highly respected and endowed with an incontestable authority.[38]

The behavior displayed by the Ponts-et-chaussées did not limit itself to this belief that technology was likely to solve any problem. It also reveals the disregard in which they held elected authorities (city council and mayor), not to mention their fellow citizens. Not only did the Ponts-et-chaussées not inform them of what was happening, or did so only at the very last minute giving them no choice, but they consciously lied to them. The city authorities were usually not fooled. Information always leaked one way or another. But as it was not official, the city found it more difficult to defend itself.

Let's take an example. As I explained above, a railway crossed the city right in its middle, and the train passage through the city center was cut by twenty-three level crossings that made the traffic extremely difficult and dangerous. Rerouting the train around the north of the city was an emergency measure, and the decision had been formally made in 1920. But the Ponts-et-chaussées needed arguments to convince the ministry (whose agreement they needed) that their project for Nantes was the best. In spite of the agreement reached in 1920, for which the city had struggled hard, they began to hint that the filling of the hospital branch could kill two birds with one stone: a lot of sand would be taken out from the main river branches, the Madeleine and Pirmil branches (to "improve" them and make them "tide basins"); the shipping of all this sand would be very costly; using it to fill another branch just beside the place of extraction would be an excellent solution, they said. It reduced the transportation cost to a minimum. Furthermore, filling the Hospital branch would

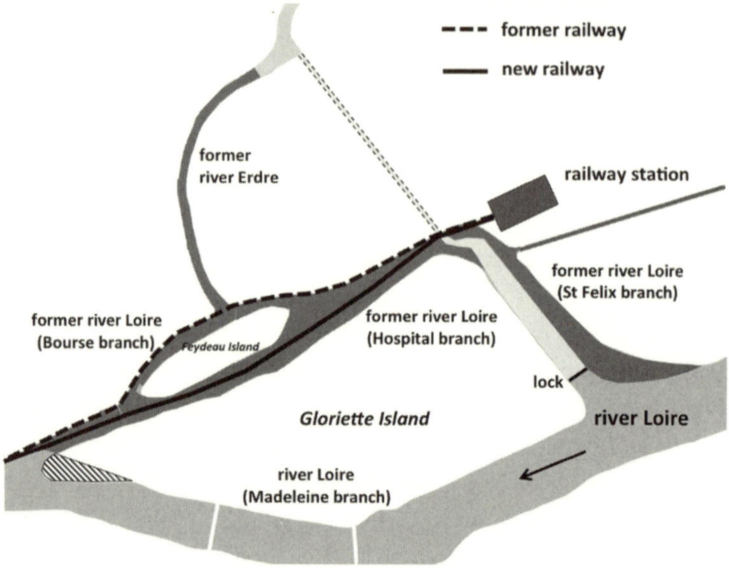

FIGURE 4.13. Former and new railway outlines.

open a free space through the city, from east to west—just what was needed for the new railway route! This would save costly expropriations north of the city. The space created by the filling of the river belonged de jure to the State, which could use it for this purpose. The only problem was that if the filling of the Bourse branch could, at most, be justified (it would supply the city with a welcome open space), that of the Hospital branch was definitely contestable: it deprived the city center of *any* river, as filling the Hospital branch also meant filling the Erdre. And as for the passing of the train on the former Hospital branch, it was in fact totally silly: it freed the Bourse branch quays, indeed, but the new railway would pass only one hundred meters or so from the previous one! This would not improve the traffic in any way, nor reduce the number of accidents, as many level crossings would obviously remain (fig. 4.13).

This plan was so contestable that it was not made public immediately. This is why in 1921, the Ponts-et-chaussées suggested filling only the Bourse branch, not the Hospital branch (fig. 4.13). But the records (internal letters, not supposed to be read by the city officers) attest that they already planned to fill the two branches. The city engineer, who opposed the project, let the mayor know what the Ponts-et-chaussées had in mind, explaining that their official plan hid another, an undisclosable and shameful one.[39] The long report written by the committee in charge of the survey regarding the Pont de Pirmil, also entrusted with the study of the competing propositions concerning the fillings of the north branch(es), also stated this: the filling of the Hospital branch would make it possible to move the railway onto the reclaimed ground, which was excellent, *but this argument should never be mentioned publicly.* No need to say that the ministry in charge would be most responsive to this unexpected saving!

The railway would finally pass there, on the reclaimed ground. As this decision was silly (the new track was as awkward for the urban traffic as the previous one) it was soon modified: the train would pass there, but in a tunnel. But to get down to the tunnel, it needed a distance, as the slope could not be too steep. And the train could not begin its descent before it had left the railway station (see the location of the station in fig. 4.13). It finally passed in a trench in which the train slowly descended from the railway station to the tunnel. This trench divides the city like a scar. Still today, inhabitants of the city who live south of this deep trench have to walk a long detour to reach the old city center, the castle, and cathedral that are only a stone's throw away! The urban landscape that has resulted from the conglomerate of "improvement works" was never designed by a city architect or even by an engineer. It is the result of a succession of small, shortsighted decisions taken independently from one another, on the basis of purely functional reasons (the river had to be deepened and the current increased, the reclaimed ground was cheap, the level-crossings were inconvenient and dangerous, etc.) that did not allow any space for aesthetics or for the comfort of citizens' everyday life. The latter not only were deprived of an appealing cityscape, they were granted a terrible one in exchange.

The environmental catastrophe that struck Nantes was in no way, as the Nantais still massively believe today, the only possible answer to a "natural" problem (the "wickedness of the river" destroying their city). It was a man-made catastrophe, due to the way the decisions were taken: piecemeal, without any overall strategy, distant from all democratic consultation. These decisions were made by an engineering body whose perspectives were strictly economic and technical, engineers who dismissed the elected authorities and even more so the ordinary citizens, who never had to answer either for their decisions or for their consequences. It is remarkable that such an important decision as the filling of three rivers and/or branches of a river crossing a large city could be taken, in the middle of the twentieth century and in a democratic country, without even the local elected authorities at least being consulted.

Today, what was done during the second half of the nineteenth century and first half of the twentieth is being undone (at least, partly undone, because the river branches that were filled cannot be reopened). The lost branches of the river will not be regained. But engineers are working to help the river come back to its previous balance. On the pages devoted to the "Plan Loire Grandeur Nature" (River Loire Improvement Scheme) of the regional environmental department (DREAL Center Val de Loire) we can read: "What is wrong [with the river]?"[40] The answer includes the works done in the nineteenth century to reduce the size of the bed and the massive extractions done in the river until the 1990s, both resulting in (among other things) the lowering of the low water level and a worsening of the flooding. The restoration of the low water level and of the dynamic balance of the river are among the main features of the scheme.

This clearly shows that a fundamental change has occurred in the perception of the rivers. But it also shows how wrong a decision can be made by experts who do not give any consideration to the opinion of their compatriots on the ground that they are no specialists. Undoing the wrong done in the past is certainly a good thing. But it would also be valuable to question why this was done and prevent, as far as possible, the repetition of wrong decisions by changing the way they are made.

Chapter 5

The New Cuyahoga

Straightening Cleveland's Crooked River

David Stradling

In January 1917, a committee composed of Cleveland's most influential engineers published a remarkable map entitled "Realignment of the Lower Cuyahoga River" (see fig. 5.1). Narrow and rectangular, it shows the terribly crooked Cuyahoga winding its way toward Lake Erie through the city's industrial flats, separating the regular grid of downtown to the east from the irregular neighborhoods of the West Side. Clustered along the lakefront and throughout the river valley are the arching lines of the city's many railroads, which at several spots fan into switchyards and terminals. Another set of thin arching lines marks changes in elevation, which in several places cluster to indicate the steep grades separating the city level of downtown and the West Side neighborhoods from the industrial flats below. With its detailed coverage of more than three square miles of the city's core, the map provides a graphic representation of Cleveland's physical complexity, as everywhere the straight lines of roadways meet the curving obstacles of railroads, riverbed, and bluff. The map's tangled and conflicting lines perfectly capture the creative and competitive energy at work in the city, as well as the persistent role of nature in shaping industrial Cleveland.[1]

By far the most striking feature of the map is the thick red line that runs more or less straight through the center. This shaded path represents the realignment plan, an estimation of where the new Cuyahoga would run. As proposed by Major Paul S. Bond of the Army Corps of Engineers; William A. Stinchcomb, Cuyahoga County's engineer; Robert Hoffman, the city's engineer; and Edgar B. Thomas, a consulting engineer who had served as the city's river and harbor engineer, the straightened river would occupy almost none of its original bed, and the river's famed bends would be no more. The

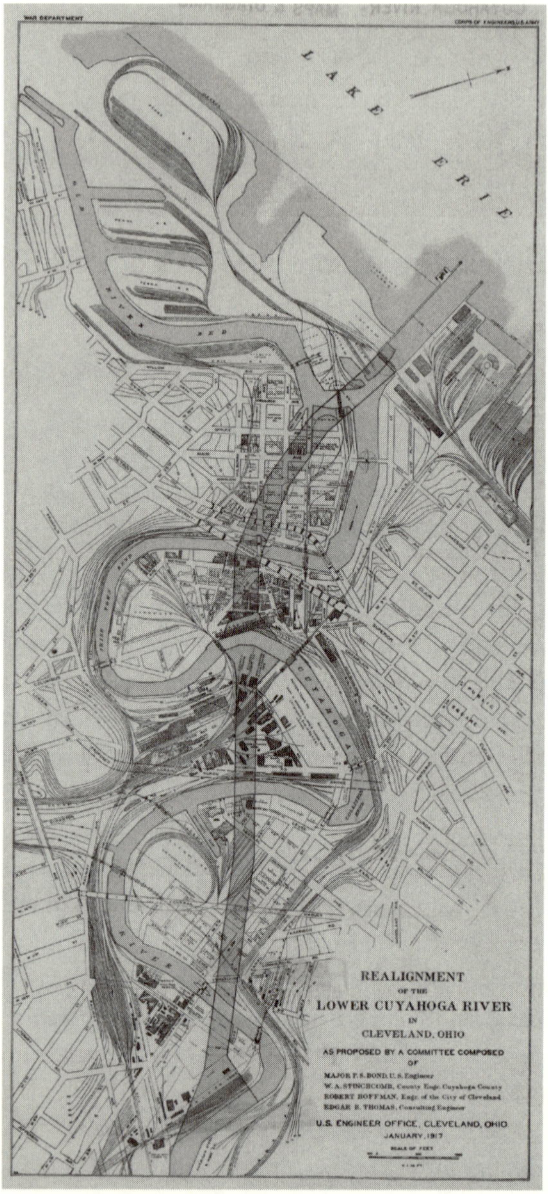

FIGURE 5.1. This 1917 map captures the most dramatic straightening plan, which proposed to replace nearly the entirety of the natural channel with a more efficient path to the giant mills upstream. Cleveland Chamber of Commerce, *The New Cuyahoga: River Straightening Recommendations by the Committee on River and Harbor Improvement of the Cleveland Chamber of Commerce* (9 Feb. 1917). Courtesy of Cleveland State University Special Collections.

plan was a remarkable bit of chutzpah—a straight line drawn in defiance of a crooked world—but it was well within the realm of possibility for the era's engineers.[2]

The engineers who drew up the plan were intimately familiar with the river and the city it had helped create. They were the leading experts on local engineering concerns, and they represented the most important constituencies engaged in transportation problem solving. In other words, this was not a fanciful endeavor by a group of dreamers and boosters but a serious proposal to solve one of Cleveland's most serious transportation problems. The collaboration of these four engineers had been suggested by the Chamber of Commerce Committee on River and Harbor Improvement, which had kept the straightening issue before city and federal officials for years. It had also kept the problem of the crooked river in the press. The committee's membership shifted over time, but year after year it included representatives from the city's large shipping firms, such as Pickands Mather and Company and the Pittsburgh Steamship Company, a subsidiary of U.S. Steel. Other members came from the city's heavy industry, including M. A. Hanna and Company and American Steel and Wire, another subsidiary of U.S. Steel. Altogether the committee had roughly two dozen members, some of whom served for many years, including George M. Steinbrenner, who represented the shipping firm started by his father.[3] The committee worked on a variety of transportation issues in the interest of facilitating industrial growth, even attending to seemingly mundane concerns such as dredging, dock repair, and bridge replacement. It also worked to improve the harbor through the extension of the federally funded breakwater in Lake Erie. But in the first three decades of the twentieth century the possibility of straightening the Cuyahoga gained the committee's most concerted attention, and the 1917 realignment plan was only one in a series backed by the Chamber of Commerce. Just a month after the committee of engineers had drawn up the new straightening plan, the Chamber published it, map and all, in a pamphlet hopefully titled *The New Cuyahoga.*

The study of a river, especially one that runs through an industrial city, provides a wonderful opportunity for understanding how complicated landscapes require the creation of complicated ideas about nature, natural resources, and industrial spaces.[4] In reading the words of Chamber of Commerce members as they contemplated the crooked river, one cannot help but be struck by the blurring of natural and industrial landscapes, the blending of natural and social resources. The Chamber's 1910 annual report, for example, simply defines the Cuyahoga as "the main industrial artery of Cleveland." Two years later the Chamber called the underutilized flats along the Cuyahoga above the navigable channel "waste land," while arguing that it should be seen as a "vast potential workshop."[5] Despite language such as "industrial artery" and "potential workshop," which clearly positioned the Cuyahoga and its banks outside nature, it remained a natural river, too, as Chamber members well knew, disappointingly crooked and unpredictable, prone to flooding and silting. Chamber members had to think of the Cuyahoga all at once as a critical component of the industrial landscape, as a natural resource, and as a force of nature.[6]

Historians have paid particularly close attention to how industrialization occurred inside workshops and mills, how capital intensification brought technological innovation, new labor arrangements, more complex corporations, etc. But industrialization took place outside factory walls as well. Urban historians might define industrialization as the struggle to remake the landscape to suit the needs of production, to increase efficiency and profit.[7] In a process described by William Cronon in *Nature's Metropolis*, economic elites and government entities created a second nature that more closely aligned with the needs of capital than had first nature. Through this long, uneven process, however, urbanites rarely fully distinguished between first and second nature, instead thinking of the world as mostly a mix of the two. As the story of the Cuyahoga makes clear, the struggle to make a suitable second nature was filled with conflict, compromise, and imperfection. Accomplishments were temporary, the obstacles posed by nature omnipresent. Try as they might, Cleveland's captains of industry simply couldn't reduce the Cuyahoga to merely an industrial artery.[8]

In the 1910s and 1920s, Cleveland's Chamber of Commerce was a collection of the city's economic elite engaged in typical boosterism. The organization was conservative on the topics of labor and taxation but progressive when it came to government activism, as was typical in that era. Indeed, Chamber members tended to be urban progressives, confident in government's ability to facilitate and (sometimes) regulate growth, and optimistic about the consequences of continued industrial development. Chamber members consistently conflated the interests of their businesses with those of the commonweal, assuming that the interests of industry aligned wholly with the interests of the community. Tellingly, the River and Harbor Improvement Committee did not concern itself with the other pressing problem on the river: pollution. For decades the Cuyahoga had served as an open sewer, with both industry and the city discharging all types of wastes into the river and its tributaries. Perhaps most famously, Kingsbury Run, which empties into the Cuyahoga at the spot where Standard Oil operated Refinery No. 1, had carried untreated waste from city sewer pipes, a slaughterhouse, a paint factory, and several refineries and oil works for at least thirty years before the Chamber of Commerce began its straightening crusade. Indeed, pollution discharged into the Cuyahoga had repeatedly spoiled the city's water supply, even though Cleveland's water came from Lake Erie rather than the river. Four times from 1874 to 1904, the city had to extend its intake pipe, from three hundred feet to four miles into the lake, to avoid contamination from the Cuyahoga's discharge. Thereafter, however, the pollution of the river itself persisted, and only after the state board of health demanded action did the city finally begin to work on a sewage treatment plant in 1912. Even then progress on water purification was remarkably slow. Through all of this, the River and Harbor Improvement Committee remained aloof, thinking that transportation improvements were critical to growth, while pollution remained a mere nuisance.[9]

The Cuyahoga is a surprisingly narrow river, at some points only 130 feet wide, even near its mouth. It drains just over eight hundred square miles of mostly agricultural lands in northeast Ohio, although it passes through other cities and towns before arriving in Cleveland, most importantly the small industrial city of Akron. Still, in the early 1900s, the Cuyahoga's waters were fairly clean when they entered Cleveland, and the city's most vexing problem concerning the quality of the inherited water involved the large amounts of suspended soil and floating deadwood that flowed out of the largely rural watershed. The river's slow-moving waters dropped a considerable amount of silt along the last few miles of its hundred-mile length, right where Cleveland needed an open channel.

By the time the movement to straighten the river gained momentum, the city had already widened the river's mouth, built long jetties to keep it open, and grown accustomed to the large annual expenses for dredging the navigable channel through the industrial flats. Many property owners had built bulkheads along the banks, easing the docking of lakegoing vessels. In 1904 the city had made a significant improvement to the riverbed, the Jefferson Cut, which straightened and widened the river south of the Erie Railway bridge and extended the navigable channel by nearly two miles. Since the completion of that work, industries had invested millions of dollars in new physical plants, including at Cleveland Furnace's pig iron facilities at the new head of navigation. After the Jefferson Cut's success became obvious, the focus of river improvement shifted farther south, where even more flat land awaited development, and to the north, where two extreme turns, Irish Town Bend and Collision Bend, had become bottlenecks.[10]

By 1910, as the straightening debate began in earnest, Cleveland was a booming industrial center. With a population well over five hundred thousand, it was the United States' sixth-largest city. It had more than doubled in size in the previous twenty years, and its growth continued unabated over the next twenty. One did not have to be a member of a boosterish organization like the Chamber of Commerce to assume that the city's future was bright. Commercial and industrial uses gathered where rail and lake travel met. Materials handling occupied much of the land in the Cuyahoga flats, where firms bought and sold lumber, stone, sand, and wheat. In 1911, more than ten million tons of goods moved in and out of the harbor in the hulls of lake steamers, which docked nearly four thousand times in Cleveland during the season of navigation, from March through December. Lake transportation and the old Ohio Canal, which connected the Cuyahoga south to the Ohio River, had sparked Cleveland's industrial growth, but it was fanned by the construction of railroads. By the early 1900s, rail lines laced the Cuyahoga Valley, and of the twenty bridges that traversed the river's navigable channel, eleven carried rail traffic. In addition to local short lines, such as the Cleveland Belt Line Railroad, some of the nation's largest rail systems had a significant presence in the valley, including the Baltimore & Ohio, Big Four, Erie, New York Central, and the Pennsylvania Railroad.[11]

Industry had begun to develop along rail lines throughout the city, though primarily on the east and south sides, most evidently along the Pennsylvania lines. But the largest industrial zones remained on the lakefront and along the Cuyahoga. Some of Cleveland's most important companies occupied the riverbanks, including Standard Oil, American Ship Building, the Sherwin-Williams Paint Company, and Grasselli Chemical. By far the most important industry, however, was steel, which by the early 1910s occupied large sections of the flats. The Upson Nut Company sat on the peninsula between Irish Town Bend and Collision Bend, while American Steel and Wire, Corrigan and McKinney, and Cleveland Furnace (purchased by Otis Steel in 1919) were farther upstream. With its pivotal position between the lakeborne iron ores of the upper Midwest and the railborne coal of Pennsylvania and Ohio, Cleveland stood with Pittsburgh as a leader in steel production. In addition, the availability of local steel encouraged the growth of a number of metal-intensive industries. In 1913, Mayor Newton Baker could declare, without seeming to exaggerate, "Cleveland, by virtue of its location, is destined to be, perhaps with the exception of Chicago, the greatest lake port."[12]

Indeed, Cleveland's future as an industrial giant seemed all but assured, except for one increasingly obvious problem: the crooked river that lay at the heart of the city. The Cuyahoga's bends limited the size of passing ships to 470 feet in length, and bridge obstructions limited them to 95 feet in height. Unfortunately for Cleveland's industrialists, newly constructed Great Lakes ships were almost always larger than this. By 1913, a Cleveland City Council investigation concluded that only about half of the American freighters in the Great Lakes could pass through the river's bends. Perhaps just as damaging to industry in the valley, the turns and bridges caused extreme delays in shipping, with the larger ships taking about five hours to travel the five miles from mouth to the upper reaches of navigation near Cleveland Furnace. Every day required a choreography of conflicting transportation systems, as ships waited for drawbridges to rise or swing out of their way, and trolleys, pedestrians, and, more and more frequently, automobiles waited at river crossings while ships inched by. Trains idled on tracks awaiting their turn to pass. Regularly scheduled bridge closings made this delay more predictable, but hardly less annoying. Altogether, poor planning and rapid growth in the city's core had created a jumble of rail lines, city streets, and high and low bridges, all interspersed with industrial and commercial buildings to create a formidable challenge to those who hoped to pass through the valley, let alone to those who would straighten the river to make it more useful.[13]

In the spring of 1909, several members of the Chamber's River and Harbor Improvement Committee toured the Cuyahoga on a government tug, joining a number of other men interested in navigation on the river, including members of the Board of Engineers for Rivers and Harbors, the body Congress created in 1902 to grant final approval to Army Corps of Engineers projects. The tug went as far as the Cleveland Furnace dock at the head of navigation, but then part of the group carried on in a smaller launch. From the water, the centrality of the Cuyahoga to Cleveland's economy appeared as obvious as the bottlenecks, and such tours became routine for those eager to improve the river. Later that year, the committee affirmed its opinion that it should be "the object of our endeavors to make the Cuyahoga navigable for any boat on the lakes; only by doing this can the Cuyahoga River valley hope to compete with other ports where ore can be melted." Interestingly, this assertion came in reaction to an Erie Railroad plan to build a new lift bridge over the river, a plan that had been submitted to the city and the Army Corps, which in turn passed it on to the Chamber for its opinion. Of course, the Chamber's opinion was that the Erie should do nothing that would contribute to the obstruction of the river.[14]

The Chamber's lobbying for a comprehensive improvement plan paid off in February 1911, when Congress authorized an Army Corps survey of the lower river, specifically for the purpose of determining how the Cuyahoga might be straightened. In that same year, and at the Army Corps's suggestion, the city created the River and Harbor Commission to study the problem and report to the mayor and council.[15] All of this activity struck the Chamber of Commerce as "the real beginning of the history of a greater Cuyahoga river." It applauded the Commission's report in a statement that perfectly mixed the boosterism of urban progressives and the then-powerful philosophy of conservation: "If we, as citizens of this sixth city of the United States, have a proper appreciation of our opportunities, and bear in mind the great tasks which other cities have undertaken and are undertaking to preserve and augment smaller natural advantages than ours, we shall take the results of the river survey and proceed untiringly, with fixed purpose, working through all possible agencies of cooperation, to develop and make available the great industrial possibilities of our inner harbor and the upper valley of the Cuyahoga." The Chamber saw no irony in the proposition that "natural advantages" should require untiring work to develop.[16]

The conceptual blending of natural advantages and human improvements was a hallmark of the Progressive Era's conservationist philosophy. Chamber members' rhetoric included all the other hallmarks of the conservation movement, too.[17] First, the Chamber clearly understood the importance of study and expertise. Before advocating a particular plan, the Chamber supported a detailed survey of the valley, to be conducted by the Army Corps, the experts most suited for the work. As the Chamber's annual report summarized in 1910, "The first essential to systematic progress is the knowledge as definite and complete as can be had, of the possibilities of the river, particularly from an engineering standpoint." The goal of this study was efficiency, a second hallmark of conservationism.[18] And, third, Chamber members had high expectations for government's role in problem solving. Its 1912 annual report noted that the Chamber had "pinned its faith to the government survey of the river as necessary to determine which plan or method of river improvement is most feasible."[19] The Chamber also assumed that government would fund and undertake the actual work of straightening the river. Further, Chamber rhetoric followed the common argument that government would protect democracy by assuring the wide distribution of economic benefits derived from the nation's natural resources. Indeed, supporters of river improvements repeatedly described government involvement in straightening the Cuyahoga as necessary to improve the lives of all Clevelanders. As the city's River and Harbor Commission concluded, "It is our belief that 'the greatest good to the greatest number' demands the improvement of the Cuyahoga River."[20]

In the fall of 1911, Colonel John Millis, the Army Corps's Cleveland district officer, emphasized these ideas in a preliminary presentation of his survey findings to interested parties. He then listened to the opinions of invited attendees, including local government officials, representatives from industry and railroads, and several members of the Chamber's River and Harbor Improvement Committee. In the ensuing discussion, it became clear that if the river were straightened the greatest good would probably not go to the greatest number. One group of investors, who formed the Cuyahoga Valley Realty Company, had control of much of the undeveloped land around and beyond the head of navigation, and so stood to benefit disproportionately from improvements that increased the size of ships that could reach their property. The company's principal owners included former Ohio governor Myron T. Herrick and his law partner Willard J. Crawford, with whom he had been engaged in other real estate deals. (Both Herrick and Crawford were active in the Chamber of Commerce as well, although they did not serve on the River and Harbor Improvement Committee.) It was obvious that these men stood to earn millions through rising property values if the river were straightened.[21]

Perhaps more troubling, other business leaders suspected that river straightening might actually harm their interests, including several railroads, some of which would be greatly disadvantaged by a reconfiguration of the flats. In addition, West Side businesses feared that prioritizing straightening would slow the construction of high-level bridges over the valley. Such bridges would facilitate and speed crosstown travel and, unlike drawbridges, never interrupt traffic for the passage of boats. The Chamber of Industry, composed of West Side businessmen, pushed for the timely completion of the proposed high-level Detroit–Superior Bridge, which would greatly ease travel between the West Side and the business and retail center of downtown. Created in 1907 to protect the interests of the West Side, the Chamber of Industry didn't have the clout of the Chamber of Commerce, but it had reasonable economic arguments against straightening. As the straightening plan moved forward, the Chamber of Industry claimed that "the sane and proper course appears to preserve the channel which nature designed," although with some modest improvements. Such an approach would be much less expensive and would not interfere with the high-level bridge.[22]

After considering local opinions and completing his survey, in June 1912 Millis issued a report outlining three potential straightening plans. He also estimated the cost of each approach. The first plan, which Millis recommended to the Board of Engineers, proposed a radical realignment of the river, cutting off all three major bends in the river and creating a new channel through the center of the narrow valley. Under this Plan No. 1, almost none of the natural bed would be used, much like the similarly radical plan that would be proposed by the local engineers in 1917. Millis's second plan was less radical in its design, though remarkably expensive. Its central feature was a new channel from Irish Town Bend that would reconnect with the river upstream from Collision Bend. The extreme cost of the plan derived from the necessity of taking a substantial property from the Erie Railroad and pushing back the bluff on the west side of the valley. The third plan, which gained almost no attention for years, proposed the widening and bulkheading of the current channel so as to make navigation possible for ships up to six hundred feet long. Indeed, this plan was more or less a continuation and acceleration of the city's established practice of periodically dredging and widening trouble spots.

Despite railroad resistance and Chamber of Industry opposition, by mid-1913, the Army Corps, the city's River and Harbor Commission, and the Chamber of Commerce's River and Harbor Improvement Committee had all issued reports supporting Plan No. 1. The following year, however, this plan failed to earn the support of the Federal Board of Engineers of Rivers and Harbors, meaning that the project would not receive federal dollars. The Board of Engineers expressed sympathy for the goals of straightening, even arguing that "the formerly despised lowlands toward the head of the valley are urgently required for factory sites." But the board recognized that each plan contained significant flaws, and so it determined that the issue required further study. Indeed, each of the plans introduced complications, including conflict with bridges or railroads, and had sparked local opposition. Given the lack of consensus in Cleveland, the board considered each plan merely "tentative." More ominously, it recommended that when the city finalized a satisfactory plan, the federal government should pay only the cost of dredging, which would be a small percentage of any radical straightening.[23]

In response to this setback, the Chamber supported the creation of the special committee of engineers that eventually developed the 1917 plan. Their effort was slowed by the shifting reality created by the building of the Detroit–

Superior Bridge, which was under construction from 1914 through 1917. At the same time, the city considered another high bridge, the proposed Huron–Lorain Viaduct, which would greatly affect straightening plans. Uncertainty about that bridge, the inability to arrange financing for any straightening plan, and undoubtedly the European war all helped create a three-year hiatus in progress toward the creation of a new Cuyahoga.

In 1917, even before the special committee of engineers issued its plan, the Chamber pursued a new path forward, attempting to use heightened concern about flooding to reinvigorate the movement toward a straightened river. Statewide flooding in the spring of 1913 had caused extensive damage to several cities, including Dayton, where hundreds of residents died. In Cleveland, the crookedness of the river and the many bridge piers worsened the flooding; lumber swept off the banks jammed behind abutments and low bridge decks, damming the river. Flood waters covered 2,740 acres of the Cuyahoga valley—including the "potential workshop" above navigation. In partial response to the flood, the state legislature passed the Conservancy Act of Ohio in early 1914, designed to "prevent floods, to protect cities, villages, farms and highways from inundation, and to authorize the organization of drainage and conservation districts." The law allowed the creation of conservancy districts via petitions, signed by five hundred freeholders (or a majority of freeholders) in the proposed district, and filed with the court of common pleas. Alternatively, a city or a combination of cities could request that the courts form a district, which turned out to be the path taken in Cleveland. An incorporated conservancy district could prepare plans for improvements and exert broad powers within the district, including eminent domain. Funding would come largely through the floating of bonds, to be repaid by special assessments on property within the district. In 1917, with progress stalled, the Chamber of Commerce encouraged the City of Cleveland and Cuyahoga County to request the creation of a conservancy district in an effort to secure funding for river improvements.[24]

The *Cleveland Plain Dealer*, the city's morning newspaper, had linked the straightening project to flood prevention while the floodwaters were still receding in 1913, but six years later the Chamber of Commerce was still making the argument in a pamphlet entitled *Conservancy and the Cuyahoga*. Adding flood prevention to its familiar list of reasons to fix the river, the Chamber now argued that merely protecting the valley from flooding justified the expense of rechanneling the river. Unfortunately for the Chamber, support for a Cuyahoga conservancy district was not particularly strong, especially for a district that encompassed the entire county and promised to spread the cost of the project well beyond the flats. What's more, the Chamber of Industry renewed its objection to straightening, arguing that since the plan wasn't really about flood prevention the use of the Conservancy Act was inappropriate. In addition, the West Side organization resolved, "We are opposed to the creation of a conservancy district because we do not believe, at this time, in the creation of an additional taxing body in Cuyahoga County, with unlimited powers." As always, the Chamber of Industry prioritized the building of high-level bridges, although it did not need to press that argument in the effort to prevent the creation of the conservancy district.[25]

As the conservancy issue made its way through the courts, in 1919 the city changed its focus to Plan No. 2, due to new realities created by the proposed Huron–Lorain Bridge. This plan then gained the support of the City Plan Commission and the Chamber, but it still required a county bond issue or the creation of a conservancy district to provide funding. The constant shifting of straightening plans, along with the shifting explanations for why Cleveland should straighten its river, didn't help the cause, and eventually the attempt to create a conservancy district failed when it died in the courts in 1921. Subsequently, all straightening plans sat unused through the 1920s.

The long uncertainty about what land might be taken by eminent domain for straightening purposes undoubtedly led to a lag in investment in the valley. *Cleveland Press* reporter John Love wrote a series of articles on the Cuyahoga in early 1925, warning that the city had been "slowly growing hollow at the center." In a bit of hyperbole, Love claimed, "the valley floor along the banks of the meandering Cuyahoga is nearer its original pastoral state than it was forty years ago." Love even seemed to blame the improvement plans themselves. "Somebody laid a curse on the valley when he drew the first hypothetical lines for straightening," he wrote. But more than the fear of future eminent domain takings had slowed investment along the river. Love noted that newer establishments preferred to be "on the level," out of the valley, where urban transportation, including walking, was easier. And, more important, Love made the same argument the Chamber of Commerce had been making for nearly twenty years: the bends in the river hampered the city's growth. Under the headline "Where the City's Future Is Threatened," Love and the *Press* pointed the finger directly at Irish Town Bend, through which modern ore carriers could not pass.[26]

After the failure of the conservancy district idea, the Chamber sought yet another way forward. Industrial leaders worked with the City Plan Com-

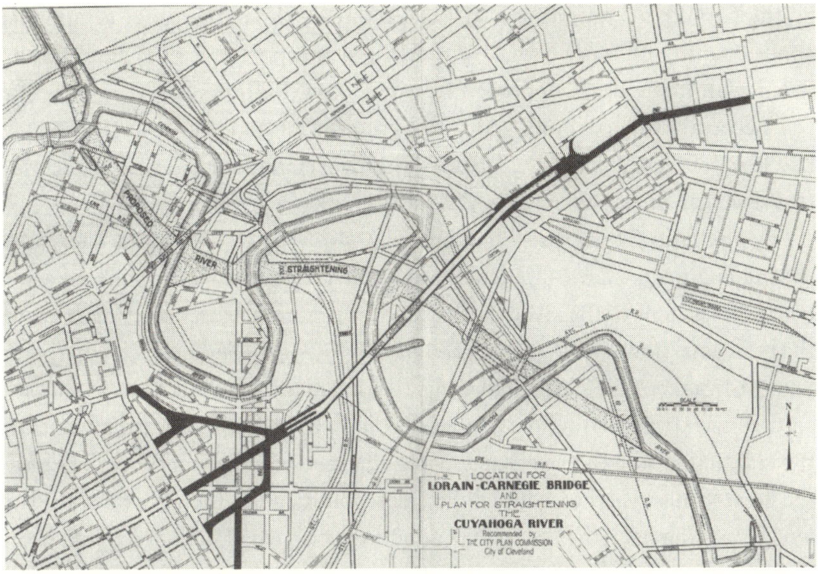

FIGURE 5.2. Uncertainty about when and where the city would construct high-level bridges complicated planning for dramatic river improvements. This 1926 map captured plans for the Lorain–Carnegie Bridge, which the city completed in 1932, and the straightened river, which never materialized. City Plan Commission, *Lorain–Carnegie Bridge and River Improvement* (City of Cleveland, 1926). Courtesy of Western Reserve Historical Society.

mission as it took up the Lorain–Carnegie Bridge, which replaced the earlier Huron-Carnegie proposal and which the people of Cleveland supported through the passage of a 1921 bond issue. The City Plan Commission solicited opinions on what should be done about the Cuyahoga so that it might coordinate plans for the bridge and the river. Some respondents said, "We ought to pave the river," while others expressed the opinion "that the community should immediately set about the straightening of the channel of this river to ample width to provide for the mooring of the largest lake vessels at the docks on each side of the river." Clearly the idea that the river should be primarily a transportation route remained dominant, but the commission concluded that the "chaos of opinion has had a very disastrous effect on land values within the valley and has absolutely prevented the normal growth of industry which might be expected in such a location." Edgar B. Thomas, who had served on the special committee of engineers in 1917 and still served on the Chamber of Commerce River and Harbor Improvement Committee, helped draw up straightening plans in conjunction with the new bridge. In a report published by the City Plan Commission in 1926, he made the old familiar arguments about opening up more land for industrial development. As the new coordinated plan made clear, improved transportation both up and over the river was necessary for continued industrial growth. Tellingly, however, the bridge had been designed and completed by 1932, carrying four lanes of traffic and two streetcar lines 180 feet above the still-crooked Cuyahoga.[27]

By the fall of 1929, the Chamber of Commerce had hired another consulting engineer, Charles Keller, to study the river again. Keller offered a sober reassessment of the river's utility and future prospects, featuring none of the bold assertions of the earlier realignment plans. Keller used a comparison with Milwaukee and Chicago to argue that dramatic improvements to the river, through straightening particularly, would provide only limited benefits to the city. Extending the river's navigable channel would probably benefit only the steel industry, he concluded. So, he recommended the adoption of Plan No. 3, pushing back the riverbanks at bottlenecks, which had heretofore been rarely discussed. It was the least expensive and least radical of the plans, and, Keller argued, good enough to accomplish more modest goals. Keller's report concluded, "In these days of heavy land traffic, a large river in the heart of a great city is no longer the benefit that it was in the earlier days of such places as Cleveland, Chicago and Milwaukee."[28] By 1929, the Cuyahoga had become as much an obstacle as a highway, and Keller argued that widening to improve travel *on* the river should not be made at the expense of travel *over* the river.

So the matter stood until 1937, when a river improvement project was finally under way, with the goal of increasing the length of ships that might use the river to six hundred feet. The work focused on widening, dredging, and "easing" the Cuyahoga, the latter being the word used by the Chamber of Commerce to describe the work. Improvements also involved removing and replacing several low and narrow bridges, while Works Progress Administration workers took a bend out of the river at the head of navigation, straightening the river in front of the expanding Otis Steel Mill. Ironically, in 1938, just as the city began widening bends in the river with the financing augmented by the Public Works Administration, Republic Steel opened the largest continuous strip mill in the world, more than a mile above the head of navigation, on the land planners had long slated for heavy industry.[29] Clearly industrial expansion was not predicated on the creation of a new Cuyahoga. Still, by 1947, navigation had indeed been eased, with significant improvements at several bottlenecks, including Collision Bend.[30]

As the city seriously debated a straightening plan for the final time in 1928, the *Plain Dealer* asked "Whose River?" The editorial noted that the steel industry, led by American Steel and Wire, had expressed opposition to the revived plan because it feared growing competition for the river's water. Expansion of the mills would require access to water, which the industry consumed in huge quantities for cooling metal. In essence, American Steel and Wire said it would rather have clear access to a free water supply than the cheaper ore costs promised by a straightened river, a reminder that industry used the Cuyahoga for more than just transportation. But like the Chamber of Commerce, the *Plain Dealer* was undeterred by the steel industry's opposition. The river should be straightened for the sake of the entire community, the newspaper argued, claiming that the Cuyahoga Valley was the best place in the United States for industrial growth. "Any industry calling for joint rail and water transportation will find here, in the heart of America's greatest consuming area, a location suited to its needs," the *Plain Dealer* claimed. Still, the valley was the best place for industrial development *only* if access to deep water could be assured through the accommodation of larger ships. Despite the ongoing debate and growing opposition to the straightening plan, the editorial ended with an assurance: "Cleveland's river will be straightened for Cleveland's sake. The interest it promotes is the interest of Cleveland."[31]

Of course, the *Plain Dealer* was wrong in its prediction that the river would be straightened and in the assumption that the city had a unified interest. Today the Cuyahoga still bends its way through Cleveland's industrial valley, its rough corners merely softened through piecemeal improvements. The Cuyahoga had been "streamlined," in the words of Mayor Harold Burton, who in 1937 finally admitted to the people of Cleveland that their river would never be straight. The "New Cuyahoga" promised by the Chamber of Commerce in 1917 would never materialize. Instead, the crooked Cuyahoga still marks the limits of the economic elite's ability to remake its environment to suit its needs, and serves as an object lesson in the city's conflicting interests and complicated politics.[32]

The "streamlined" Cuyahoga was just a part of the broader remaking of the Cleveland environment to suit the needs of industrial development, and the river was now straight enough to facilitate massive investments in the steel industry in the 1940s through the 1960s. New mills filled the flats south of the end of navigation, the former "waste land," just as the Chamber of Commerce had predicted fifty years earlier. New factories, including automobile plants, sprang up in surrounding suburbs. But even as capital investment continued, a series of river fires in the 1940s and 1950s, punctuated by a major blaze in 1952, foreshadowed problems to come, when a shifting economic geography would diminish Cleveland's "natural" advantages. Industrial growth with essentially no regulatory oversight had created a terribly polluted river, as sewage, oils, phenols, acids, heat, and tons of iron filings poured into the biologically dead and physically dangerous Cuyahoga. In the 1960s, as pollution flowed into the river, jobs, people, and capital flowed out of the city. This flight intensified in the 1970s, leaving behind the shell of an industrial city, and a much manipulated and terribly polluted environment.[33]

Some observers smelled trouble well before the 1960s and 1970s. In the spring of 1947, the urban economist Mabel Walker gave a remarkably prescient speech in New York to the Municipal Finance Officers Association titled "The American City Is Obsolescent," in which she used census numbers to describe the shrinking of central cities. "I think the simplest and briefest explanation of the population trend," she said, "is that people did not come to the city in the first place because they liked living in the city, but because the city offered them a means of livelihood." Walker noted that residential and employment opportunities were decentralizing, and that planners and politicians should not work against the will of the people. She concluded that the "creaking old structure must be readjusted to modern needs." The city would have to be rebuilt. Although Walker did not imagine just how far the decentralization would proceed, at least not in this speech, she described perfectly the severity of situation. "A new type of city adapted to present and future needs is struggling to evolve, but it is held back by the dead carcass of the past."[34]

In industrializing metropolitan Cleveland, the needs of production had often trumped the needs of the people, a dangerous imbalance that became all the more clear as investment and production slowed. The shell of the industrial city—factories, rail lines, inadequate housing, polluted and inaccessible waterfronts—would take decades to remake. As part of this process, after the river caught fire one more time in 1969, it became clear to most Clevelanders that the city needed yet another new Cuyahoga, this one less polluted and less prone to burning. And the many bends in the river? They could stay.

Chapter 6

A "Slum River"

The Unequal Urbanization of Bogotá (Colombia) and the Transformation of the Tunjuelo River in the Twentieth Century

Vladimir Sánchez-Calderón

Between May and June 2002 the amount of water carried by the Tunjuelo River (also called Tunjuelito) in southern Bogotá (Colombia) increased by one hundred cubic meters per second, an amount twenty to thirty times greater than its mean stream flow. These flow levels exceeded the storage capacity of La Regadera and Chisacá dams—located in the upper areas of the Tunjuelo basin—and caused floods downstream, which affected both Bogotá's main source of aggregates—a material of construction needed for reinforced concrete—and four low-income neighborhoods located downstream from the mining zone (see fig. 6.1).[1] These were not the first flood-related disasters to occur in this part of the city,[2] yet they made evident the active participation of actors not considered in the previous events. Until then, most official and news reports had directly blamed the river for causing the damage and indirectly blamed the neighborhoods' mainly poor inhabitants for settling there.[3] The 2002 floods made visible the participation that the aggregates mining companies and the Aqueduct Company had had in the modification of the river course and riverbanks for more than four decades. Taking the 2002 floods as a starting point, this chapter aims to analyze the interaction between Bogotá and the Tunjuelo River throughout the twentieth century, a period characterized by the urbanization of the river basin.

The Tunjuelo River urbanization has been part of Bogotá's overall urbanization, a process characterized by a highly sociospatially segregated structure, a feature common to most Latin American contemporary cities.[4] This

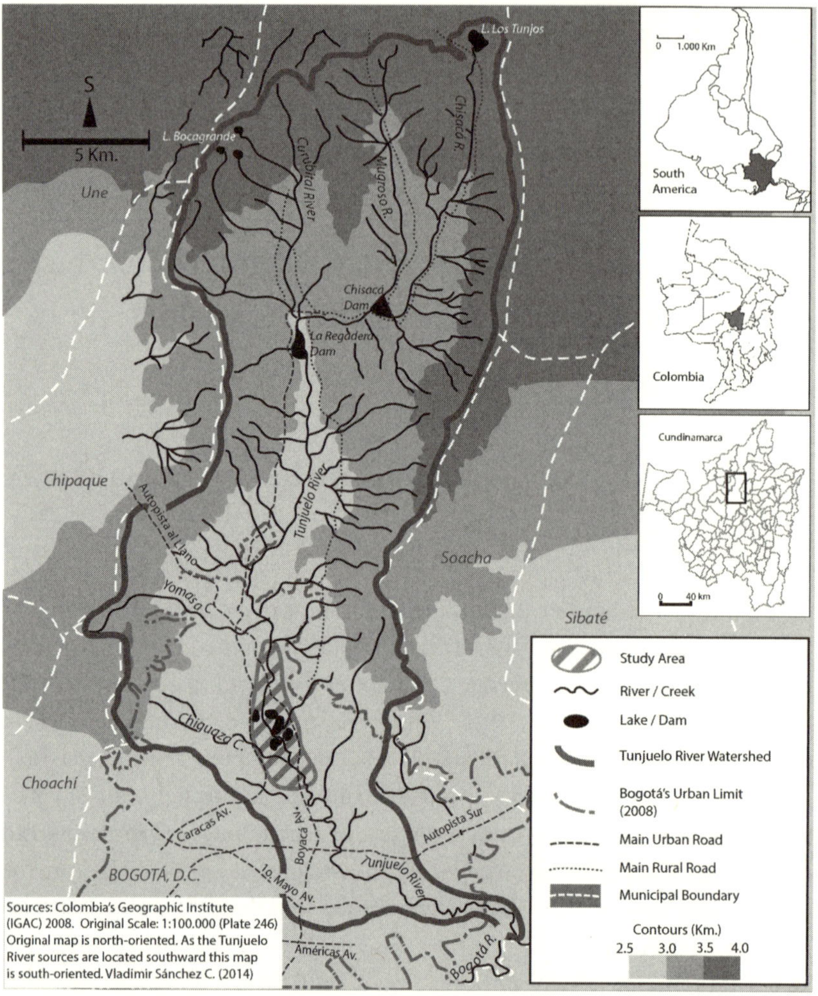

FIGURE 6.1. Location of the Tunjuelo River watershed and area of study.

segregated urban landscape was shaped, on the one hand, by the concentration of planning efforts in the northern part of the city, where the elites moved consistently since the beginnings of the century, and on the other, the predominance of "informal barrios" in the south (where the Tunjuelo River is located), settlements occupied mainly by poor people and characterized by little control and direct participation in state institutions and developed without observing the planning normative, a kind of urban settlement that can also be termed a slum.[5] In that way, this work shares the idea that urbanization is a socioecological and historic process that is never socially or ecologically neutral, and that it is necessary to consider "the question of who gains and who pays" for the environmental change that the urbanization of the river entails.[6]

The Tunjuelo River and Bogotá show, in a certain way, a typical situation between a city and a river, in which the economic and demographic growth of the first implies the ecological degradation of the second.[7] However, it emphasizes how that river degradation entails social inequality as well. In this way, it stresses how urban inequalities are not only "human" but "socionatural." Historic relations between Latin American cities and its rivers have not been much explored in historiography. However, the cases of Santiago (Chile), São Paulo (Brazil), Medellín (Colombia), and Buenos Aires (Argentina), which have been studied in recent years, suggest that links between social exclusion and a river's degradation are a common situation in the region's main urban centers. The importance of "self-building" in solving housing demand did not diminish over the twentieth century in the region, as was the case in North America or Europe. As a result, environmental problems derived from urban rivers transformation, such as floods, did not alleviate poverty after World War II, but have continued to be a threat up to now.[8]

In this chapter the periodical occurrence of disasters related to floods caused by the Tunjuelo River is analyzed, based on the use of oral, written, and cartographic primary sources, as a historical product stemming from the socioecological relation between Bogotá and the river during the twentieth century. In doing so, three main roles the river played throughout this period are discussed: the main water source between 1938 and the 1950s, a source of land suitable for low-income inhabitants beginning in the mid-1940s, and an aggregates source for the construction of the city from the early 1950s. This chapter contributes to the analysis of the historical interactions between rivers and cities in two different ways.[9] On one hand, it illustrates the relevance that the inclusion of other "natural" elements, beyond water and the river course itself (in this case aggregates and land), has for gaining a deeper understanding of the historic relations between the city and the river. On the other hand, the Tunjuelo case illustrates how an urbanization process characterized by a spatially segregated pattern has led to unequal environmental transformations in terms of: (1) The distribution of risks related to floods, affecting mainly "informal" (slum) settlements; and (2) A biased valuation of those risks by blaming the river's natural conditions and the informal urbanization for the recurrent floods, while at the same time ignoring/neglecting the effects that formal private and public companies had in those events.

Water for a Thirsty City: A River Gained for the Capital

Bogotá is the capital of Colombia and its biggest city with almost eight million people in 2010.[10] Unlike most of the biggest cities in other continents, Bogotá is located inland, in the eastern branch of Cordillera de los Andes at 2,600 meters above sea level (8,500 feet) and about 1,000 kilometers (658 miles) from the country's main seaports. Its Spanish foundation (1538) on the foothills of the Cerros Orientales (Eastern Mountains) was made near the rivers San Francisco and San Agustín, but relatively far from the most densely populated indigenous area, west of the Spanish settlement and very near the Bogotá River, the region's main river.[11] Despite great variations between its foundation and the end of the Colonial period (1810–1819), the demographic and physical size of Bogotá can be considered modest compared to other Latin American cities like Mexico City or Lima. Thus, in 1801, 21,394 inhabitants lived in an area of almost four hundred hectares. During the nineteenth century, Bogotá experienced a huge demographic growth; between 1801 and 1912 the population increased fivefold, although it did not experience the same physical expansion. The size of the city increased by only approximately one hundred hectares by the end of the nineteenth century.[12] This enormous densification entailed problems related to public and private health and the depletion of resources nearby, especially the forests in the hills, which in turn affected the supply and quality of the water that rivers could provide to the city. Thus, by the end of the nineteenth century, the amount and quality of the water distributed in Bogotá had significantly decreased.[13]

This impoverished water supply led to a search for other water sources that extended until the 1930s, when, after several technical reports from geologists and engineers, the mayor of Bogotá joined with the city council and Colombia's government in deciding to support the Tunjuelo River project.[14] Located at approximately 10 kilometers (6.2 mi) from Bogotá's historical center, with a total length of 73 kilometers and with its sources located nearly 3,800 meters (12,500 ft.) above sea level in the Páramo de Sumapaz region (fig. 6.1), planners believed the Tunjuelo would provide enough water for Bogotá's increasing population, which they thought would exceed 1.5 million people by the end of the twentieth century.[15]

The main element of the Acueducto Nuevo (New Aqueduct), as the Tunjuelo River project was called, was the La Regadera dam, located in the contact zone between the upper and medium basins, nearly three thousand meters above sea level. This dam and its related works, like the pipes for transporting water that extended twenty-two kilometers, meant not only the greatest physical intervention since the Spanish arrival but also the control of the river under the city's rule. This physical intervention was made possible by the political and administrative control the city enacted through several laws during the 1920s and reached a turning point in 1955, when the whole Tunjuelo River basin became part of Bogotá's jurisdiction by the annexation of the rural municipalities of Usme and Bosa.[16]

In 1938, as part of Bogotá's fourth centennial celebration, the Acueducto Nuevo was inaugurated with great jubilation by Colombia and Bogotá's authorities.[17] However, less than eighteen months later, this happiness vanished. Due to a strong dry season linked with an El Niño period, by early 1940 the water provided by the Tunjuelo had become insufficient to fulfill Bogotá's increasing demand.[18] The solution provided by Arthur Casagrande, one of the best-known US engineers in the dam-building field and who was contracted by the Bogotá government, was to build another dam, upstream from La Regadera.[19] On August 6, 1951, the date of Bogotá's anniversary, the Chisacá dam, a five-millon-cubic-meter water reservoir (one millon cubic meters more than La Regadera), was inaugurated. However, city planners and engineers had lost their faith in the Tunjuelo, especially because it could not sustain the increasing demand in a city that had surpassed the demographic estimations made when La Regadera was designed. So, while Chisacá dam was being built, another one, called Neusa, eleven times bigger than both dams of the Tunjuelo combined, was under construction.[20] Thus, by the end of 1950s the main sources for Bogotá's aqueduct were located in the northern part of the region. Since then, the Tunjuelo aqueduct system represented an ever-decreasing portion of Bogotá's water supply. In fact, by 2004, the Tunjuelo River accounted for only 4 percent of the water supplied to the city.[21] However, the dams built on the Tunjuelo meant that the city had physical and legal control over the whole watershed since the 1920s, a prelude of what happened in the following decades.

Aggregates for the "Formal City": A Gain for Bogotá, a Loss for the Tunjuelo

La Regadera and Chisacá dams point toward two distinctive features of Bogotá that reshaped its relationship with the Tunjuelo River since the second half of the twentieth century. On one hand, the city's physical and demographic growth demanded increasing resources over rural areas, in terms of water and, as is explained below, land and construction materials. In fact, Bogotá, like most of the largest cities in Latin America, experienced enormous physical and demographic expansion during the twentieth century.[22] Thus, in 1905 a little more than one hundred thousand people inhabited the city and occupied nearly five hundred hectares (1,235 acres), while by 1999 there were more than six million people living in almost thirty thousand hectares (almost 75,000 acres).[23] On the other hand, there was a technological change in the materials and techniques of construction, which were initially used for big infrastructure projects such as dams but soon afterward became extended to roads, pipes, housing, and building construction. This technological change is epitomized by reinforced concrete, a new material that mixed water, cement, iron poles, and a variable mix of different particles of rock called aggregates. Reinforced concrete increased both the resistance and malleability of the materials, making it possible to build taller buildings, longer pipe systems, and bigger and stronger dams.[24]

While both dams on the Tunjuelo River were built entirely of concrete, these were not the first projects to use this material in the city or its surroundings. In 1910, during the National Exposition that celebrated the one hundredth anniversary of Colombia's independence, the Compañía de Cemento Samper made a formal delivery of the first work made totally with Colombian cement, an event that was interpreted as the arrival of a new era of national and urban development. In fact, it was interpreted as a signal for the "Entrance to Modernity."[25] Although the speech made by the Samper Company exalted Colombian cement, the building they donated to the city would not be possible without access to another raw material, which was also local and played a prominent role in the wide use of reinforced concrete. Aggregates represent almost 80 percent of the total weight in a concrete-made work, and include all of the rock fragments between 0.06 and 150 millimeters long (0.002–6 in.).[26] So, the quality of concrete depends greatly on the quality of the aggregates used. According to Smith and Collis, "the essential requirement of an aggregate for concrete is that it remains stable within the concrete and in the particular environment throughout the designed life of the concrete."[27] Among other features, the mechanical properties are a key factor for estimating aggregates' stability. Until the 1940s most of the aggregates used in Bogotá's construction sites achieved a 2,000-psi resistance, an indicator of its mechanical properties considered to be insufficient according to international standards.[28] These aggregates came from two types of sources: quarries located on the surrounding hills of the city and pits located in some of the creeks nearby.[29]

As the cement companies needed to promote cement consumption, they realized that they needed to use better aggregates and bigger deposits in order to be profitable. In the case of Cemento Samper, they had cofounded an aggregates mining company named Central de Mezclas, which set forth to discover better aggregates deposits. By the early 1940s, Central de Mezclas had discovered a huge aggregates deposit suitable for exploitation (almost one hundred meters deep and one thousand hectares in area) in the Tunjuelo River, then located in the municipality of Usme, ten kilometers from Bogotá's downtown.[30] After some laboratory tests by US and Colombian laboratories, the estimated resistance of the concrete made with these new aggregates increased to 3,350 psi.[31] These characteristics, the extent of the deposit and the resistance of the materials, were considered unique in the surroundings of Bogotá, just at a moment when there was a debate in the city around the mining extraction in the nearby mountains.[32] Consequently, by the early 1950s, according to interpretations made during this research of aerial photographs, the extraction started in a site about one kilometer from the river's current course.

The Tunjuelo had, besides the mechanical properties of the aggregates, another advantage in terms of the economic and technical conditions of this deposit. Due to its geological history and hydrological dynamic, the Tunjuelo River made the production of aggregates easier for the companies. As previously noted, the river has its sources above 3,800 meters above sea level in the Páramo de Sumapaz region.[33] For that reason, the river and its main tributaries have been directly influenced by the alternation between glacial and interglacial periods over the last three million years.[34] During the glacial periods ice layers accumulated and acted like an enormous rock crusher, cracking and crushing stones, shaping long areas of material easily transportable by the water during the interglacial periods, when the stream flow was bigger because the ice melted and merged. So, the Tunjuelo River and its main tributaries had more charge capacity in those periods, transporting considerable amounts of material from the highest parts and accumulating it in lower and flatter zones where the stream flow loses its strength. The result is a cone-shaped geoform, known as a fluvioglaciar cone, thirty-nine kilometers from the river sources and composed of a complex mix of different rounded rock particles that vary from clay to cobbles and boulders (rock particles greater than sixty-four millimeters).[35]

These bigger particles were ideal for producing the aggregates needed for making reinforced concrete, because cobbles and boulders were big enough for crushing into smaller and irregular gravel-size particles. The gravel and smaller-sized particles, which were also produced by the river itself in the eroding and transporting process described above, had a disadvantage: their shape. According to a technical report made by a private company in 1968, most of the natural gravel deposited by the Tunjuelo was rounded or angular, the most preferred shapes for aggregates.[36] It estimated that, in that particular plot, "natural aggregates" made up 16 percent of the total production. That implies that the Tunjuelo had saved a lot of energy and labor for the private company.[37] In other quarries—located in the mountains near Bogotá's downtown, not in the flat zones—these kinds of particles were also produced, but the process was more expensive because it was necessary to use more human work and more energy.[38] In sum, the Tunjuelo River not only produced a huge deposit of better aggregates but made exploitation easier and more profitable in comparison to other deposits located in the city or in its surroundings.

Another feature of the Tunjuelo aggregates deposit, shared with other mineral exploitation sites far from the city's historical center, is the big land plot in which the extraction was made. Although fragmented and ostensibly smaller than the colonial haciendas, most of the fluvioglaciar cone was divided among few owners—nearly half a dozen—a fact that facilitated bigger explorations, more production, and profits.[39] Furthermore, only three big private companies exploited most of the deposit since the late 1950s: Central de Mezclas, Cementos Diamante (both with the participation of Cemento Samper), and Ingeniesa. These three main companies shared the exploitation of the Tunjuelo deposit with three other smaller companies: Carlos Madrid, Manuel Rey, and Sánchez y González, which also initiated the exploitation in the 1950s on a big land site owned by Fundación San Antonio, a subsidiary of the Catholic church.[40]

Research based on aerial photographs demonstrates that the exploitation of aggregates was more intense between the late 1960s and 1990s,[41] the more intense period in demographic growth and also a period of great urban changes on account of the consolidation of urban master plans, the expansion of architecture as a new profession, and the boost given to construction as a key economic sector for national development.[42] And the Tunjuelo's fluvioglaciar cone has played a fundamental role: according to several sources, this site supplied at least 80 percent of Bogotá's demand for aggregates in the period mentioned,[43] and several of the city's most significant buildings and urban sites have been built with these, including Bogotá's two highest buildings (Avianca and Colpatria), the International Airport (Eldorado), and the main highway in northern Bogotá (Autopista Norte).[44]

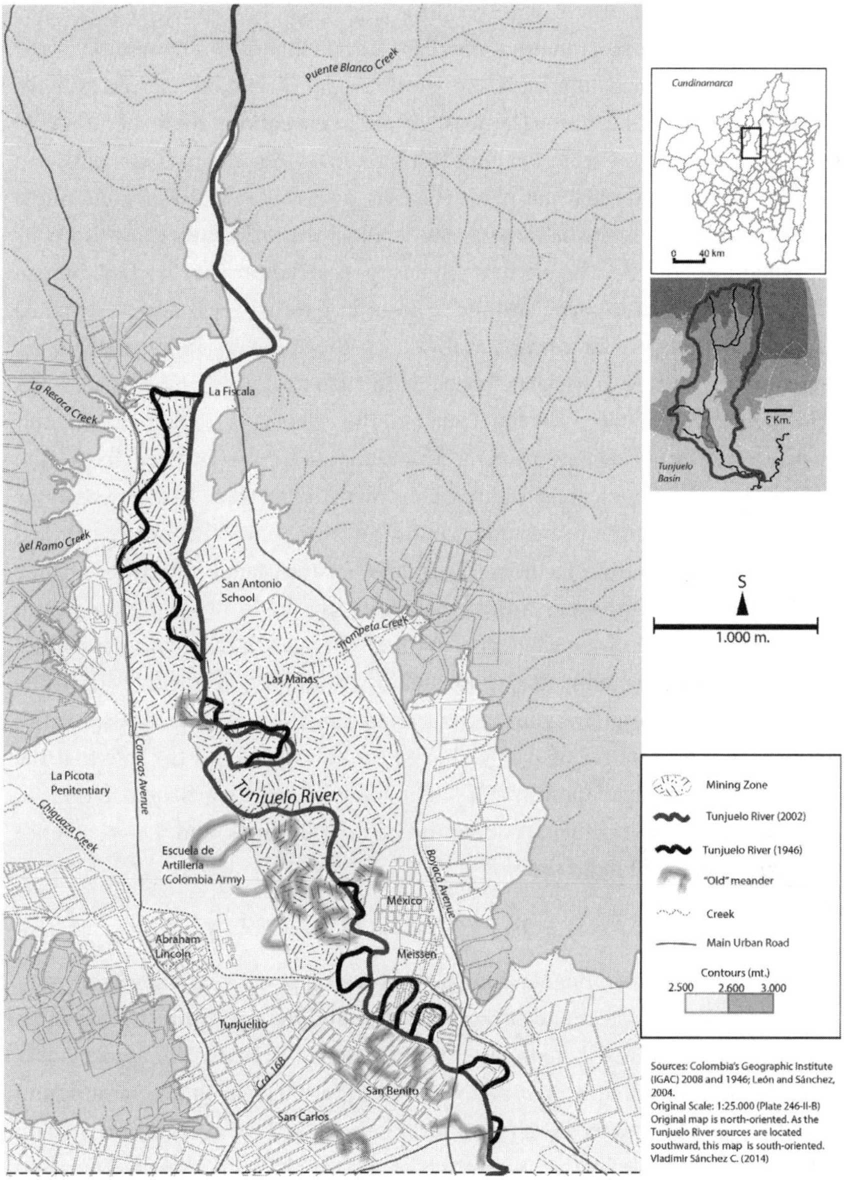

FIGURE 6.2. Tunjuelo River's course modifications in the aggregates mining zone (1969–1998).

The extraction of stones to produce aggregates has been made possible through a complex socionatural process that has included intervention in the Tunjuelo and some of its tributaries. The photointerpretation referred to above enables the identification of at least seven interventions over the Tunjuelo course made in the extraction zone during 1969–1998, the period mentioned (fig. 6.2). These interventions made the river linear in a transition zone where the river adopts a sinuous form (meanders), which previously diminished its speed and energy. As a result, more linearity means more speed, which in turn means more eroding capacity. In fact, according to the Secretaría Distrital de Ambiente (Bogotá's Environmental Office), the sinuosity index of the river diminished in the mining area from 1.84 in 1940 to 1.36 in 2005.[45] Also, when the meanders are cut, as in the Tunjuelo, the river has to adapt and modify its course by building new areas for accumulation of material and dissipation of energy. These new areas are located downstream from the interventions and tend to be prone to periodic floods. So, these interventions meant less probability of flooding in the mining zone but translated it to lower areas.[46] As a result, the aggregates mining has been "the main modeling process in the lower Tunjuelo watershed as a result of changes in the course, increases in the amounts of sediments, and instability of dikes and slopes."[47] Thus, the technological change in construction related to the adoption and spread of reinforced concrete resulted in a change in the uses of and the role that the Tunjuelo River played in the history of Bogotá—changes translated, in turn, to the modification of the river course and its surroundings, and the occurrence of disasters related to floods.

Land for the "Informal City"

As mentioned above, Bogotá, like most of the biggest Latin American cities, experienced the fastest and biggest physical and demographic growth of its history in the twentieth century, especially since the 1950s. This process was the result of several factors: rural to urban migrations related to economic and political displacements, the city's natural growth, and the availability of urban land made possible by the annexation of the municipalities located between Bogotá city and the Bogotá River in 1955. Thus, between 1951 and 1973 the city grew almost fourfold in population, from 717,250 to 2,855,065 inhabitants, and almost twofold in area, from 8,000 to 13,985 hectares.[48] Nevertheless, this extraordinary demographic and physical growth was uneven in sociospatial terms. Again as with most Latin America cities, this growth was characterized by a significant sociospatial segregation. Social groups, defined by economic income more than by ethnic, racial, regional, or religious differences, tend to live in different parts of the city and they participate differently in the city's production and reproduction process.[49] In general, this differential participation is expressed in unequal living conditions that can be expressed in degrees of access to different basic and specialized urban services.[50]

In Bogotá, this spatial pattern consolidated since the 1930s and followed a north-south axis. The wealthiest groups settled in to live in the northern sector of the city, and the poorest occupied primarily the western and southern peripheries.[51] Although there has been a great diversity in the modalities of housing access and in the ways of producing urban space in the Tunjuelo's watershed,[52] the main form of satisfying the housing needs has been through a process currently called "informal urbanization."[53] In the case of Latin America, this term comprises a multitude of ways of producing urban space that share, at least, two main features: (1) an incomplete fulfillment of the legal urban planning requirements, and (2) a stronger role played by the inhabitants in the physical building of their homes and neighborhoods, commonly known as "self-building."[54]

In the case of Bogotá, both the relative abundance of land suitable for urban expansion in the southern areas and the spatial concentration of the elites in the northern sector of the city acted as trigger factors for informal urbanization of rural sections, shaping the spatial pattern mentioned above. Some calculations estimate that between 1950 and 2000, informal urbanization represented 33 percent of Bogotá's physical growth, a proportion that increases up to 60 percent in the case of the Tunjuelo's watershed, being especially important in the *localidades* of Ciudad Bolívar, Usme, Rafael Uribe, and Tunjuelito.[55] In fact, the first barrios in the Tunjuelo's watershed, dating from the mid-1940s, had informal origins.[56] All of these settlements are located in a flat, though not homogeneous, terrain downstream from the mining areas described in the previous section, and preceded those extraction activities. Some of these settlements were established in the surroundings of the Camino a Usme (road to Usme) a road that was built during the construction of La Regadera dam and followed a pathway that existed since precolonial times.[57] These barrios contrasted with the urbanization of the surrounding hills (current *localidades* of Ciudad Bolívar and Rafael Uribe), a process that accelerated since the 1960s and was significantly shaped by small sand and clay mines.[58] Areas upstream of the aggregates mining zone (*localidades* of Ciudad Bolívar and Usme) also started to be occupied in the 1960s in a process that continues to the present.[59]

In contrast to formal settlements, in informal neighborhoods private developers, called *urbanizadores*, associated with landowners, promote the sale of unbuilt land plots. The decision of moving into distant and peripheral locations, as has been the case of most of the Tunjuelo's inhabitants, has been supported by confidence in its eventual "formalization" or its incorporation as a "formal" settlement in terms of planning requirements and property rights.[60] But formalization is a process that usually takes several years and implies long negotiations with the city's authorities. Therefore, the dwellers of a *barrio informal* usually experience a long period without access to basic public and private urban services such as roads, aqueduct and sewage pipes, electricity, education, health, and recreation facilities. In the case of the barrios included in this study, the lack of these services lasted around ten years and was characterized by the emergence of a wide range of communal responses that varied from the self-construction of pipes, roads, and churches to de facto acts such as strikes, legal complaints, and clientelist agreements in election periods.[61]

According to oral history inquiry conducted in the study area, one of the most critical issues that people experienced in the beginnings of the barrio was the lack of water supply.[62] Here, the Tunjuelo River and its tributaries in the current urban zone were fundamental for supplying that service.[63] For example, one of the founders of the barrio México said, "It was only grass over there. The river passed farther from where it is nowadays, but it was very clear. In those days [ca. 1968] there were no pits; the water was pure. We went there for laundry and body cleansing for children. We found some *piedritas* [stones] and on Sundays everybody washed their clothes."[64] For cooking, home cleaning, and adult bathing, people had to pick water up from public tanks, where the water (provided by the Aqueduct Company) was available one or two days a week. The same woman from the México settlement said that "for eating, there was a *pila* [water tank] in the [barrio] Meissen, and sometimes we could take water from it. My two older kids were in school, and after school they changed their clothes and filled up some cans we had and sold water to the neighbors."[65] In contrast to the urgency with which people demanded clean and potable water, disposal of domestic wastewater was solved in a more pragmatic way—here the Tunjuelo River and its tributaries also played a fundamental role, in this case by acting as an open sewage channel. In the early stages of the settlement, with few people living nearby, sewage use was not necessarily considered in opposition to the other uses mentioned above. Nevertheless, when new neighborhoods were built and the old ones had consolidated and continued growing in population, the quality of the water impeded those alternative uses.[66] Furthermore, as the concrete demand increased in the rest of Bogotá, more sediments were added to the river in the aggregates mining area, contributing to the river pollution.

Floods in the Tunjuelo: A River Lost in an Unequal City

In addition to the water supply and sewage disposal, the Tunjuelo River and its watershed have also offered another important resource for Bogotá's urban inhabitants. It has been the terrain itself, the material piece of earth over which the houses and the neighborhood have been built. Some years before the extraction of aggregates started, several *barrios informales* began to be built in the section downstream from the extraction area in a lower section of the fluvioglaciar cone. This geoform does not exhibit great topographic differences and it can be viewed as a flat or almost flat terrain where the Tunjuelo River periodically overflowed during the rainy seasons (May through November).

Although it was not the most important aspect to be considered when buying a land plot, interviewees usually stated that they (or their parents) preferred a flat terrain over a hilly one, because it was easier and less expensive to build on than the former. Nevertheless, as they soon learned, building houses on these terrains ended up having some serious issues. Another woman, who has lived in the Abraham Lincoln settlement since 1948, said that "shortly after we moved here, it started to rain, and the first rainfalls made our *lote* (land plot) waterlogged. There was no cement on the ground, only mud for several days. Every rainy season was the same until we could make a landfill and several years passed before the Aqueduct Company built the sewage pipes."[67] On a closer inspection, the fluvioglaciar cone is not as flat and even as expected at first. In general, areas that are higher and consequently less prone to flooding are located south of the river course. The upper mining zone is about twenty-five meters above the lower housing areas (San Benito and Meissen Barrios). But lower areas were also produced by the constant course change that the river made over the last few thousand years in the fluvioglaciar cone. As a way of diminishing speed and energy the river adopted a sinuous course and as a result of that process, several wetlands and old meanders (oxbows) existed prior to the urbanization. Those areas were lower sectors—between 0.5 and 3 meters—that also became flooded with heavy rains or with the river outflows. Registers of flooding for this part of Bogotá in press archives can be found since 1949, although since the 1960s this kind of event never happened again in most of the neighborhoods located downstream of the aggregates mining zone. In Venecia and El Carmen settlements, for example, located three to four kilometers downstream from the area of study, these wetlands were not directly connected to the Tunjuelo River. In those cases, draining works were made over several years, by landowners first, then by inhabitants, and finally by the Bogotá's Aqueduct Company in the 1970s, the decade when the First Aqueduct and Sewage Master Plan was developed.[68]

However, floods continued to be very common in four neighborhoods—Tunjuelito, San Benito, México, Meissen—until 2002 (see fig. 6.2). These neighborhoods were located immediately downstream from the aggregates mines and very near to the Chiguaza Creek, Tunjuelo's main tributary in the urban zone, and also built on the lowest terrain in the area. In these cases, landfilling, instead of drainage works, was preferred for adapting the earth. In contrast to drainage, landfilling, a task that sought to increase the level of the *lotes*, was made mostly on an individual basis. As a result, the level of land lifted varied from house to house, which in turn represented more or less vulnerability to the periodic floods. One of the local leaders of Meissen stated, "We had to raise our house, and some other houses have kept a lower base level. They became inundated when it rains a lot.... I think we landfilled almost one meter at least... we use any kind of material, especially construction debris."[69] So, even though a lot of landfilling work was done by the settlement's inhabitants for increasing their house level, it did not prevent them from flooding during the rainy season. People from this settlement described landfillings of above 1.5 meters (5 ft.), a task that demanded a lot of work, material, and money, and yet the water table during floods could exceed that level.[70] Indeed, in some floods, like one in May 1996 that affected the San Benito and Tunjuelito settlements, the water level reached almost two meters above the houses' level.[71] The works made by the inhabitants to deal with floods were complemented by the river deviations, linearization, and channelization works and course control by artificial dikes made by the Aqueduct Company since the mid-1970s. For example, according to aerial photographs, at least three meanders were cut between 1977 and 1986 in this area, works that were complemented with artificial levees in both of the riverbanks (fig 6.2).[72] However, as the floods of 1996 and 2002 showed, these engineered works were not enough for completely averting the disasters related to the river dynamic.

From the mid-1950s until 2002, the persistence of floods in these four settlements was recurrently evaluated both in terms of the settlements' informal origins and location on inadequate physical areas and the river's variable water level.[73] Indeed, the Tunjuelo River exhibits a great variability in its flow, which in an average year ranges from 3–8 m^3/s during the dry months (December through March) to 70 m^3/s during the wet ones (especially June through July and October through November).[74] In fact, La Regadera and Chisacá dams were built calculating that maximum transport capacity (70 m^3/s). Nevertheless, as noted in the introduction, between May and June 2002 the amount of water carried by the Tunjuelo River increased to 100 m^3/s, reaching a maximum of 180 m^3/s on June 9, 2002, demonstrating once more its great variability.[75]

What is interesting about the historical relationship between Bogotá and the Tunjuelo River is that only from 2002 did it become evident to "all the city" that the current river condition and the recurrent floods were not only consequences of the informal settlements and the river's natural dynamics but also associated with the mining activities and the Aqueduct Company's interventions on the river. In spite of having technical evidence of their impacts since the late 1960s, mining, dams, and works on the river course were not taken as factors that could contribute to the periodic occurrence of floods.[76]

The reasons behind the invisibilization of the Aqueduct and the aggregates companies in the explanation of the recurrent floods have not been explored, but I would like to suggest one possible reason. Both the public and the private companies are part of the "formal city," in contrast to the "informal settlements." The latter have been seen as opposed to planned and rational urbanism as well as to the market-oriented formal economy. Thus, most of the terms used for characterizing informal settlements lead to a negative valuation of these spaces and its inhabitants. "Clandestines," "illegal," "spontaneous," "subnormal," and *piratas* (pirates) are terms that appear both in press reports and in Bogotá's institutional reports since the 1950s.[77] Although there have been great changes in the city's public policy toward informal urbanization, it continues to be perceived as an "incomplete" or "inadequate" urbanization.[78] Under this perception the incomplete urbanization is assumed to be the cause of recurrent floods and, in contrast, the interventions over the river made by the Aqueduct and the aggregates companies are assumed to be "good" due to their intentions of controlling the river (in the case of the Aqueduct), and making it more profitable (in the case of mining activities). In other words, in this view, "good interventions" cannot be responsible for the floods, and consequently Aqueduct and mining activities are not assumed as possible causes in the reports on floods. Furthermore, in this reasoning there is almost a parallel between the river and the informal settlements (or slums). Both are seen as causes for the floods because they were not controlled or produced under the "formal city" model, and what was needed to deal with the recurrent disasters was, in the case of the river, its complete channelization in the urban zone,[79] and in the case of settlements their "formalization."

It is also important to note that inhabitants of the *barrios informales* were not just passive victims of the Aqueduct Company and the aggregates companies. They also helped to consolidate this perception. For example, in July 1968, after a great flood in the Tunjuelito and Meissen neighborhoods, a local leader urged the following list of demands: construction of a new dam down from Usme settlement, linearization of the river so as to avoid "so many curves," and the closure of the aggregates extraction companies [*areneras*].[80] In sum, the assumption of the informal city as opposed to the formal city led to a misrepresentation and undervaluation of the roles that formal actors could have played in the recurrent floods that affected some informal neighborhoods on the Tunjuelo's riverbanks. Each flood event, moreover, reinforced the "formal view" about the neighborhoods and the river. This view also obscured the active process that inhabitants had in the transformation of the river and its surroundings, not only in making "inadequate" decisions as has been stressed in this "formal view," but modifying and using the river resources for building a living space.

The Production of a Slum River

The Tunjuelo River has been associated for a long time in Bogotá with the recurrent disasters related to floods. The most affected have been inhabitants of informal neighborhoods, some of whom have lived there for more than fifty years. Most interpretations have blamed the river and the settlements for the disasters. This article has shown that the urbanization of the Tunjuelo River during the twentieth century has entailed a profound transformation of the river made not only by its inhabitants but also by formal agents, both private and public. This transformation has been part of Bogotá's urbanization process, intensified since the 1940s. This process, common in the Latin American context, has been characterized by an unequal and segregated spatial pattern, where elite groups have settled in very specific sectors of the city, in this case in northern Bogotá, and influenced the focus of planning initiatives in those sectors. Less powerful groups, on the other hand, have tended to settle in peripheral areas such as the Tunjuelo watershed, where informal urbanization has predominated, with the building of slums. The interaction between Bogotá and the Tunjuelo River shows how the segregation of the city has resulted from several connections between different actors and elements.

Water and especially aggregates have been resources located in (and partially produced by) the Tunjuelo that have conditioned not only the transformation of the river but also the urbanization of Bogotá itself. Without such a big deposit suitable for exploitation, for example, the city would have had to turn to less adequate and more distant aggregates sources, causing construction to be more expensive.

The particularity of the Tunjuelo River urbanization shows the possibilities that including other resources beyond the water has for the historical analysis of city-river interactions. Bogotá not only demanded water from the river; the geologic history and hydrologic dynamics of the Tunjuelo offered a unique opportunity for the construction of the city. At the same time, this same area offered land for poorer groups who were very active in the adaptation of the riverine areas. The analysis of how these two resources, aggregates and land, conditioned the urbanization and the transformation of the Tunjuelo River reveals other ways of investigating the links between cities and rivers. In this way, not only the inhabitants or its barrios were labeled as informal, illegal or uncontrolled, the Tunjuelo River was also characterized with those adjectives. At the end of the century, the Tunjuelo was not merely *a* river, it was indeed a "Slum River."

Chapter 7

Urbanizing a River in a Bicultural Border Region

Strasbourg and the Upper Rhine on the Way to Water Modernity, 1789–1925

Christoph Bernhardt

Strasbourg is a French city that in its long history since Roman times has drawn much profit from its geographical position at the Upper Rhine. Serving as a traffic junction of major axes of communication and commerce from the Atlantic to middle Europe and from northern Germany to Italy, the city was a head of navigation on the Rhine for centuries. It also became a strong political actor as a member of the powerful Rhine municipal confederation beginning in the thirteenth century and an intellectual center during the Renaissance. Lucien Febvre has celebrated Strasbourg's role as a node of communication and as a cultural and commercial metropolis in the Middle Ages and the early modern period.[1]

But the city's golden age came to an end in the days of the French Revolution of 1789. This chapter retraces the major lines of a fundamental economic and environmental crisis that mainly resulted from the decline of shipping together with other water-related problems and would last more than a century. The main functions that the city's two rivers Ill and Rhine had for Strasbourg became critical during the nineteenth century and culminated in three major problems:

Floods increasingly endangered and destroyed parts of the municipal area.

A massive lowering of the regional water table caused by large-scale river regulation schemes put an end to shipping.

The water quality of the urban river Ill and of the regional water resources increasingly declined in the second half of the nineteenth century as a result of population growth and insufficient clearing of waste waters.

This chapter analyzes how these water problems were perceived and dealt with by the local administration. Shifts in urban governance as well as in the balance of power between the state and the city shaped water management—and in return changes in the waterscape challenged the political actors and triggered political transformation. In a politically and culturally divided border region between France and Germany the footprint of national policies on the urban and regional waterscape became especially evident in the two periods of national polarization after the Franco-Prussian War of 1870–1871 and World War I.[2] In these periods strong conflicts between parts of the urban population and the newly established German (1871) and French (1918) public administration arose, as will be shown at the end of the chapter.

Economic Decline and Reforms in Regional Water Management (1789–1840)

The crisis of the nineteenth century was not the first that Strasbourg had to overcome since early modern times. A long period of stagnation that lasted several decades was caused by the Thirty Years' War (1618–1648). It was only from the beginning of the eighteenth century onward that Strasbourg slowly regained its position as the major commercial and cultural center along the Rhine, ahead of Frankfurt and Mainz. Strasbourg's population grew from 32,500 (1709) up to about 50,000 (1789). At the end of the eighteenth century Strasbourg was again able to play its former role as "the" city and "metropolis" in the Upper Rhine region, which by that time was legally divided into numerous small independent political territories.[3]

The French Revolution put an end to this period of prosperity, as it deprived Strasbourg of its century-old privileges as a free harbor and as a transit node for navigation.[4] On the right bank of the river the Grand Duchy of Baden was established in 1803–1815 as a new state and strong political player. At the same time the regular works for flood prevention along the Upper Rhine were stopped as a consequence of the military conflicts between Napoleon's armies and the German states. The riparian municipalities increasingly resisted financially supporting projects of flood defense, and Strasbourg completely stopped its contributions.[5] In 1801–1802 the region was hit by a major flood that forced the villagers around Strasbourg "to continuously fight against the river."[6] As a consequence, numerous local water wars escalated. When in 1802 French water engineers started provisional works for a cut through a bend of the river upstream of the city the inhabitants of Strasbourg's neighboring village of Auenheim were afraid of losing a large area of land. They defended themselves by violently attacking the workers so that military forces had to be mobilized and the project was abandoned.[7]

In the first half of the nineteenth century economic decline and physical damages to the landscape caused by the Rhine fatally coincided for Strasbourg. While public and private revenues drawn from navigation were declining, floods increasingly attacked the city's communal land. This was especially true for the municipal forests. The town council estimated that during the two decades from 1818 to 1838 about one hundred hectares of precious timber land was taken away by the river.[8] This loss seriously hit the municipality as well, which drew about 15 percent of the municipal budget from its forests, as the citizens used the forests for many purposes.[9] Other parts of the municipal property could be only provisionally defended against the attacks of the river by large investment. The Rhine also repeatedly flooded a branch of the river—the so-called *bras Mabile*, or alter Rhein—which passed Strasbourg's military citadel. The French State was very concerned that the citadel would be damaged, so the government set up several projects for river regulation in the following years that concentrated on the bras Mabile and the area around the bridge between the cities of Strasbourg and Kehl.[10]

Flood Protection or Navigation? The Ambivalent Effects of River Regulation for Strasbourg

The Vienna Treaty of 1815, which defined the layout of the post-Napoleonic political order in Europe, legally fixed new territorial boundaries along the Upper Rhine. The *Thalweg* (main channel) of the river was confirmed as the borderline between France and the Grand Duchy of Baden. In the following years the corps of engineers of the two states developed new projects of river regulation in the Strasbourg area. Between 1821 and 1829 a series of costly dams and other measures of flood prevention was realized between Strasbourg and its German twin city Kehl on the opposite side of the river. As a result, the Rhine was partly moved eastward and forced into a new bed, and the bras Mabile was provisionally consolidated.[11] But the consequences of the project were highly ambivalent for Strasbourg: On the one hand, moving the river eastward meant a better protection of the city and the citadel against floods. But on the other hand, in the long run this intervention caused a great and irreversible lowering of the regional groundwater table. This fundamental change in the region's hydrological profile not only increasingly complicated the reclamation of drinking water for the city; even more problematic were the harmful effects on shipping. As the unintended consequences of the river regulation scheme

FIGURE 7.1. Regulation Scheme for the Upper Rhine region around Strasbourg (1851). Source: *Carte über den Lauf des Rheins von Basel bis Lauterburg* (Karlsruhe, 1851).

for the water table only slowly became apparent, the French water engineers and Strasbourg's municipal council did not realize in time that the city was slowly being disconnected from the river. Water authorities only became aware of the necessity to once again reorganize the bras Mabile in order to maintain its function as a junction for navigation between the Rhine and Strasbourg.[12]

As flood prevention was still given absolute priority, the water authorities launched another large-scale French-Badian river regulation project in 1840, which was completed in 1876. While the project succeeded in straightening the river, giving it a firm bed and minimizing the danger of flooding, it worsened the conditions for navigation. As a result, by 1862 the water table of the Rhine near Strasbourg was lowered by 0.6 meters compared to the depth around 1800.[13] The consequences for the city were "disastrous," as Strasbourg historian Pierre Buntz puts it.[14] The regulation scheme not only moved the river eastward away from Strasbourg and lowered the water table but also increased the *speed* of the current. When in 1861 the technical experts of the Central Commission for Navigation on the Rhine traveled down the river by ship to evaluate its physical state and the obstacles for navigation they stated that large-scale shipping to Strasbourg had become practically impossible.[15]

Massive financial support from the French government to stimulate navigation to Strasbourg by constructing the Rhine–Rhone Canal (opened in 1832) and the Marne–Rhine Canal (1853) were of limited value and only made Strasbourg "a canal harbor" (*port de canal*).[16] In their recent studies on this issue, Mark Cioc, David Blackbourn, and others have highlighted the ambivalent effects of the Rhine regulation, and there is a large body of German literature that praises the intervention as a historical milestone of water engineering, but they do not mention that the project put an end to the glorious history of Strasbourg as a Rhine port city, at least for some time. In 1864 upstream navigation was completely abandoned, and so was downstream navigation four years later.[17] As a consequence, local and economic historians have defined the first half of the nineteenth century as an economically disastrous period ("marasme économique") for Strasbourg.[18] The rise of the railways beginning in the 1840s contributed considerably to this decline and made things even more difficult for commercial navigation.

Modernization by Infrastructure: Strasbourg and Its Rivers under the German Empire (1871–1913)

In 1871, as a result of the Franco-Prussian War, Strasbourg and the Alsace became part of the newly established German Empire. Within this new legal context the relation between the city and its rivers was fundamentally rearranged in general political terms as well as in terms of water management. Directed by a German municipal and state government, a culturally divided urban society struggled massively on three key projects of modernizing the regional water infrastructures: constructing a new side channel along the Rhine from Strasbourg down to the area of Mannheim, creating new large-scale harbor facilities, and introducing a central water provision and discharge. The difficulties of these three projects reflect the fundamental cultural clash between the urban society, the majority of which wanted to preserve traditional French visions of water engineering and sanitation, and the German administration, which tried to reorganize Strasbourg's water culture along German models and strengthen its affiliation to Germany.

As early as May 1871 the Strasbourg Chamber of Commerce submitted to the German chancellor Bismarck a proposal for a canal from Strasbourg to Ludwigshafen (opposite the city of Mannheim).[19] Some weeks later it launched a canal committee that publicly referred to the arguments of an older "comité pour la défense de la navigation" and built on a broad consensus among the French urban population.[20] The committee proclaimed that Strasbourg should again be connected to large-scale shipping on the Rhine and become the new head of navigation instead of Mannheim. Within two years some sixty other municipalities along the Rhine, from Karlsruhe down to Cologne, and many chambers of commerce joined the committee and promoted the project of the side channel. If at the beginning the idea was rooted in "French traditions" and had originally been promoted by the French community of Strasbourg, it was in fact soon transformed into a joint project of a large number of German municipalities. The powerful association of the cities indicated that large-scale shipping was regarded as a key question for urban development at the time. It also demonstrated the will of the city councils to build a broad interurban coalition against their governments.

But the state of Baden strictly rejected the idea of the canal because it threatened fundamental interests of Mannheim and of the Badian railway. Government experts also claimed that the channel would be much costlier and more harmful in environmental terms than the regulation of the existing riverbed, especially for the groundwater table. After around two decades of campaigns and struggles the canal project was finally abandoned, even though the committee had been able to convince the German provincial government, which presented a detailed proposal to Bismarck in 1883.[21] In the public debate on the two concepts of regulating the existing riverbed or building a side channel, scientific expertise and the reputation of the renowned Badian engineer Max Honsell, who promoted the regulation concept, played a decisive role.[22] In 1906 a contract on a joint regulation project was made by the governments of Baden, Alsace, and the city of Strasbourg. The city contributed one million Reichsmarks. Works started in 1906 and were finished in 1913. Right from the start the new waterway strongly stimulated large-scale shipping to Strasbourg.[23]

Investment in the Future: Building Urban Harbors in Times of Depression

But the river regulation would not have been able to reintroduce large-scale navigation to Strasbourg if the city's harbor facilities had not been modernized. Around 1890, as a result of the failing of the canal project, large-scale shipping upstream of Mannheim still remained difficult. Nevertheless, the municipal administration of Strasbourg in the 1880s reconsidered the chances to reintroduce navigation to the Alsatian capital. Steamships had become more powerful and were increasingly able to overcome the strong current of the Upper Rhine, and the prospect of profits from trading food and building materials from the Upper Rhine region in exchange for coal from the Ruhr valley was very promising. This calculation seemed to be confirmed by the rise of the small port city of Lauterbourg at the northern edge of the Alsace, where conditions were not so different.[24] Strasbourg's municipal administration carefully observed these trends and made detailed comparative calculations on the profits in the transport of mass goods by railways and by shipping. As a result, municipal officers in Strasbourg—as well as those in other towns along the Rhine up to Bale—realized that intelligent infrastructural solutions for large-scale shipping, which included modern transfer sites to railways, promised large profits.[25]

Such a modern infrastructure was completely lacking in Strasbourg, however, so in 1886 the municipal administration took up a proposal presented by the Chamber of Commerce to develop a new port outside the city wall near the gate Porte d'Austerlitz (which the Germans called Metzgerthor).[26] The city was willing to invest around 1.2 million Marks in the project, which was competing with another concept promoted by the regional water authorities for the same place.[27] Municipal engagement was additionally stimulated by projects for an extended harbor area that the Badian railway company was developing for the city of Kehl on the opposite side of the Rhine.[28] The Kehl port was opened in 1902 and soon started to compete with Strasbourg's interests by introducing low taxes for transports to Switzerland and Italy.[29]

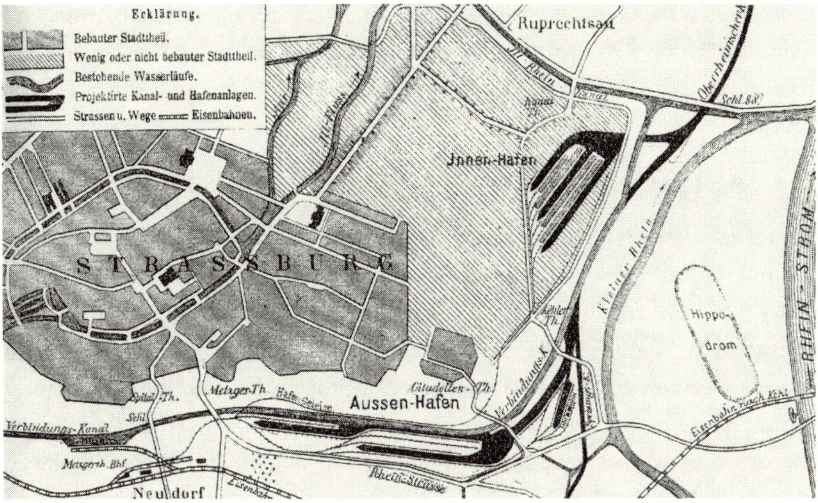

FIGURE 7.2. Plan for the new ports of Strasbourg (1889). Source: Paullsen, "Entwässerung der Stadt Strassburg," *Archiv für öffentliche Gesundheitspflege in Elsaß-Lothringen* 15, no. 3 (1893): 97–141.

Even if around 1890 Strasbourg's projected harbor represented a speculation on a future that still seemed uncertain with regard to the poor state of the riverbed, the large coal-trading companies from the Ruhr valley shared the optimistic vision of the municipal administration. They booked all the storage capacities at the new harbor even before construction works had begun.[30] The municipal administration was even more encouraged to go ahead as the petrol industry showed strong interest in constructing large iron vessels for the storage of petrol (which still today plays an important role for Strasbourg's harbors).[31] The city reacted to the lively demand and rapidly expanded their planning for the harbor, which was constructed from summer 1890 onward and opened in 1892 under the name of Hafen vor dem Metzgerthor.[32] On June 11, 1892, a high water level on the Rhine carried the first large steamship to the new port, and "Strasbourg regained the status of a Rhine port city."[33]

In the following two decades the port area expanded rapidly. As early as 1901 another new "Port du Rhin" with two large basins for industry and commerce was constructed on an area of 128 hectares parallel to the Rhine (see fig. 7.2).[34] As a result of the river regulation and the new harbor facilities, the amount of goods passing through Strasbourg's harbors was increased from 11,500 metric tons in 1892 to 635,000 in 1906 and up to 2,000,000 metric tons in 1913.[35] But even if the decision to locate the new harbors south and west of the city seemed rational in spatiopolitical terms and proved economically successful, it had a mixed effect for the regional culture: it isolated Strasbourg from the Rhine River in terms of communication and social interaction, so that the city in fact "turned its back to Germany" throughout the twentieth century.

Sanitary Modernization as "Germanization"?

If river regulation and navigation transformed parts of the Upper Rhine region around the Alsatian capital into an industrial landscape, the modernization of water provision and discharge during the last decades of the nineteenth century turned Strasbourg into a "sanitary city." After the city had been integrated into the German Reich in 1871 it soon became the field test for some of the most advanced German municipal reformers. The local administration, led by town mayor Back and other leading reformers like Alexander Dominicus, developed pioneering concepts in the fields of housing reform, slum clearance, and so on.[36] But all these projects were massively contested in the context of a strong confrontation between the new German administrative reform elites and the French local society. The implementation of a new water infrastructure was to become one major point of issue, so that Vivian Claude has analyzed their introduction as a prominent example of the "Germanization" of Strasbourg.[37]

Around 1871 the traditional system of water provision and discharge in Strasbourg was not as disputed as in London, Paris, or most other German cities. Water for drinking and for private households was still taken from public wells. This system of decentralized, free public water provision found broad acceptance among the urban population, while critique from hygienists did not become well known outside of expert discourse. The introduction of a new system of piped drinking water was started with the first modern waterworks in 1879, which was expanded in 1899 and again in 1905. But the publicly financed new pipe system was for a long time limited to only a few quarters *intra muros* and was only slowly expanding. Strasbourg stayed, in sharp contrast to the intentions of the municipal administration and to the surrounding Reichsland of Alsace-Lorraine, a latecomer in this field of hygienic modernization.[38]

One major reason for this situation was that the demand for and consumption of the piped water seriously lagged behind the expectations of the planners. The rate of consumption was calculated around 1879 as 150 liters per day per inhabitant, but reached only 67.3 liters in 1897 and 136 liters in 1919, although it was massively increased by uses for several public purposes after 1900. It was only around 1918 that the last traditional wells were closed. The acceptance of the new system was marked by strong social and cultural differences, because the French population in particular was strongly attached to the old system.[39]

The limited use of piped water also meant that the physical and social pressure to discharge large quantities of wastewater was not very strong for a long time. Thus, the sewage system maintained its traditional form for even longer than the drinking water system had. Around 1871 Strasbourg's rainwater was evacuated by two major canals into which twenty-two kilometers of secondary canals (the so-called *Dohlen*) discharged their water. The two main canals and a number of more or less irregular minor *Dohlen* discharged their wastewater directly into the river Ill. Moreover, the network of *Dohlen* illegally received considerable quantities of feces and only drained one-third of the urban area, while two-thirds remained unconnected. The majority of the human waste was transported by private entrepreneurs outside the city and used as fertilizer for agricultural purposes. In the early 1880s between seventy and one hundred kilograms per inhabitant per year were exported outside the urban area.[40] The system of *Dohlen* caused considerable environmental problems, as they silted up very rapidly and many of them leaked, thereby polluting the urban ground. They had to be cleaned quite often, and the mud that was removed had to be transported by car through the streets—a difficult and dirty business. Although this system caused considerable sanitary problems, the old main canal was still expanded between 1881 and 1895 by 3.4 kilometers and the secondary canals by around 12 kilometers.[41]

Toward a New Path in Urban Water Infrastructure

The traditional system was in a way stabilized by its broad acceptance among the French population. While the German-dominated municipal board of health seriously called for a new "rational" system as early as 1873,[42] this was not introduced until more than twenty years later. The municipal administration intending to overcome the serious resistance against sanitary modernization developed a strategy of three steps: weakening the traditional system by cutting its resources, creating a "showroom" for the new technology in the modern quarters of the city, and mobilizing as many institutional and financial resources for the new system as possible.

In 1889 the municipality started two interventions against the old system: in that year the powerful German Association for Public Health (Deutscher Verein für oeffentliche Gesundheitspflege, DVfoeG) was invited to hold its annual conference in Strasbourg. For the conference, a Medical Officer Krieger from the Ministry for Alsace-Lorraine published a broad and very critical survey on the state of the regional water courses and the sanitary conditions in Strasbourg.[43] At the same time, the local authorities took over the service of removing the mud from the canals outside the town and fixed a sum to be paid for this service by every homeowner.[44] Making a municipal service out of this previously private business would prevent, if the system were to be abandoned, legal action by private entrepreneurs for financial compensation, so that a potential pressure group was eliminated. The public intervention was also facilitated by the decline of profit and private interest in the business in waste, which was more and more replaced by chemical fertilizers for agriculture. In the same period, the construction of traditional *Dohlen* was stopped because of the uncertain perspective of the whole system. This policy caused serious reluctance in several quarters as it also prevented the paving of a number of streets.[45]

So while from the early 1890s onward any financial support for the old system was stopped, huge investment was made in the new quarters of Neustadt, which was constructed primarily in order to house German immigrants. From the beginning of the planning the commission for the extension of Strasbourg had asked for a modern sanitary system in this area. Consequently, in this part of the town the majority of the streets were financed, because of their large size, by the municipality, and no construction fees had to be paid by the homeowners.[46] The new quarters, with their broad streets, huge public buildings, and modern sanitary equipment, became the model for the modernization of the old city.

Changing the Role Model: A Farewell to Paris

It was only in 1891 that the municipal department for construction presented its plan for a modern sewerage (*Entwässerungsplan*) that the board of public health had asked for nearly twenty years before. Its main principle was to separate the wastewater from the natural water courses in the urban area and discharge them without any special cleaning downstream into the Ill river. All rainwater as well as domestic and commercial wastewater should be evacuated by a mixed-water system through a network of canals (a new type of *Dohlen*). This decision called for the construction of seventy kilometers of new canals (among them twelve kilometers of old canals to be restored) for an amount of 5,520,000 Marks.[47]

The controversial public debate provoked by the municipal project generated several arguments that were well known in sanitary discussions of the time. Among them comparisons between the role model of Paris and the

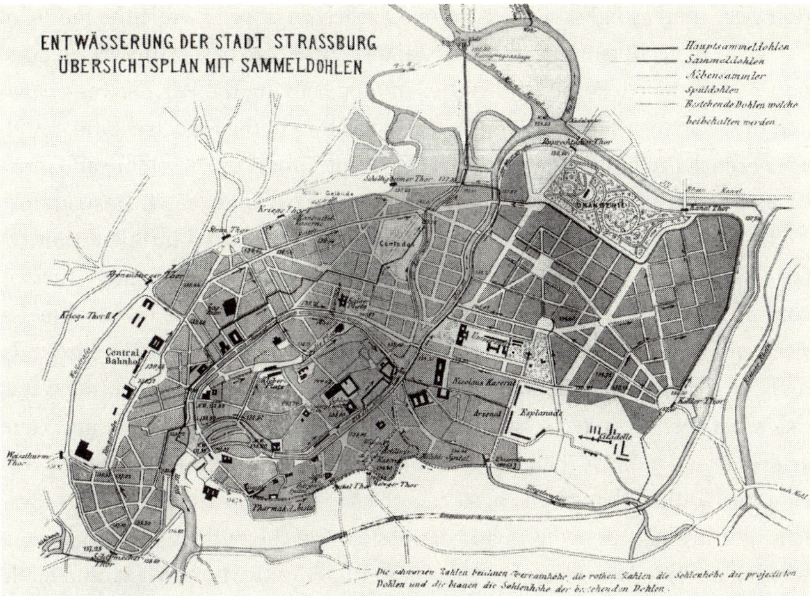

FIGURE 7.3. Plan for the new sewerage of Strasbourg (1893). Source: Krieger, *Topographie der Stadt Strassburg nach ärztlich-hygienischen Gesichtspunkten bearbeitet* (Strasbourg, 1889).

differing solutions for London and several German cities found broad attention. The city's medical officer Goldschmidt called for the introduction of a mixed water carriage system by pointing to the poor sanitary conditions in Strasbourg and the considerably lower rates of typhoid fever in London compared to Paris.[48] In contrast, the homeowners association, represented by its vice-president Franz Fastinger, strictly denied the need for a new system: "Why give a healthy man medicine?" he asked in a keynote speech given to the meeting of his association in 1892 and published the same year. According to Fastinger, cities where the new system had already been introduced did not show lower rates of infectious diseases. In contrast, the modernization of the streets and the houses would increase the annual costs for sewerage from 30 to 125 Marks per house. Furthermore, Fastinger underlined the problem of polluting the river Ill by discharging wastewater downstream of the city. This argument referred to a similar debate in Paris where a commission directed by Pasteur massively criticized the projects for the French capital.[49]

The resistance of the residents could seriously endanger municipal planning, as had been shown by the stoppage of a sewerage project by public vote in Bale in 1876.[50] But at that time German concepts of urbanism and sanitation were increasingly accepted among French engineers, while the model of Paris was in decline in the international discourse of experts. This was in large part due to the enormous difficulties in modernizing the Paris sewer system and the relative weakness of municipal legislation in the French capital as well as in French provincial towns.[51] The former role model of Paris more and more became the symbol of a stagnant situation in the field of urban infrastructure.

In contrast to the Strasbourg homeowners, the local administration referred exclusively to German debates and the example of German cities. A commission for the study of the municipal project decided in 1893 to conclusively abandon the old system because it would not be possible to efficiently manage the traditional *Dohlen*.[52] In a second step, a "critical decision" was made between a mixed system and four types of separated systems. Here again a broad public debate showed the strong French culture of sanitary orientations within the urban society, as the separated pneumatic "Lienur System" got strong support in public discussion. But the municipal commission, which had visited the German cities of Berlin, Frankfurt-am-Main, and Halle (Saale), made its decision in favor of the mixed system.[53]

In contrast, the commission voted, after having undertaken extended studies on several clearing technologies, against the Berlin model of irrigation. The commission's report referred to geological problems together with financial arguments and the contemporary "self-cleansing theory" when justifying the decision to discharge the wastewater downstream into the river Ill. As a result, the system that was finally adopted represented a kind of bicultural compromise, as it showed some elements of the Berlin model but shared other features with other German and French cities. While, for instance, in the technical regard the Berlin mixed system was realized, the irrigation model of Berlin was not copied, and the financial contributions of the homeowners were split up into two bills, for the discharge of feces and of the domestic wastewater, respectively.[54] In the debate arguments drawn from the French or German cultural background were amalgamated with common "rational" arguments like cost, efficiency, or geophysical conditions. In the lively public disputes the general confrontation between the two cultures spread into a variety of culturally shaped positions, interests, and arguments.

Once the decision was made the new system was constructed, which would have remarkable consequences between 1900 and 1910. By that time in the inner quarters 85.1 percent of all built-up parcels and 81 percent of all privies were connected to the new system, and this number was increased by around four thousand houses per year. The value of the new network was estimated to represent 10,145,000 Marks in 1911.[55] As a result, the incorporation of Strasbourg into the German Reich and its role as a field test for sanitary reform had triggered modernization and established a new path in the city's urban infrastructure.

Nationalizing Urban Infrastructures after World War I: Strasbourg's Harbors and the Reinforcement of the French State (1918–1925)

During World War I Strasbourg and the Upper Rhine became one of the most contested areas of national policies in Europe. In the course of the war the geography of cities and regions rapidly gained significance as much in scientific as in geopolitical terms. When in 1917 the French government established a scientific committee (*comité d'étude*) of renowned geographers and historians such as Lavisse, Seignobos, Martonne, and others, it intended to develop scientific arguments to justify national political claims and expand the French territory.[56] Other nations, like the United States, followed the same strategy. In the negotiations for the Versailles treaty, the geographer Martonne and his team served as a think tank and as counselors for the government, and helped to elaborate the paragraphs that reintegrated Alsace and Strasbourg into France.[57] One of Martonne's main arguments since the early days of his career as a state counselor was that holding Strasbourg and developing it as a driver for regional economic growth and head of navigation at the Rhine would considerably strengthen the French state.[58]

After the war this strategy was very rapidly realized by using the new "hinterland" (*arrière-pays*) concept of economic geography as an instrument to stimulate regional economic growth in the Strasbourg region and integrate it into the national state.[59] The French government considered the urban harbors and navigation as the key instrument of a top-down strategy by which the city of Strasbourg as well as the German state and private German owners were in fact expropriated from the shipping infrastructure they had owned in the prewar time.

The economic benefits of this strategy promised to be high, as trade volumes in Strasbourg's harbors were increasing from 1913 to 1924 by 38 percent, driven by the rapidly growing production of potash in the Alsace region.[60] But creating a "French hegemony" over shipping on the Rhine had to overcome major challenges, like a lack of steamships in French ownership, a deficit of French mariners licensed for the Rhine, and a dominance of German owners in Strasbourg's harbor facilities, where companies from the Ruhr valley held storage capacities, cranes, etc.[61] But with the help of legislation drawn from the Versailles treaty, sophisticated surveys of newly founded associations like the Société d'Etudes pour la Navigation Rhenane, and public propaganda in journals like *La Navigation du Rhin*, the political and economic dominance of the French central state in the local economy of navigation was consolidated step by step.[62]

It was especially the new associations mentioned above that carefully studied and calculated the economic value of navigation on the Rhine and organized expropriation and redistribution of the harbor facilities. In addition, during the first decade after World War I Strasbourg profited from control over the neighboring German harbor of Kehl, which the Versailles treaty subordinated to French authorities until 1928.[63] The most important project of the new French government was a rapid expansion and modernization of the harbor facilities. Besides modernization of the entrance to the existing harbors and renovation of railway junctions, a new large harbor called a "Port autonome" was built, which comprised an extended basin for the trading and storage of petrol.[64]

Nationalizing and expropriating urban infrastructures for shipping provoked serious conflicts not only with the Germans but also between the local administration and the central state. Starting with the foundation of the powerful Société d'Etudes pour la Navigation Rhenane in 1919, the City of Strasbourg as well as the local Chamber of Commerce found itself deprived of control over the shipping infrastructure and its benefits. It needed strong protests to achieve for the local administration some representation on the boards and a share of capital of this Société.[65] The new Port autonome, too, was created as a legally independent organization that initially was directed by the Commissariat Général of the central state. Strasbourg's mayor called this construction explicitly a de facto expropriation of the municipality. Although this conflict was solved by mutual agreement, local representatives were dominated in the boards of Port autonome by actors from the regional and the national scale.[66]

Turning Strasbourg into a French industrial town not only massively changed the character of the city in economic and cultural terms, it also caused environmental decline. The trading of petrol polluted the riparian landscape as well as the river itself, and Strasbourg also became a center of other harmful infrastructures such as incineration plants for the combustion of garbage.[67] Serious environmental damage was also progressing on the regional scale of the Alsace. There the construction of the Canal d'Alsace, which was opened in 1928 in the south of Strasbourg, reinforced erosion and massively lowered the water table. Consequently, environmental historians of the Rhine consider the twentieth century and the Canal d'Alsace in a long-term perspective as the peak of environmental damage of the river and its waters.[68] But in a long-term perspective from the French Revolution onward, industrial pollution of the twentieth century only completed environmental degradation that had been started in the fields of river regulation and transformation of the urban riversides as early as the late eighteenth century.

Toward a European Water Modernity

The case of Strasbourg during the nineteenth century shows that the very different functions of rivers and regional water systems for cities caused contradicting challenges and conflicts of interests. Being located near the river offered considerable economic chances for navigation for cities but at the same time increased the dangers and damages of flooding. In a broader study that I have published in spring 2016 I have shown that cities principally tended to highly estimate the vicinity to rivers and their economic potential while rural communities preferred to keep a certain spatial distance to rivers which promised to minimize the danger of flooding.[69]

As this chapter has shown, the nineteenth century was a period of large-scale water management that had mixed and partly unintended consequences and raised complicated problems of governance. This was especially true for a border region like the one at the Upper Rhine, where the municipality of Strasbourg was confronted with the national water authorities and strategies of the foreign policy of the states of France, Baden, and the German Reich.

In the long run of the nineteenth century the interventions of water authorities increasingly succeeded in controlling the rivers and minimizing their potential for destruction so that the profile of risk changed, and citizens' awareness of risk itself declined in riparian communities. The introduction of modern sanitary infrastructure in the culturally divided city of Strasbourg demonstrated the strong impact of cultural attitudes and social constructions in the field of urban water management. But in a broader comparative perspective the role of bicultural confrontation on water issues was slowly diminishing and dissolved into a multifaceted, transnational modern "European" consensus on the governance of rivers and urban waters in the period of "water modernity."

Chapter 8

Path Dependencies Managing the River Elbe and the Requirements of Hamburg's Open Tidal Seaport

Dirk Schubert

Rivers form barriers but they also create opportunities for settlements and urban development. In cities with seaports, which are often located on a river, there is a close interrelationship between the development of port and city. The course of development in the type and design of port expansion measures was, in general, established at the beginning of the nineteenth century, when industrialization and an increase in world trade created new demands. Decisions taken at that time are still significant today in determining the options for future development in cities and ports.

The topographies of river ports and seaports around the world are different—no two are alike. All exhibit unique phenomena based on their history, culture, location, and transport connections to the hinterland. Port cities on riverfronts offer special opportunities compared to other cities. Hamburg is situated on the Elbe River as well as several smaller rivers, lakes, and artificial canals, and so has been an amphibious city from the beginning. Above all, its connection to the North Sea, other ocean networks, and seaport cities worldwide via the Elbe is an essential feature of Hamburg's port history. Therefore, the development of the Elbe estuary has to be included in any long-term narrative of Hamburg and its port—it describes a fascinating tale of environmental history, shaped equally by nature and humankind. Drawing on a framework of path dependency as a guiding tool in historical analyses, this contribution explores the history and interdependencies of the city of Hamburg, its port, and the river Elbe over the last two centuries.

The concept of path dependence can be used both as an analytic tool to interpret historic development and as a vehicle for economic and spatial planning. It is not about banal explanations like "the past influences the future," but about well-founded historical analysis.[1] Currently, the concept is mainly used by economists and social scientists with a historical focus. Path dependence seeks to describe and explain long-term lines of development. These trends are heavily dependent on the course of direction set by stakeholders at certain points in time and prove to be mostly irreversible. Path dependence can refer to dimensions such as institutions and governance, to economics, politics, and technologies, as well as ideas and principles, but it is generally based on the interaction of these factors in pursuit of a common goal. After all, path dependence is also referred to in science disciplines; that is, in the context of lines of research and traditions in an evolutionary development that make way for shifts in paradigm. It also relates to a historic process that is preceded by decisions for a course of development in a similar direction. This is associated with self-amplifying effects, and their embeddedness implies considerable costs incurred by a change of strategy.[2] Paths follow an intrinsic reproduced logic. As a rule, this promotes persistence with chosen development paths and makes a change of course difficult. Situations requiring complex decisions tend to rely on "approved" strategies and lead to a reduction in complexity.

Local decisions on path determination rely strongly on external factors.[3] Developments like those in transport technology, the launch of high-speed trains, the fall of the iron curtain, and the containerization of cargo can influence the locational advantages or disadvantages of cities. They become prevalent everywhere. Such new technologies can undermine a city's economy—the stronger the specialization and uniformity of the local economy, the greater the region's vulnerability.

The crossover points (critical junctures) at which the course of direction is determined are a relevant aspect in path analysis. This throws up the "what if" question: What if a different decision had been taken and a different developmental path had been pursued? This is where we run the risk of speculation, reflecting possible but unused paths, while these counterfactual questions could help to analyze the past and explore alternative (future) paths.

A Brief History of the Interactions between Hamburg and the River Elbe

The Elbe River is the third-longest river in Germany and has a total length of about 1,100 kilometers. The river was made navigable by commercial boats and ships after 1842, straightened and dredged in subareas. The Elbe provides important trade links in the hinterland of Hamburg out to Czechia. Several tributaries such as Havel and Saale and, later, built canals (Elbe Lateral Canal, Elbe–Havel Canal, Mittelland Canal) improved waterway networks and linked the Elbe to other German industrial cities such as Pirna, Magdeburg, Dresden, Boizenburg, and Berlin. The Kiel Canal (North Sea Baltic Canal), whose western entrance is near the mouth of the Elbe, provides a connection for ships between the port of Hamburg and the Baltic Sea. Within Hamburg the Elbe has a number of tributaries, some of which are cut off from the main stream by dikes.[4]

The Free and Hanseatic City of Hamburg is both a river port and an open tidal seaport. As a city-state it has repeatedly altered the course of the Elbe River and throughout history has also deepened its riverbed in order to improve access to the North Sea and to countries and seaports worldwide. The main shipping channel of the Elbe had already been moved in the Middle Ages from the southern branch of the Elbe in Harburg to the northern branch nearer Hamburg itself.[5]

Hamburg's harbor was sited where the Alster River joins the Elbe, and offered space and shelter for many ships. In the thirteenth century, the city center developed around the harbor on the estuary Nicolaifleet, with a town hall, law courts, stock exchange, and customs and weighing facilities. At the time, the city's most important export product was beer. During the fifteenth century, larger ships were anchored outside the city and loading and unloading was done manually and with the use of handling equipment on board.

In 1816, the first steamships on the Elbe ushered in a new era. Rail links from the Prussian city Altona on Hamburg's western periphery to Kiel and from Hanover to Harburg were completed in 1844 and 1847, respectively. This in turn brought a rapid intensification of local transit shipping between Harburg and Hamburg. At the beginning of the nineteenth century, the handling of goods was predominantly conducted on the river. Ships were either anchored on the Elbe or moored on piles, and the cargo was transported to the warehouses on barges. Goods were loaded from ships onto smaller boats and taken to the shore and to merchants' buildings where they were stored, sometimes processed, and then transported farther on land or water. The merchants' buildings were used for housing and office space as well as storage, and were accessible from both the water side and land side.

Goods could also be brought directly from the ships to the shore. Special heavy-duty cranes were operated by a crane master and his crane hands, employed by the city. Loading and unloading on the river was undertaken by private companies; later, stretches of quays were also leased to private companies. The river and canals were lined by shore merchants' buildings and warehouses. Hamburg did not have a separate harbor area with roads parallel to the riverbanks and public access to the water. The "free port" area instead extended over the entire city of Hamburg.

In the mid-nineteenth century it seemed reasonable to look toward London, the world's largest port, as a model. Hamburg had already employed experts in urban planning, harbor construction, and infrastructure development from England for many years. Thus, the first proposals were for a dock harbor with locks, following the example of London. These initial designs were primarily proposed by the English engineers Lindley and Walker and presented jointly with Hübbe, the director of water construction at the time.[6] This was only the starting point for the transformation, however, as Hübbe later withdrew his support for the project. It was subsequently proposed that Hamburg should instead be a tidal seaport with open access for all ships at all times, not blocked by locks like most British ports. Hübbe's successor Dalmann took a clear position in favor of an open tidal port without locks,[7] and the first phase of port expansion, until 1880, indeed followed the fundamental decision for an open tidal harbor and the necessity of building enlarged facilities for the handling of goods. The construction of a tidal seaport was possible because the mean tidal range in Hamburg is "only" about 3.6 meters, whereas in London it is about 6 meters. This meant that goods could be transported more easily and ships wasted no time passing through locks.

In 1866 the number of sailing ships along the Elbe reached its peak and subsequently decreased, both absolutely and relatively. The expansion of steam shipping allowed for the development of schedules of departure and the prediction of approximate arrival times, facilitating calculation of, as well as openings for, speculative profits. The change in seafaring traffic brought about new and risk-reducing patterns for trading companies by improving predictability. Shipowners' companies, for example, developed into a new and independent, purely profit-oriented economic field. The increase in trade forced Hamburg to contemplate the expansion of its harbor once again.

Sandtorkai, an artificial basin constructed between 1859 and 1866, was the first modern quay facility following the decision to create a tidal seaport. Improvements on the quay walls allowed direct loading and unloading from ships to land with adjacent rail links. Sandtorkai provided quayside mooring for the largest seafaring ships, as well as single-story warehouses for the temporary storage of goods, movable cranes for loading and unloading operations, and links to the inland transport system. Barges transported goods on the water side.[8] All of Hamburg's basins constructed since the decision in favor of a tidal port, even up to the present day, have been designed in the same fashion.[9]

It was decided among the city's elites that harbor development measures should be defined as infrastructure development and therefore carried out at the cost of the state. Handling of goods on the quays was also organized exclusively by the city; however, the leasing of stretches of quay was later permitted. The storage business was conducted both in communal warehouses that were rented or leased and in private warehouses. Leased quays could be used freely for loading and unloading by the shipowners' companies, whereas on the city-operated quays an impartial allocation of moorings was meant to ensure the fair use of the facility. The state-controlled quay management organized handling operations at the quays, the loading of goods onto trains, vehicles, and inland-waterway boats and barges, and the supervision of cranes, warehouses, and quays; additionally, it collected fees for the use of quays and storage and weighing facilities. Municipal employees managed the port operations.

Hamburg was also responsible for securing a safe journey along the Elbe from the city to its mouth until this was taken over by the Reichsregierung in Berlin in the 1920s. In 1872, the bridges crossing the Elbe (*Elbbrücken*) were completed, providing Hamburg with a direct link to the rail network of the German Reich. The location of the *Elbbrücken* determined the extent of the future harbor and quay facilities to the west of the bridges. They still mark the dividing line between sea shipping and river shipping.

The next phase was triggered by Hamburg joining the customs territory of the German Reich in 1881. For centuries Hamburg had been a free port and was able to store and process duty-free goods. Duty had to be paid when goods left Hamburg, which mainly involved small quantities. This situation was advantageous to merchants and shipowners but put trading companies, which, like foreign businesses, had to pay duty on goods intended for sale outside Hamburg, at a considerable disadvantage. With the founding of the German Empire in 1871, Hamburg became a member on the political level but not in terms of its tariffs and customs duties. This ambiguous situation was resented by the imperial government and considerable pressure was put on the Hamburg government to resolve the situation. Eventually a solution was found that did not maintain the whole city as a free port but instead established a small free port zone. The size and location of the free port were controversial, and several alternatives were considered.

Finally, it was agreed to designate a specific free port area in which no housing would be allowed. The contract between Hamburg and the Reich laid down that the customs authority remained with Hamburg, that duty-free access along the river Elbe was secured, and that businesses for processing duty-free raw materials or semifinished goods could be established in the free port zone. The Reich granted financial aid for the construction of the redevelopment measures. Existing buildings had to be demolished to make room for new warehouses. About twenty thousand people had to be rehoused for the construction of the warehouse district (Speicherstadt). This additional population pressure on neighborhoods close to the port further aggravated existing hygiene problems that likely contributed to a subsequent cholera epidemic. Hamburg had to master gigantic building operations between 1882 and 1888, when it joined the customs territory.[10] Initially, the free port incorporated large disused areas of land, mainly south of the Elbe.

The Speicherstadt, a part of the free port, was designed for the storage of high-quality goods. Direct unloading from seafaring ships into warehouses was not possible and goods were transported on barges. On the land side these could be delivered or collected by vehicles or rail. Housing was not permitted within the free port area and port workers had to "move closer together" in central areas or were forced to search for new accommodation in districts farther away from the harbor.

Apart from trading, shipyards were the main hub of economic activity in the port. Since the nineteenth century, shipyards were located on the southern bank of the Elbe. In 1870 fifteen yards were located on Steinwerder and Reiherstieg. In 1877 the shipyard Blohm & Voss, soon of worldwide reputation, was established, and in 1907 another large shipyard, a branch of the Stettiner Vulkan-Werft, settled there. Gigantic ocean liners were built and about two million emigrants moved to the New World, departing from Hamburg and sailing down the Elbe on their voyage to—they hoped—a better future.

Up to the twentieth century, the Elbe River provided food (fish) and drinking water, but also served as a waste dump and sewer. When in 1892 a cholera epidemic erupted in Hamburg, claiming around eight thousand lives, the seaport city was criticized all over Germany. The disease spread during a hot summer when the Elbe carried little water. Drinking water was taken directly from the river and left untreated prior to use. The outbreak was further amplified by unhealthy living and housing conditions. It was a disaster for Hamburg, not only because of the death toll but also because of the closing

Abb. 12.
Nicht weiterverfolgter „Groß-Hamburg"-Gedanke aus dem Jahre 1919.

FIGURE 8.1. Plans for Greater Hamburg in 1918 including the estuary zones of the river Elbe, proposed for the Workers' and Soldiers' Council in 1918. Source: Fritz Schumacher, *Das Gebiet Unterelbe-Hamburg im Rahmen der Neugliederung des Reiches* (Hamburg: Broschek, 1932), 27.

of the port and all traffic along the Elbe for many weeks. Immediate improvements to the supply of drinking water and the sewers were required and slums next to the Elbe had to be cleared.[11] But the Elbe remained the lifeblood of the city, for making a living as well as for growth and prosperity. Environmental issues were deemed irrelevant, and the river as a natural resource remained exploited.

At the beginning of the twentieth century, Hamburg reclaimed more space from surrounding Prussia for further port extensions. Altona's port was located west of the port of Hamburg, while the port of Harburg lay to the south, across the Elbe. Further extensions became necessary due to rapidly expanding trade and the national importance of the port not only for the city and the region but for the whole of the German Empire. In 1919, during the chaotic situation following World War I, a plan was put forward to amalgamate all the littoral zones between Hamburg and the North Sea into "Greater Hamburg."[12] While this plan was drawn up and supported by Hamburg, the neighboring states and cities, especially Prussia, objected.

During the 1920s the port expanded south and westward, with shipbuilding industries and modern quays and warehouses. Artificial extensions were built step by step, all connected with the river Elbe. Canals and waterways were deepened and quays and warehouses constructed. The city of Hamburg took over the responsibility for planning, building, and leasing land and quays in the port area that were owned by the city-state of Hamburg. Hamburg had always needed to control traffic and free trade along the Elbe. As an outpost and outer harbor, the city of Cuxhaven at the mouth of Elbe was made part of the city-state of Hamburg in 1937.[13]

The opportunities for expansion were limited, however, because Hamburg was surrounded by Prussian cities including Altona, Harburg, and Wandsbek, and they posed considerable competition. The restrictive administrative boundaries of Hamburg were also a hindrance, and agreements with Prussia concerning the extension became necessary. But Hamburg had the advantage of the best topographical location compared to Altona and Harburg, which was improved and assured by far-sighted land policies for port extensions. In 1937, the competing ports and cities were finally amalgamated by the National Socialist government into Greater Hamburg.

New Challenges after World War II

During World War II, the port and city were bombed heavily and most of the infrastructure was destroyed. Sunken ships lay in the Elbe River and the port. The shipbuilding industries were also not allowed to build new ships during the British occupation. Due to the partition of Germany and Europe after 1945, the port of Hamburg lost most of its hinterland in the eastern part of Germany, in Czechoslovakia, and in the Baltic countries, including Latvia and Lithuania. The loss opened up opportunities for competing ports like Antwerp and Rotterdam, as well as for Rostock and other Baltic seaports. This had disastrous consequences for Hamburg, as it lost most of its established trade connections. Modernization required many improvements to the port infrastructure.

Hopes for growth and an increase in trade after the expensive optimization and modernization measures of port facilities were challenged by new technologies in shipbuilding and cargo handling. Beginning in the 1960s, many of the seaport industries were relocated—shipbuilding to Asia and (dangerous) oil handling to deepwater ports like Wilhelmshaven are examples. In 1973, Deutsche Werft in Finkenwerder, the largest shipyard, was closed. This resulted in a heavy loss of jobs and reduced capacities to deal with modern advanced technologies. But most important have been the changes in logistics. From the late 1960s, the amount of goods handled in containers increased, which meant a loss of tough, dangerous, traditional dockers' jobs. Since Hamburg initially refused to jump on the "containeritis" train, the first container ship was unloaded in Bremen, and Hamburg could do little to stop the revolution in cargo handling.[14] When the new, updated Port Development Plan was decided upon in 1976, around forty thousand jobs were located in the port area while around one hundred thousand jobs were related indirectly to the port.

The 1962 Flood Disaster in Hamburg

In 1962, 120 years after the Big Fire in 1842 and 90 years after the cholera epidemic in 1882, growth was interrupted by another unexpected catastrophe.[15] Although there had been floods with some damage over the previous decades, the dikes along the Elbe River and in Hamburg seemed to be adequate. But overnight, from February 16 to 17, 1962, a storm front pushed water into the German Bight and the mouth of the river. The disaster that followed affected mainly the coastal regions along the North Sea and in particular the city and the port of Hamburg. Breaches along the coast and the rivers Elbe and Weser led to widespread flooding of huge areas. In Hamburg, more than a hundred kilometers away from the North Sea, the residential area of Wilhelmsburg—Europe's largest river island—was most seriously affected.

Storm warnings for northern Germany on February 15 had predicted a higher flood. But the flood in the evening of Friday, February 16, affected the dikes more than anticipated and led to some fifty breaches between the North Sea and Hamburg. When more details about the breaches were circulated, an alarm for Hamburg was also raised. But notification of the affected population was not sufficient. Foresighted civil protection plans had not yet been prepared.[16] Due to disrupted telephone lines, warnings and information about the breaches could not be sent onward from the coastal cities to the hinterland.[17] The alarm system was not well connected and electric lines were demolished by the storm. By around midnight breaches of several dikes happened also in Hamburg. Dikes could no longer withstand the rising flood levels. The level of the dikes was not high enough and the landside slope of the dikes had not yet been reinforced, so that first water destroyed the dikes from behind and offered the way for the flood.

In the end the homes of around sixty thousand people were destroyed or damaged and the death toll amounted to 315 in Hamburg. Already during the night the rescue operations were coordinated by Helmut Schmidt, *Innensena-*

tor (senator of the interior) of Hamburg, later to become chancellor of the Federal Republic of Germany. The help of the army (Bundeswehr), particularly engineers, was made possible. Schmidt established his image as a "doer" and "decision maker" and also ensured help from other NATO states, who sent helicopters, termed *Fliegende Engel* (flying angels) by the local population. Some days later the extent of the catastrophe became clear: 120 square kilometers along the river Elbe, or a sixth of the city of Hamburg, was flooded, streets were impassable, and railway operations had to be suspended, leaving Hamburg temporarily without supplies.

The extent of damage caused by this disaster was clearly a result of the decision for an open tidal seaport and the path dependence of not providing an effective closable flood barrier. As a consequence of the catastrophe, Hamburg changed its port and urban development strategies. After the catastrophe the western part of the island of Wilhelmsburg was designated for port and industrial uses, while only the eastern part was zoned for residential use. This involved the relocation of many homes from the western part of the river island to large housing estates north of the river. Emergency plans were implemented after the flood, and dikes were shortened and strengthened. More sustainable dike constructions were adapted also, severing some branches of the river and creeks from the sea.

Setting the Course for the Twenty-First Century

Only three years after the flood disaster a new regional development strategy was decided upon, based on the improved economy and infrastructure of the lower Elbe region. A new growth model involving the import of primary materials was implemented: nuclear power plus deep water access for ships plus industrialization of the regions north and south along the river. Three nuclear power stations were constructed, as well as a state-subsidized aluminum mill and a steelworks plant in Hamburg. The opening of the first nuclear power station was to mark the transformation of the lower Elbe region into the "Ruhr area" of the north. The strategy of aggressive "industrialization of the lower Elbe region" was based on the concept of Fordistic "catch-up" modernization. It was developed by Hamburg's Minister for Economics and Transport, Helmuth Kern, and based on radial axes along the main transport corridors with Hamburg at its natural center.[18]

The city of Brunsbüttel on the north bank of the Elbe River was to become a chemical center ("Bayer Town"—named after the biggest chemical company), while a nuclear power plant, chemical industries, and an aluminum plant were

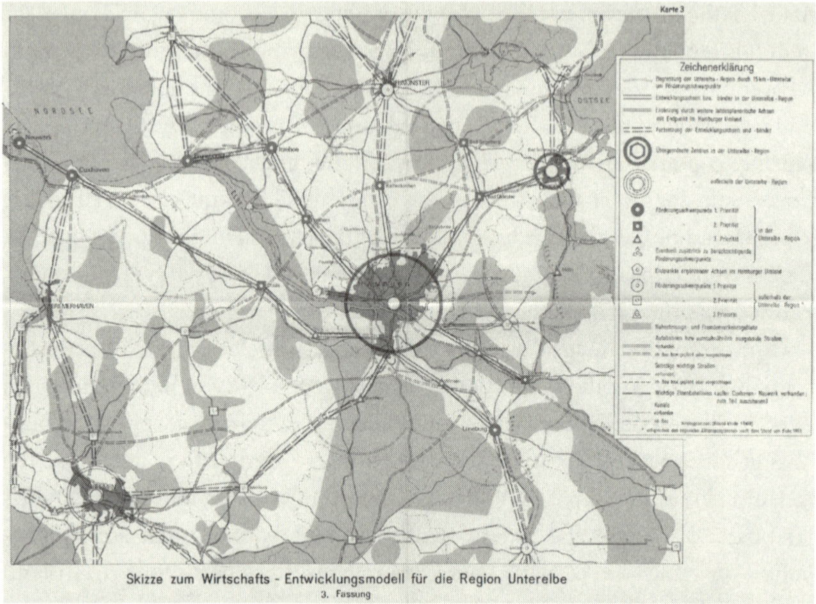

FIGURE 8.2. Regional Development Model for the Elbe estuary (with new deep water port Neuwerk Scharhörn). Source: Helmut Kern, *Ein Modell für die wirtschaftliche Entwicklung der Region Unterelbe* (Hamburg: Freie und Hansestadt Hamburg, 1972), Maps.

built in Drochtersen near Stade. The offer of cheap energy, using the river for the cooling of power stations, and an available workforce attracted only a few globally oriented companies like Dow Chemical. The strategy was intended to overcome the gap between urban and rural in the region as well as Germany's north-south divide by exploiting the lower Elbe region just for this purpose. Hamburg, already the most important player, kept in the background while the territorial neighboring states hoped for modernization, new company branches, and jobs.[19] The strategy was based on a short-term perspective that focused on those industries facing rapid internal restructuring, as well as global challenges. Hopes were pinned on the Elbe to be an artery for prosperous development in the region.[20] The Elbe became polluted, however, and its fish stocks were destroyed. In the mid-1970s the environmental movement emerged—thanks to the influence of Rachel Carson and others—as well as the anti-nuclear power movement, and stricter environmental laws were implemented.

There was debate about immediately constructing a new sea harbor on the coast next to the city of Cuxhaven, similar to what Bremen had done in Bremerhaven in the nineteenth century. It would have been costly, caused environmental problems, and created an isolated port in the mouth of the river without any service industries.[21] Hamburg acquired nine thousand hectares of land from Lower Saxony on the estuary. The area of mudflats was intended as an investment in the future of a growing new port, where the largest vessels could be berthed.[22] In 1971, plans for the new "Europort," modeled on Rotterdam, were completed. It was to be connected with the mainland by a dam.[23] However, Hamburg finally canceled these plans due to environmental concerns in 1979.[24]

While flood-protected land was enhanced, new flood barriers were built on tributaries, the road and rail infrastructure was improved, and newly solicited companies were subsidized, meaning unemployment in the lower Elbe region increased. The structural difficulties and weakness of the lower Elbe region shifted the focus toward "hard" location factors, like deep water access, spaces for industries, and cheap energy supply available in the region. The ideology of looking only at "more tonnage" and economies of scale implied the neglect of other modern technologies, high-tech industries that would generate a more diverse regional economy. Finally, in 1980, Hamburg founded a technical university, the only one in northern Germany that addressed the problems of the Elbe region and diversification of its economy, and proffered a new perspective for jobs in areas such as offshore technologies.[25]

The politics of "river and coastal industrialization" failed because of global changes, international division of business, the rising costs of raw materials, and the strategies of developing countries for processing their resources themselves. Internal restructuring of companies, fewer jobs than anticipated, environmental risks, and growing environmental awareness stopped further settlement of company branches. While Hamburg was proactive, looking for better coordination and pushing the implementation of a spatial vision, the territorial neighboring states were rather passive. This led to an erosion of the industrial base, high rates of unemployment in the region, and heated debate about more innovative strategies.[26] In 1982, the mayor of Hamburg, Klaus von Dohnanyi, declared, "Hamburg will stay a port city; however, the future perspective is not based on water but on land." The port development plan of 1989 set out that Hamburg would also become a logistics center with attached service industries.

In the mid-1980s, the decreasing attention being paid to regional planning was acknowledged, and a new start was made to overcome harmful competition and promote cooperation with the surrounding municipalities.[27] This resulted in a new Regional Development Concept (REK) in 1994, after Ger-

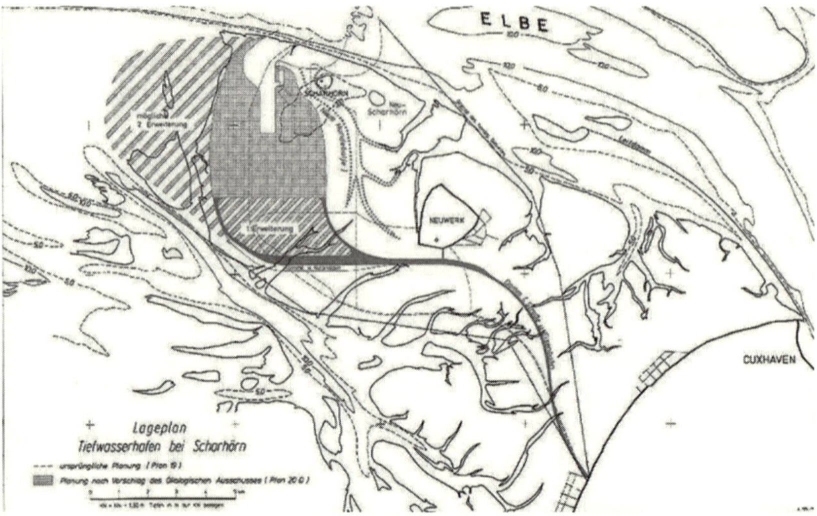

FIGURE 8.3. Deep water port Neuwerk Scharhörn "Europort" (1979). Source: http://www.cuxpedia.de/index.php/Datei:Tiefwasserhafen.jpg.

man reunification, which completely changed the economic geography of Hamburg's hinterland. This concept is still in force today. General principles concerning economics, housing, water supply, transport, leisure, nature, and the environment as well as waste management were determined.[28]

After the fall of the iron curtain in 1989, massive growth of the port was predicted. New connections to the east would make Hamburg the westernmost port of the Baltic, although it was situated on a river connected "only" to the North Sea.[29] Its location in the hinterland, about 120 kilometers from the North Sea, seemed to be an advantage because carrying goods on ships was the cheapest and fastest means of transportation for long distances. Formerly located in the periphery in the European Union, Hamburg now reclaimed a central position with a vast hinterland.

River Port, Deep Water Port, and Complex Environmental Problems

The Elbe River has been dredged nine times since the nineteenth century to secure access for ever-larger ships to the port of Hamburg. In April 2012, after four years of discussions, Hamburg's neighboring states approved a further deepening of the Elbe. In the meantime, environmental groups (Bund für Umwelt und Naturschutz Deutschland BUND and NABU Naturschutzbund in Hamburg) have lodged a protest, taking recourse to legal means, leaving

Germany's highest court still to decide about their complaint. The dredging of the river had been scheduled for 2008 but cannot be started pending the final decision. Although an institution in Kiel (Wasser- und Schifffahrtsdirektion Nord ASD Nord) is responsible for the application, the Elbe is a navigable waterway that is under the jurisdiction of the federal government. Hamburg has an interest in the immediate dredging of the river as the intended depth of 14.5 meters would allow (nearly) all ships—so states the argument of the Hamburg Port Authority—to reach the port, even at low tide. There is, however, a complex case to be made both for and against dredging.

The following concerns regarding the environmental impact of the dredging are shared by environmental groups and farmers along the riverbanks alike. The dredging of the river will move the saltwater zone twenty-five kilometers upriver. The amount of sediment removed will equal the total of the three previous dredges. Farmers are worried that their land, some of central Europe's most productive, will become saline, and their fears are echoed by environmentalists. At more than ten thousand hectares, northern Germany's *Altes Land* is one of the biggest fruit-growing regions in central Europe. It borders the western bank of the mighty Elbe River, which provides fruit farmers with the water they need to grow their produce.[30] How far saltwater would intrude inland is still a matter of dispute.

Dredging would also increase the Elbe's velocity, causing problems for many fish species as well as endangered plants. The environmental impact would be devastating for the fauna. The shallow waters along the riverbanks that act as nurseries for Elbe fish and habitats for other species would be disturbed by increased light levels caused by the dredging. If these shallow water margins were to decrease further, the system could collapse. Environmental groups warn that dredging the river also threatens some of the region's endemic marine and plant life. The state of Lower Saxony has offered some fruit farmers in *Altes Land* financial aid to adapt their irrigation systems, but the region around the village of Jork, for example, is too far upriver to benefit. Although only the higher elevations ("hunchbacks") on the bottom of the Elbe will be excavated, about thirty-eight million cubic meters of sand and mud will have to be carried off. The costs are estimated at around five hundred million Euros, most of which will be borne by the federal government. The environmental groups' case makes reference to recent strong European environmental law which requires spaces for compensation and zones that can be flooded.

The reasoning of the powerful Hamburg Port Authority (HPA) is quite different, of course. After Rotterdam, which has a depth of around eighteen meters, the port of Hamburg is the second-largest in Europe, providing jobs for an estimated 150,000 people in the region. The Hamburg Port Authority endorses the deepening, as it would enable them to compete in the global shipping market. Hamburg port officials argue that the measure is essential to keep up with the ever-increasing size of oceangoing container vessels. They want to deepen the Elbe to allow ships with a draft of 13.5 meters to reach the port of Hamburg, extending to 14.5 meters at high tide. The changes would amount to a meter below the present depth.

If large oceangoing vessels cannot reach the port because they have outgrown the river, it will have adverse effects on the entire region. If the port cannot handle these large vessels they will go elsewhere and Hamburg will lose its significance as a world port and a hub. About 3,200 large container vessels navigate the river each year and around 400 of them need a high tide to reach or to leave the port. Even larger container vessels that can carry about twenty thousand container units are now under construction. Since the world financial crisis, container traffic in Hamburg has been decreasing or stagnating and the growth scenarios from ten years ago never materialized.

Moreover, most of the logistics companies' packing and unpacking containers are located not in Hamburg but in Schleswig-Holstein in the north and Lower Saxony in the south. This means that containers are transported through Hamburg to the quay and then immediately loaded onto container vessels that use the Elbe as a transport corridor to Asia, in particular. This implies that fewer jobs (in Hamburg) are involved and that the added value is created in the region. The port is becoming a mere container-lock that bears little relation to its existing urban surroundings, as already executed in many other seaport cities. The path dependence that leads toward the perpetual dominance of maritime stakeholders in economies in the Elbe region seems almost to be irreversible. If a more diverse regional economic strategy is not implemented it is likely that the continuity of the maritime focus will persist.

In 1993, plans for a new deep water port in Wilhelmshaven, Lower Saxony, were initiated. In Wilhelmshaven even the largest vessels could be accommodated without any problems. This strategy would generate labor division among the German seaports Bremerhaven and Hamburg on the North Sea coast and the four involved states (Hamburg, Bremen, Lower Saxony, and Schleswig-Holstein). Hamburg had initially decided to participate in the construction and development of a new terminal around the year 2000 but later withdrew its support. Consequently, the dredging of the Elbe became more urgent in order to maintain a competitive position. Hamburg's decision marks the end of its cooperation with other German seaports, compelling environmentally problematic dredging and more costly competition for attracting container lines in the future.

Meanwhile, Hamburg's most important employer is no longer the port but Airbus Industries. Parts for the construction and assembly of airplanes are delivered to Hamburg by plane, not by ship. Because other German seaports like Bremerhaven and Wilhelmshaven already offer the water depths required for the largest container vessels, the question arises whether the Elbe River should really be dredged again and again to improve access to the port of Hamburg, given the potential environmental impacts. Although it is difficult to estimate exactly what impacts the deepening of the river would have, it is clear that they would affect the ecosystem, fish stocks, the speed of tidal surges, the quality of the water, and the neighboring fruit farms, which need to irrigate their orchards with water that is not too brackish.

New challenges brought about by global warming and rising sea levels will make it more difficult to predict all of the possible, interrelated impacts of deepening at local, regional, and national levels, as well as the manifold environmental consequences. While rivers and waterfronts were categorized as devalued land and dangerous areas for decades, this image has changed over the years and now they are becoming "spaces of opportunity" and "spaces of promise." This leads to a reinvention and redefinition of the interdependencies between water and land, river and city. HafenCity Hamburg is the most important transformation project in Hamburg and one of the largest developments of its kind in Europe. The HafenCity reestablishes the connection between the River Elbe and the city center, giving Hamburg a new direction for growth: down and along the river.

The dredging of the Elbe and widening of its fairway will accelerate tides and increase the flow of water. The consequence of past dredging has been summed up in the catchphrase "higher floods require higher dikes." Recent, more flexible flood control policies at a regional level involve a cascading system of controlled flooding and innovative redevelopment projects in HafenCity and Wilhelmsburg.

Interdependencies between Economy and Ecology

The port of Hamburg has always dominated the local economy, above other economic sectors. "What serves the port serves the city" was the unquestioned ideology and the unscrutinized development path of the open tidal seaport decision. However, since 2008 the construction of new port terminals has been delayed or canceled. A perspective focusing more on the unique qualities of the port and the lower Elbe region rather than on container handling may be more sustainable.

The central question for the future of the river Elbe and the port of Hamburg is the issue of dredging. It is not only the river that needs to be dredged continuously to avoid the danger of silting up but also the port basins. There is no single irrefutable answer; it depends on what is considered more important in the long run: the environmental or the economic perspective.

The quality of water has improved as pollution from the former German Democratic Republic and its factories has been cleaned up, and more species of fish are now found in the river. The Elbe and its shores are increasingly used for leisure purposes like sailing, rowing, jogging, and even swimming. The mudflats in the mouth of the river have been designated a world heritage site. However, globalization and port logistics have severed nearly all river ports and seaports functionally, geographically, and visually from cities. Today, cities are growing more alike while ports and shipping have become more specialized. Port facilities such as Maasvlakte 2 in Rotterdam and Eurogate Container Terminal in Wilhelmshaven are turning into secluded worlds that are separated from their urban context. They have been spatially and mentally severed from the city, with their own employment, operators, and administration structures. They are segregated for different reasons, such as noise, security, demand for huge spaces, and deep water access. While older port areas near the city center have been (re)integrated into the urban fabric, the new port infrastructures are separated from major city networks and situated in areas where deep water ports and large areas of land are available. Seaport cities and local port authorities will gradually lose the ability to determine the course of "their" ports, while logistics firms operating globally will be setting the agenda.[31]

Seaports with direct access to seas and oceans nowadays have clear advantages over river ports, which have, and always will have, the problem of dredging. For many centuries, however, the comparative advantage had been with river ports, which offered opportunities for safe storage, easier cargo handling, and protection against enemies. Over the last few decades we have seen how those parts of the port dating from the Middle Ages to the nineteenth century, which are located next to the city center, have been transformed and integrated into the urban fabric. On the one hand, the decision-making processes of global logistics companies follow top-down strategies, with manifold impacts at the local level, and on the other hand, local urban planning in sea and river ports follow different paradigms from more (or less) democratic bottom-up schemes. The result is a scenario of skepticism: the city needs the port and dredging of the Elbe River for jobs and taxes, but the port (more precisely the container port) no longer needs a location next to the city.

Part III.

Cultural Dimensions of Urban Rivers

CHAPTER 9

RIVERS AS PRISMS OF URBAN IMAGINING

Eastern Sichuan Work Songs

Igor Iwo Chabrowski

In this chapter, I shall discuss how understanding of the city-river relationship in China changed in the twentieth century, through the example of the southwestern province of Sichuan. To demonstrate the important shift in the way this relationship was forged and comprehended, I first focus on the case study of the Sichuan junkmen, a working-class group, who in their life and work constantly traversed the borders between the cities and the rivers. Boatmen's perception of the relationship that connected these two environments is accessible through their work songs, called Sichuan junkmen *haozi* 號子 or *chuanjiang haozi* 川江號子—a unique form of singing with an attested history of at least three hundred years. Over the last quarter century, the same songs and the very image of the Sichuan boatmen was appropriated by the authorities of the Chongqing Municipality (administrative entity of the provincial status formed in 1997) in order to legitimize the Three Gorges Dam project. The reasons why junkmen suddenly became important for the Communist state and the manner in which their representation and their traditions were used form a second topic of this chapter.

SICHUAN BOATMEN, THEIR WORK SONGS, AND THEIR AESTHETICS

Sichuan boatmen were the main transportation workers of the Sichuan Province before the introduction of railways (1952).[1] They existed on the merging of the two spatially and socially distinct environments: on the one side, the organized, governed, and geographically urban space; on the other, the wild, unadministered, and floating world of the waterways. The concurrent engagement with both realities allowed junkmen to traverse the cultural and spatial boundary dividing these two worlds. It was also a source of their unique perception of the links that connected rivers to cities and cities to rivers in the predominantly rural and parochial country that Qing (1644–1912) and Republican (1912–1949) China was. In fact, both in their daily practices (work, movement, residence, etc.) and in their cultural imagery, boatmen managed to integrate urban space in the broader river world of Sichuan. Or, in other words, they dissolved the walled and sheltered cities and towns in the broth of personal, commercial, and religious links that traversed borders and converged with the riverways.

Boatmen's representations of these dual but intermingling realities found expression in their work songs. These songs were boatmen's tools of work, helping them perform backbreaking labor, their main sources of entertainment, as well as their main means for circulating and inculcating knowledge and ideas. In this chapter I focus on the work songs' texts, which from the perspective of the junkmen obliterated the borders dividing the physical space of cities and rivers and traded them for a uniform social and cultural space. In this way, imagination and perception of the city was firmly tied to the representations of the river; conversely, the river provided a vantage point from which the city was seen and understood. *Haozi* also expressed the boatmen's sensitivity to geographical, social, and historical realities—to cities of various sizes and their cultural and economic significance. Using the example of the local metropolis Chongqing 重慶 and the smaller trading center Hechuan 合川, we shall explore the diverse limits to boat workers' vision of the relationship between the river and the city.

Who were the Sichuan boatmen? Chinese preindustrial regional economy, which lasted until the Second Sino-Japanese War (1937–1945), was reliant on specialized labor groups. This was especially pronounced in the field of transportation, where steamers, railways, and paved highways were either nonexistent or had only limited significance. The shipment of goods was largely conducted by small guild-based companies employing traditional wooden junks and large groups of specialized wage laborers who towed, sailed, and rowed the boats. Multiple rivers in the region, each with its unique morphology and its own set of dangers and difficulties, contributed to the creation of some extremely specialized transportation communities and cultures. Here I will focus exclusively on one of the most significant, the one that managed the two biggest waterways of Sichuan: the Yangzi and the Jialing 嘉陵.

Although they had strong roots in earlier centuries, the Sichuan boatmen's fast growth to prominence occurred during the eighteenth century. The number of people engaged in this profession increased dramatically throughout the next century as a result of the accelerated parallel processes of rural population growth, urbanization, and surging volume of interprovincial trade. According to Paul J. Smith, the same period experienced a population shift within the province. Its eastern part lining the valley of the Yangzi River became much more populous, which in effect turned its main city, Chongqing (capital of Ba County, Baxian 巴縣) into one of the most important cities in southwestern China.[2] The consequences of these demographic changes have been described by Wang Di, who outlines this multistage socioeconomic formative process. Wang states that one of the most important elements of Qing Sichuan's history was an inflow of a migrant population from central China, which not only took land under cultivation but also flowed into towns. Moreover, the cities that saw the biggest surge in population were those that existed along the waterways and contributed the trade networks exploiting Sichuan's natural riches. From the nineteenth century onward, this province was progressively more integrated in the cross-imperial web of commerce propelled by the immigrant groups dominating its cities. These trends did not abate either in the nineteenth or in the early twentieth century; on the contrary, as pointed out by Wang, eastern Sichuan cities, particularly Chongqing, grew rapidly (for example, Chongqing's population increased tenfold between 1812 and 1890, from thirty thousand to three hundred thousand) and the volume and variety of goods they produced and traded gradually became more sophisticated.[3]

One of the results of these processes was a growth in the number of people employed as boatmen, especially on the most important Yangzi route, as well as in the number of boats braving the inland waterways. From the early 1920s traditional junks had to compete with the expanding network of steamship companies, yet in the early decades of the twentieth century these were more the general economic conditions than technological advances that posed a threat to traditional shipping. Multiple accounts claimed (though often quite inaccurately) that in Sichuan's main port there were more than ten thousand boats moored at any time of the year. From trade surveys conducted and published in 1946, we know that most local river routes were serviced by such traditional boatmen.[4] Moreover, on a majority of routes, wooden boats successfully resisted competition from the steamship companies up until the 1960s.[5]

The minute division of labor was characteristic of the profession of Sichuan boatmen, which divided tasks among rowers, boat pullers (so-called trackers), swimmers (who untied ropes when they were caught between the rocks), headmen (who led the way for the tracker gangs), workers on the boat (who used long poles to push the boat away from the rocks or the shore), cooks, helmsmen, pilots, captains, and so on. This division also implied social inequalities, roughly splitting the boatmen into one group of wage workers and another of entrepreneurs, who had a financial stake in each trip and/or co-owned the boat (this is especially characteristic of the latter three professions listed above). The workers themselves were divided into two groups—those who were permanently employed and seasonal workers. The former seem to have obtained their jobs through personal or secret society associations such as the Gowned Brothers (Gelaohui 哥佬會 / Paoge 袍哥), which was ubiquitous in Sichuan. Seasonal workers were simply hired only for a short time in front of more daunting rapids. Despite these status divisions, the boatmen seem to have shared the same culture, based largely on the oral transition of ideas and stories within largely masculine, river-based communities occupying fringes of the settled world of cities and villages.[6]

Sichuan boatmen's work songs were not only one of the main forms of cultural production of these workers but also an indispensable tool of their work. Ted Gioia claims that we cannot separate the work from the work song, since this art form was integral to the very performance of particular tasks.[7] Indeed, to state it in the shortest possible way, *chuanjiang haozi* were, apart from their musical and entertaining quality, a means for communication, harmonization, and organization of work. They allowed for managing the pace of the trackers' work gangs, which often spread across a very large distance—keeping in touch with the boat they towed was essential for the security of the tracking enterprise, as the rivers of Sichuan are known for their fast currents and multiple gorges and eddies. Singing together also provided a way of controlling the breath and movement of the large groups of trackers, alleviating physical stress and protecting against accidents. Moreover, *haozi* marked different sections of work—their pace sped up or slowed down according to the river conditions and required movements—becoming the sonic guides for each stretch of the waterway.[8] At the same time, they were venues for the construction, transmission, and assertion of cultural images within the boatmen communities. These images will be the focus of the later parts of this chapter. Finally, it needs to be mentioned that it is highly improbable that the *haozi* circulated outside the boatmen communities before they were turned into an exemplary people's art in the early years of the People's Republic. The folklorization of boatmen's songs coincided with the demise of traditional shipping and thus with the disappearance of the primary condition for these work chants' existence.[9]

Although I have extensively treated the following topic elsewhere, a few points need to be made here.[10] First, Sichuan junkmen *haozi* did not constitute an entirely unique genre existing solely in Sichuan. For example, as demonstrated by Gioia, work songs of various kinds were an integral part of the historical labor culture in the United States.[11] They were also common in other regions of the world, and certain kinds, such as sailors' shanties, became globally recognized and formed a part of the entertainment culture even after they stopped serving their primary, work-aiding function.[12] Some forms of river transportation in Europe and North Africa somewhat resembled the Sichuan junkmen's, such as raftsmen (*flisacy*) in the Polish-Lithuanian Commonwealth, *burlaks* (Бурла́к) in Russia, or boatmen on the Mississippi. Each of these groups created some particular forms of professional knowledge and produced a set of cultural traditions.[13] There are, however, quite strong differences between the traditions produced by the above-mentioned groups and the Sichuan boatmen. Junkmen's *haozi* were characteristic for their flexible adjustment to the changing river morphology and their textual richness. That is, not only did they operate with the use of calls and short words, they included whole passages of texts, sometimes extracts from operas, though more often descriptions of places, feelings, conditions of life and work. Second, an important difference between Sichuan boatmen and other similar groups lies in the accessibility of their cultural tradition to the current researchers. A large number of the songs were collected and published in the 1980s and later in the 2000s, as a reaction to the aging of the last people to have been actively engaged in performing traditional shipping in Sichuan. The published anthologies, through which we can access the culture of the boatmen, such as *Zhongguo geyao jicheng Chongqingshi juan* 中國歌謠集成重慶市卷 (Grand compendium of Chinese folksongs, volume on Chongqing Municipality) and *Chuanjiang haozi* 川江號子 (Work songs of the Sichuan Rivers), are obviously not ideal sources. They are selections and, to a degree, standardizations of the popular traditions and not mirror images of the oral arts of the boatmen. That said, much of the collectors' work was based on very thorough research and left more than three hundred texts as well as musical notations, which allow for a fairly thorough reconstruction of these laborers' life and mentality. Moreover, in the 1980s, 1990s, and even in the early 2000s, there were still living junkmen who worked before the Communist takeover and who had learned to sing *haozi* from the old masters—that is, the tradition of singing

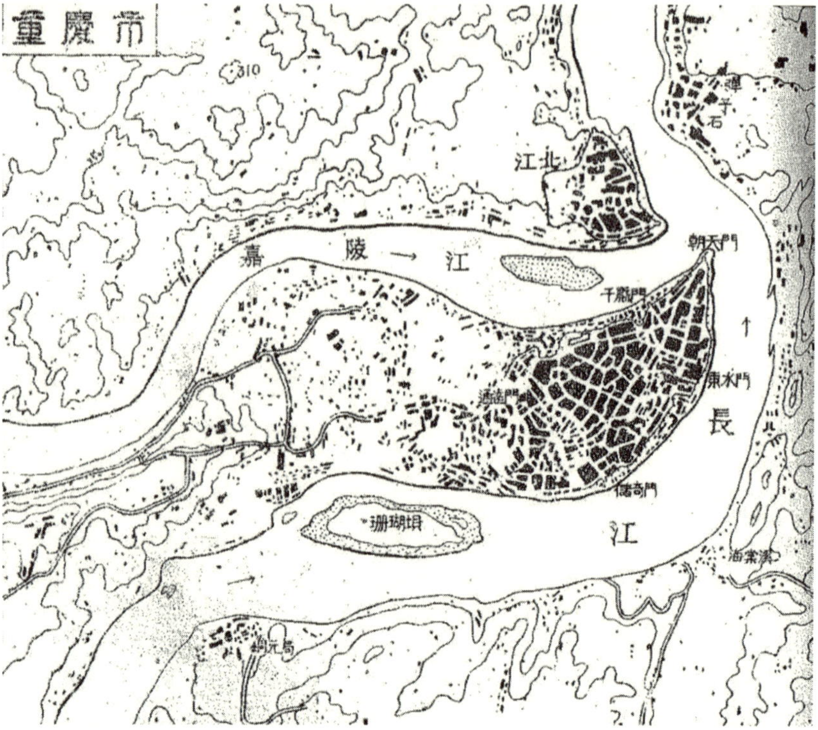

FIGURE 9.1. Chongqing and Jiangbei 江北. Source: Zhou Lisan 周立三, Hou Xuetao 侯學燾, and Chen Siqiao 陳泗橋, *Sichuan jingji dituji* 四川經濟地圖集 [Collected economic maps of Sichuan] (Beibei: Zhongguo dili yanjiusuo, 1946), n.p.

work songs was not "fossilized" in the form of the written word. This factor determines a fairly high authenticity of the material presented in the anthologies. These publications, however, are interesting not only as collections of documents but also as evidence of nostalgia for the old world whose social and natural features underwent irreversible changes. Taking this into consideration, we can state that restoring the memories of the Sichuan boatmen and of their culture was a form of public memory cast against the sweeping "modernization" of the Chinese social and natural landscape.

Finally, it is worth mentioning the cultural aesthetic context in which the junkmen's *haozi* functioned. At least from the fall of the Han dynasty (220 CE), Chinese scholars and poets developed an appreciation for mountains and rivers. Mark Edward Lewis ascribes this new aesthetic movement to the shift of the Han population south of the Yangzi River and their consequent enchantment with the local landscape that was so drastically different from their

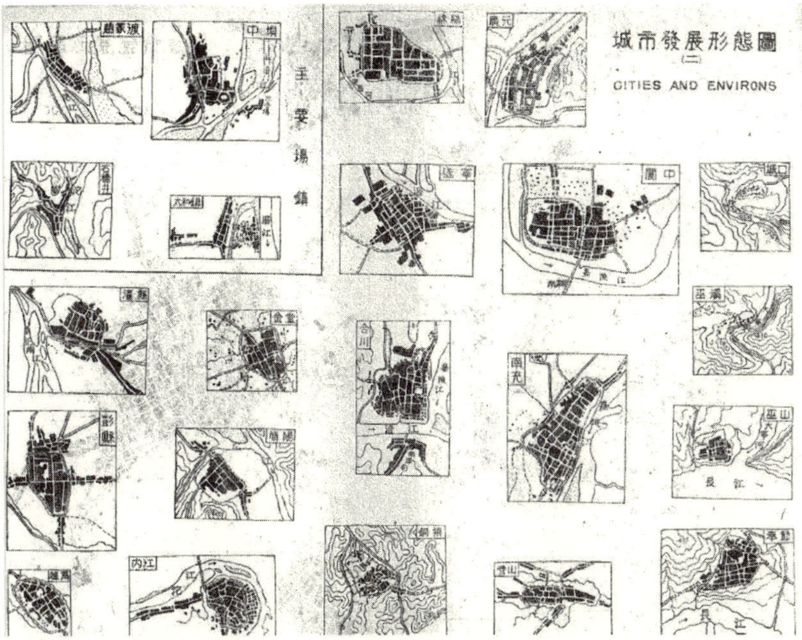

FIGURE 9.2. Shape and development of the cities—the main urban centers of Sichuan. See the characteristic orientation of most cities toward the rivers. Exactly in the middle there is a map of Hechuan, a city discussed here in some detail. Source: Zhou Lisan 周立三, Hou Xuetao 侯學燾, and Chen Siqiao 陳泗橋, *Sichuan jingji dituji* 四川經濟地圖集 [Collected economic maps of Sichuan] (Beibei: Zhongguo dili yanjiusuo, 1946), n.p.

home environment in the north.[14] Simply speaking, the dramatic beauty of the new provinces became synonymous with beauty as defined by the idiom *shanshui* 山水—mountains and water. More often than not, however, this beauty was enjoyed within an urban environment, reproduced in gardens, enjoyed on viewing platforms and towers, etc. Until the end of the classical education and artistic taste, roughly during the Republican decades, city descriptions by the literati (as found, for example, in county gazetteers, *xianzhi*) abounded in detailed accounts of local mountains and rivers—essential features of the local environment and indispensable elements of the landscape of any respectable city.[15]

This literati culture rarely produced a picture in which a river and a city would be integrated or in which such a connection was central to the narrative. Rather, two magnificent rivers, the Yellow River and the Yangzi, became topics of their own, represented in paintings and lauded in poems.[16] Yet most literati separated the natural environment from that of a city. To take an example from eastern Sichuan, multiple poems written between the Ming and Republican periods spoke of terrifying rapids and currents, boatmen exerting themselves like beasts but jolly in their leisure hours, and sights of fame and splendor. Nature (even in the form of a garden) stood in opposition to human activities epitomized in the busy world of the cities.[17] It is in this context that the boatmen narratives and their point of view become especially interesting.

Picturing the Spaces—Boatmen Singing about the Cities

Whether depicting Sichuan cities or river valleys, work songs projected an environment bustling with activity. Two urban centers, the provincial capital of Chongqing and the regional port of Hechuan, attracted extensive attention and numerous *haozi* were created to describe and guide one through them. Both cities were viewed from their gates, which in the case of eastern Sichuan is indivisible from the outlying river docks. Under the reign of the Qing dynasty as much as during the Republic, the gates divided the space that was within and outside the city walls—they were places of inspection and of collection of the custom duties.[18] Furthermore, they were places where goods were unloaded and stored, where companies and guilds congregated, and where people thronged to buy and sell incoming products shipped from the surrounding counties and even distant foreign locations. For the boatmen the gates functioned not as borders but as places of access to the urban space while still being integral parts of the river world they lived in. This is why junkmen's depictions of the cities are blind to the monumental buildings, which distinguished the urban world, and instead are very sensitive to the chains of contact points that linked the town to the province and to the river—its main highway for commerce.

To illustrate this point, we should look at one of the most recognizable boatmen's songs, titled "Among Sichuan harbors Chongqing is the most important" (Sichuansheng shuimatou yao shu Chongqing 四川省水碼頭要數重慶).[19]

四川省水碼頭要數重慶，	*Among Sichuan harbors Chongqing is the most important,*
開九門閉八門十七道門：	*Nine gates open, eight gates closed, together seventeen gates:*
朝天門大碼頭迎官接聖，	*Facing Heaven Gate* [Chaotianmen], *great dock, invites officials, receives holies,*

千厮門花包子雪白如銀，	*Thousand Servants Gate* [Qiansimen], *snow-white bundles are just like silver,*
臨江門賣木材樹料齊整，	*River Overlooking Gate* [Linjiangmen] *sells wood lined up in perfect order,*
通遠門鑼鼓响抬埋死人，	*Passing to Far-Off Gate* [Tongyuanmen]—*bells bid farewell to the deceased,*
南紀門菜籃子涌出涌進，	*Gate of Regulating Southern Kingdoms* [Nanjimen] *is crossed by baskets filled with vegetables,*[20]
金紫門對着那鎮臺衙門，	*Purple Golden Gate* [Jinzimen] *faces city's yamen,*
儲奇門賣藥材供人醫病，	*Holding Marvel Gate* [Chuqimen] *is where medicines are sold to treat the sick,*
太平門賣的是海味山珍，	*Great Peace Gate* [Taipingmen] *is where rare delicacies are sold,*
東水門有一口四方古井	*Eastern Water Gate* [Dongshuimen]—*there is a square ancient well*

Tying particular gates to the trades and practices performed there created a static and timeless vision of the city. The song describes not only goods but also ritual practices, such as receiving officials, processions associated with religious holidays, and funerary celebrations. Certain indications are given of the outstanding elements of the urban landscape. The most conspicuous among them are government yamen in the vicinity of Jinzimen (Purple Golden Gate) and the old wells, or rather water tanks constructed as an element of firefighting infrastructure. Some elements in this song call for special attention, especially the choice of goods listed. Wood, medicines, cotton, and foodstuffs were all traditional trades of Chongqing confirmed in sources at least since the eighteenth century. The choice is, nevertheless, slightly misleading in the context of boatmen's *haozi* from the latter part of the twentieth century, since what is represented cannot be connected to any particular time in the city's history in the last three hundred years. What is important about this song, however, is that the river gates represented a part of the city that was geographically and socially most accessible to the junkmen. On both sides of the city gates, as far as it can be confirmed for the first half of the twentieth century, the urban and suburban wards were inhabited largely by the members of the laboring classes. The outer part belonged to the "amphibious" communities that were settled on boats and in makeshift housing, whereas the inside was a realm of petty traders and workers.[21] Boatmen not only hailed from the equally low-class backgrounds but also were bound by a variety of social networks with the people who lived there. It was in these intermediate realms of the city where they traded, obtained work contracts, found entertainment, or engaged in religious ritual.

Another work song about this city, called "Talking of Chongqing" (Shuo Chongqing 說重慶), yields an even clearer view on how the relationship between the river and urban spaces was conceived by the boatmen.[22] This song, outstanding for its level of detail, demonstrates as well the superficiality of the river laborers' engagement with the urban world; or, in other words, the limits of their access to the city life and culture. Reading "Talking of Chongqing" is like looking into the recollection of a stroll by a boatman who dared to venture into the city's streets and alleys. He starts his walk from the riverbank, crossing the various gates of Chongqing, therefore, the first lines of the song are almost identical to "Among Sichuan harbors Chongqing is the most important." The deeper he goes among the houses and shops, the more details are recorded about particular trades and urban spaces. The method of description is the same as previously—the space is mapped and organized by matching names of places with characteristic activities performed there. For example:

"較場壩"地方寬多少美景，	*Parade Field* [Jiaochangba] *is a broad place, so many spectacular views,*
談生意講買賣賺金賺銀。	*Talking business, bargaining, earning gold and silver.*
想當年春秋操好不齊整，	*Back then the military discipline was very good,*
眾教頭操壩上大練兵丁。	*All the officers were putting soldiers through their drills.*

Junkmen saw the Parade Field, the main—if not the only—city square, as a place where outstanding public events took place, where goods passed from hand to hand and money was made, and where communal memory was encoded. Talking of military drills, the junkmen recalled the Parade Field's previous function—during the Qing dynasty it bordered the Hall of Martial Arts (Yanwuting 演武廳) and was a place of regular drills for Qing troops. Moreover, as pointed out by Peng Botong, Jiaochangba was associated with some important historical figures such as Li Hualong 李化龍 (1554–1611 or 1624), a renowned Ming general, and especially Zhang Xianzhong 張獻忠 (1606–1647), an infamous late-Ming rebel who killed thousands of Sichuan people and dismembered numerous imperial officials there.[23] Since nothing commemorated this history or the previous practices during the Republic (the field was converted into a permanent market and traffic intersection),[24] we can see that for the boatmen the very name of Jiaochangba was enough to recall the past. Moreover, this was one of the very few cases where boatmen had a notion of the activities performed by the state, such as military drills or executions. The urban space thus was a contact zone with the state and by that defined what the state (and its history) was.

The social contrasts—characteristic of urban space and rather absent from the impoverished and by comparison egalitarian junkmen's social reality—formed another important element of their perception of Chongqing. The city, to their eyes, brought people from various walks of life together and the juxtaposition of them appeared shocking if not condemnable:

"三牌坊"富貴家表之不盡，	*Three Archways Ward* [Sanpaifang] [, where] *opulence is demonstrated without restraint,*
"魚市口"到冬至要殺犯人。	*Fish-Market Door* [Yushikou] *when winter comes, criminals are killed there.*

The obscenely rich lived next door to the most miserable. The privileged lavished their wealth at the same time and alongside the place where criminals were brought to a public execution. Such a striking concentration of diverse walks of life was characteristic only to the city.

Residential and government functions did not exhaust the spectacular quality of Chongqing. The city ran on trade in goods and money, and for boatmen it looked like a huge maze, as varied in its diversity as long stretches of the river valleys.[25] So, instead of the bays and rapids, the stations on one's way were formed by banks, company houses, and markets:

"朝天門"開字號兌換金銀。	*Facing Heaven Gate opens to shops changing gold and silver.*

The song pilots the reader (in the past, the singer and listener) from one market to another as if trying to show everything one could buy in Chongqing:

"道門坎"賣瓜帽還賣鉛粉，	*Ridge of the Way* [Daomenkan] *sells gourd hats and lead powder,*[26]
"雙火墻"賣氈帽又賣氈纓。	*Twin Fire Wall* [Shuanghuoqiang] *sells felt hats and felt tassels.*

[...]

"大匠街"賣銅器又賣冬筍，	*Great Artisan Street* [Dajiangjie] *sells brass ware and winter bamboo shoots,*
"新街口"賣衣綫又賣頭繩。	*New Street Door* [Xinjiekou] *sells cloth threads and ribbons.*

The goods, whether of local origin or imported, were what attracted most junkmen's attention. They were the splendid sights of the city, equal, if we compare these songs to other *haozi* that picture Sichuan, to rapids, gorges, temples, and spectacular rocks lining the province's riverbanks.

Religious and Mythical Spaces

"Talking of Chongqing" also gives us some information about the cultic affiliations of boatmen. Chongqing was a space of substantial religious significance, but there is insufficient evidence to speculate if the whole city was embedded with supernatural power or if only its various parts were considered efficacious due to the presence of gods or congregation of geomantic power. For instance, Qiansimen [Thousand Servants Gate] possessed powers particularly well disposed toward the boatmen:

"千厮門"雞毛土地靈得很，	*Thousand Servants Gate, Chicken Feather God's efficacy is great,*
還雞願燒香火要數船民。	*And wishes fly with incense smoke, innumerable crowds of boatmen come.*

Boatmen revered this local god and associated it with a city gate and a nearby dock. Apart from its protective function, the cult of Chicken Feather God was a way for the boatmen to psychologically, ritually, and physically breach the borders between the river and the city. The name of the god was derived from a ritual practice of performing chicken blood sacrifices in thanksgiving for safe arrival to Chongqing. The close relationship of the god with the junkmen was explained by a legend. According to it, the Chicken Feather God was a weasel reborn into a bodhisattva caring for the life and safety of both the boatmen traveling on the Jialing River and the people residing close to the Thousand Servants Gate. Another legend stated that this local spirit warned people of the conflagration that struck Chongqing in 1949, thus depicting him as a protector of all the workers and petty traders of the city's riverbanks.[27] That the supernatural protection was common to the boatmen and the suburban and urban populations shows that in the lower-class worldview of the time, the river and city were a continuum rather than sharply divided spaces.

Moreover, the particular quality of this earth god (*tudi* 土地), a type of deity that typically protects a clearly designated swath of agricultural land or a ward in the city, was that the borders of its efficacy were described not by the land but by its ritual community.[28] This largely meant the boatmen who went to or away from the Thousand Servants Gate and who traveled on the Jialing River. In their day-to-day religious practice, therefore, boatmen not only imagined the urban space as flexible and intermingled with the river world but also defied the division between social and natural spaces.

A further mixing of the natural and human environments can be seen in the way boatmen took seriously the geomantic vision of Chongqing. "Talking of Chongqing" demonstrates how both the buildings and knowledge of local history served for the creation of a holistic, efficacious vision of the city. Their imagination of space had a cosmological meaning, which made them see in the natural and man-made landscape and construction reflections of the auspicious animals that connoted moral fundamentals. Namely, in the eyes of the junkmen Chongqing's buildings and wells, placed according to geomantic principles, *fengshui* 風水, shaped it into the form of a giant turtle, a potent symbol of longevity and persistence.[29]

"沙井灣""鹽井坡"烏龜眼睛。	*Sand Well Bend* [Shajingwan], *Salty Well Slope* [Yanjingpo] *are tortoise eyes.*
"三元廟"不撞鐘光是敲罄;	*Temple of Three Seasons* [Sanyuanmiao], *to make an empty sound, do not strike the bell;*[30]
"龜頭山"敲了鐘不得安寧。	*Tortoise Head Mountain* [Guitoushan], *when the bell tolls one cannot remain in peace.*

The sense of the appropriate geomantic characteristics of Chongqing was strengthened by the conviction that the number of its gates was not accidental but consciously planned by the officials of the past who wished to grant good fortune to the city. Both songs quoted here repeat the following formula:

開九門閉八門十七道門	*Nine gates open, eight gates closed, together seventeen gates*

In other words, they state that the number of city gates mirrors the Nine Palaces and Eight Trigrams (*jiugong bagua* 九宮八卦), i.e., the geomantic compass. We do not know to what extent boatmen shared the local scholarly tradition, which associated the construction of city walls and gates with two officials, Peng Daya 彭大雅 (before 1241) and Dai Ding 戴鼎 (fl. 1371).[31] We can, however, say that a number of motifs that persisted in local culture and were engraved in the popular vision of the city, its shape, and the meaning of particular places it contained connected the city to the local environment and undermined the strictly human-centered image of the urban space.

The curious gathering of elements, which were included in this *haozi*, reflects not only the depth of impression the city made, but also the amount of knowledge necessary to "master" it and be able to exploit its complexity. Mapping Chongqing was like mapping the Yangzi and Jialing valley, but a city of this size, richness or diversity needed to be treated on its own and could not just be integrated into the river routes and represented only by its name. As we have observed, even in boatmen's descriptions the urban environment could obliterate their acute perception of the river and the relationship between rivers and the city, because for them, at least in the way they narrated it, a city of this size was like a whole river itself.

Provincial Spaces—Rivers above Cities

This was certainly not the case with smaller centers. If we look at the songs depicting second-ranking towns, such as Hechuan, we will see that the method of description and the focus of the singer-observer were similar, but the size and importance of the urban center affected boatmen's words and images. It is clear from the outset that gates and docks, areas connecting river and urban spaces, dominated the content of songs and rendered "inner" areas of the city invisible. The ballad "Counting city gates of Hechuan" (Shu Hechuan chengmen 數合州城門) will serve as our example.[32]

瑞映門扎的部隊多得很，	*Gate Reflecting Auspiciousness* [Ruiyingmen] *is crowed by brawling soldiers,*
盡是手拉營。	*They are all from the riflemen's camp.*
塔耳門熱鬧得很，	*At Pagoda's Ear Gate* [Taermen] *it is very lively,*
今天要殺几個人。	*Today they are going to kill a number of men.*
洛陽門道路寬套得很，	*At Luoyang Gate* [Luoyangmen], *where the road is very broad,*
又可走太和又可走遂寧。	*Some go to Taihe and others to Suining.*[33]
青龍橋楊颜婆正在粉，	*On Green Dragon Bridge* [Qinglongqiao], *lady Yang Yan is just putting on her makeup,*
周么妹正在拉人。	[and] *little sister Zhou is pulling people in.*
學昌門熱鬧得很，	*The Gate of Prosperous Learning* [Xuechangmen] *is crowded and noisy,*
住的是几條江拉船工人。	*There reside trackers that work on local rivers.*

人和門廖瞎子拉皮條得行得很，
At Harmonious Relations Gate [Renhemen], *the blind Liao is a skillful pimp,*

五個錢包你人一個進門。
With five strings of cash you enter the [brothel's] *door.*

大南門賣雜貨齊全得很，
Great Southern Gate [Dananmen] *sells groceries all in great order,*

大蒜老姜花椒麻辣死人。
Garlic, ginger, pepper, spice—[enough to] *kill people.*

小南門賣旺子相因得很，
At Small Southern Gate [Xiaonanmen], *they sell pig blood really cheap,*

几個錢就可買一大盆盆。
So for some cash one can buy a big pot.

北門裏唱川戲的多得很，
At the Northern Gate [Beimen] *they sing so much Sichuan opera,*

三撥班子都在唱駝子回門。
Three theater groups all sing "Hunchback comes back home."[34]

This song, quite similar to "Talking of Chongqing" fleshes out multiple details of urban life, which was filled with various local groups and with bustling activities of daily life. The urban representation in "Counting city gates of Hechuan" was distinctively lower class, filled with people and activities far divided from that entertained by local elites or even more established urbanites. It was a rough world on the edge of the rivers, which served both as channels for goods and for people of all walks of life. Their labor and entertainment were placed side by side, in fact, in the exact same locations in the riverside markets and shanties surrounding the city gates. Flowing as the river's current were the crowds of boatmen calling on prostitutes, listening to the din from the local opera stage, or getting away from the squabbling soldiers. The song shows a society forged on the margins of urban life and belonging to the chaotic, culture-deprived, and uncontrollable world of the rivers.

A different picture of Hechuan and the Jialing River valley appears in the verses of another song, "Departing from Hechuan" (Hechuan kaichuan 合川開船):[35]

合川開船下嘉陵江，
Embarking from Hechuan, sail down Jialing River,

八角亭修在梁梁上。
Eight Horned Pavilion [Bajiaoting] *is built on top of the ridge.*

南津街的白塔挨着火柴廠，
The white pagoda of Southern Ford Street [Nanjinjie] *is close to the match factory,*

東津沱有個渝豐紗廠。
Eastern Ford Pool [Dongjintuo] *has one Yufeng cotton mill.*

馬廟尼姑想和向，	*At the Horse Temple* [Mamiao], *Buddhist nuns are craving for monks,*
張公灘險要南倒船邦。	*South of dangerous Prince Zhang Rapid* [Zhanggongtan] *it is too difficult for the boatmen.*
枯水天甑子石在幹坡上，	[During] *low water, stones steam;* [one has to] *climb a cliff,*
釣魚城威風修在巨石梁。	*The Fishing Town* [Diaoyucheng] *is proudly built on a huge ridge.*
軍民保城把元兵抗，	*Soldiers and civilians guarded it against Yuan* [Mongol] *invaders,*
蒙哥大汗把命喪。	*Great Han Möngke lost his life* [there].

With a distinctive focus on production and factories or manufacturers established in the 1930s, the *haozi* recalls glorious elements of the past, such as decades of resistance to the Mongol invasion during the Southern Song dynasty (1127–1279), of which remained a nearby fortress of the Fishing Town (Diaoyu).[36] As in Chongqing, Hechuan tradition lauded famous military figures irrespective if they were local heroes (Bamanzi) or villains (Meng'ge—Mongol Great Khan Möngke, 1209–1259).[37] Another quite characteristic motif, perhaps borrowed from the popular literature of the last two dynasties, was the sarcastic depiction of the lewd Buddhist monks and nuns, such as the one at the Horse Temple [Mamiao].[38] Irrespective of the differences it is again a view at the city from the riverbank, and a mental construction of the urban and riverine space organized by someone moving with the flow of water. Each step along the way from Hechuan down the current is marked by human construction or some prominence or beauty. Yet the city is drowned in the descriptions of the natural features of the river environment such as pools, eddies, rapids, and rocks that gained primary importance. The human constructions were simply other points on this list marking boatmen's way along the river valley. What is important to mark here, however, is that they were integral and not distinctive points, which would divide the environment into the urban and river-dominated spaces. Although surely boatmen could draw a distinction between the two, they "humanized," by naming and historicizing, the works of nature at the same time and in the same manner that they "naturalized" the products of human hands.

Images of the Industrializing River-Cities

Haozi as well as the kinds of work on traditional wooden boats did not disappear in the blink of an eye with the introduction of new transportation technology (steamers, railways, and airplanes) or with the political revolutions of the midcentury. This genre of songs together with the profession of the Sichuan junkmen interacted with a changing reality and their texts recorded the rapidly industrializing landscape of the Sichuan cities and river valleys. For example, the *haozi* "Counting docks" (Shu matou 數碼頭) shows how boatmen saw the increasing urbanization of the Chongqing municipality that occurred between the late 1930s and early 1950s (when the city was a provisional capital of China, 1938–1945, and after the Communist takeover).[39]

朝天門，大碼頭，	*Chaotianmen, great dock,*
車來人往不停留。	*Cars came, people go in an incessant flow.*
金字門，南紀門，	*Jinzimen, Nanjimen,*
兩個碼頭是近鄰。	*Two docks are close by.*
菜園壩，是車站，	*Caiyuanba is a train station,*
火車汽車不斷綫。	*Trains and cars form an unbroken thread.*

It is curious to see that Sichuan junkmen still saw the city from the perspective of the river as a chain of gates and docks that lined up the waterway. In this song we can see that the singer was moving up the flow of Yangzi and appreciated the activity and change happening along its banks. The new elements he particularly paid attention to were the trains and cars—new means of transportation that were undermining his main source of livelihood: transporting goods and people in and out of the city. These new machines, at the same time, were slowly rendering the boatmen and also the river-oriented city space obsolete. Consequently, the boatmen's understanding of the integrated river-city space was becoming an invalid concept, not matching the novel industrialized environment. At the same time, junkmen's *haozi* show the emergence and spreading of a new vocabulary with which the riverine space could be described, such as "cargo dock" (*banyun matou* 搬運碼頭), or a "road" (*lu* 路) operated by the car traffic—for example, Hankou–Chongqing Highway (HanYu lu 漢渝路).

磁器口，井口站，	*Ciqikou—Wellhead station,*
那是化磁中心站。	*It is a central station for chemical magnets.*
北碚碼頭多靠河，	*Beibei dock along the many rivers,*
朝陽北廟和黃葛。	*Chaoyang, Beimiao, and Huangge.*
漢渝路，有土灣，	*HanYu Road is in the place called Tuwan,*
中渡口在小河邊。	*Zhongdu is close to the Small River* [Jialing].

As these examples prove, the mid-twentieth-century changes in the socioeconomic condition of boatmen's work as well as the rapid industrialization during the Second Sino-Japanese War and in the first years of the Communist rule influenced the relationship between cities and rivers in Sichuan. On the one hand, it came as an outcome of the challenge to traditional shipping posed by the development of the land transportation—trucks and railways—two important areas of investment from the late 1930s on. On the other hand, the change in the regional economy that came with the development of heavy industry (mines, steel mills, and arsenals) reshaped the traditional urban spaces of docks and city gates.[40] The way Sichuan boatmen saw, imagined, and presented their world through the work songs was permanently altered and did not correspond with the observed reality. Consequently, as one of the main functions of *haozi* texts was to create valid and useful maps of the areas where river laborers went, the songs started to change and include new places as well as new vocabularies with which they could be described. We should bear in mind, however, that the forces of change, though persistent and destructive toward the previous reality, were neither fast nor unidirectional. I have demonstrated elsewhere that the process of industrialization in Sichuan temporarily increased the demand for traditional shipping for the simple reason of the insufficiency of the modern means of transportation.[41] Moreover, irrespective of the scale of economic reconstruction of the region during the war and the Mao period (1949–1976), only the recent decades, with the construction of the Three Gorges Dam and contemporary large-scale urbanization, stamped out both physical features and social memory of the historical city-river relationship in Eastern Sichuan.[42]

Turning Tables: Boatmen as Twenty-First-Century Urban Representations

There has never been any uniform line on the reappropriation of Sichuan boatmen culture in the People's Republic of China. It is thus difficult to support the point that there has ever been any clear policy, idea, or even a view on the question of reusage of the old traditions. Until the late 1980s scholars claimed that traditional junk transportation disappeared altogether with the Communist "Liberation" in 1949. The work of trackers and rowers was seen as the highest degeneration of capitalism and the starkest form of exploitation on the side of any undesirable elements within Sichuan society. Moreover, steamships were perceived as epitomes of the "modernity" that the Communist government was supposed to bring to China, a country previously drained by the greed of foreign imperialists and local warlords.[43]

By the 1980s, however, the huge interest in fading local traditions turned the eyes and ears of ethnographers toward junkmen's folksongs. The publication of boatmen songs' lyrics in two main volumes in 1989 and later, in 2007, a separate thorough study of both lyrics and music as well as a plethora of popular and academic articles somewhat ennobled this popular tradition and cast a different, much more positive light on the boatmen themselves.[44] Finally, in the years preceding Chongqing's separation from the Sichuan province and its establishment as an independent municipality in 1997, Party-affiliated Chinese scholars started seeing boatmen chants as possible political and cultural capital. For example, according to Deng Ping 鄧平 and Li Liangpin 李良品, boatmen *haozi* confirmed the municipality's socialist credentials, recognized Chongqing's historicity, and gave future generations a bright example of a genuine working-class spirit.[45]

Here, however, I discuss something else: the visual representations of boatmen that were put up in Chongqing to celebrate local history, which reached its triumphal "finale" with the construction of the Three Gorges Dam. To commemorate this momentous event the government of Chongqing Municipality commissioned the construction of a monumental representative-cum-commercial urban space, named Three Gorges Square (Sanxia Guangchang 三峽廣場), in one of the main city districts, Shapingba 沙坪壩. The square, with its large-scale sculpture program, was finished in 2003. Apart from a modeled river passing through an almost two-hundred-meter-long promenade reproducing the shape of the Yangzi Gorges, the whole structure finished with a huge waterfall at its central point. The stone walls that represent mountains lining the riverbanks on both sides of this edifice were decorated with a sculpture program based on distinctively local myths and legends. Among them figure most prominently two heroes: the demigod emperor of China, Great Yu (Da Yu 大禹), and the ancient general Bamanzi mentioned above.[46] Both were revered in Chongqing throughout imperial times, but more recently came to epitomize local culture and particularity, as visible in a number of scholarly and general publications on the topic.[47] Especially interesting among these sculptures is a composition presenting trackers hauling a boat on the rocking waves of Yangzi (fig. 9.3).[48]

This sizable composition presents a bigger-than-life group of boat haulers exerting their muscular bodies in the excruciating effort of pulling a sail junk up stormy waters of a river. All workers except for their leader face the ground and their hands are almost touching the pavement. Yet the leader stands up-

FIGURE 9.3. Sculpture of trackers hauling a boat. Three Gorges Square, Shapingba, Chongqing. Photograph by the author.

right and has his eyes fixed on a distant aim. This aim in fact is a waterfall across the square representing the newly built Three Gorges Dam. Suffering and yet powerful and masculine figures, with faces marked by anger and determination, they resemble many other statues of workers commissioned in PRC. They emphasize pride of labor, injustice of exploitation, and hope for victory by the working class. The meaning of the trackers' depression (looking to the ground) and their leaders' pride and hope expressed through his forward-looking posture is quite clear. As the square itself was commissioned by the local government with financial participation of local businesses in order to present in a few snapshots a history of the Yangzi and Chongqing, it is hard not to read its meaning as that of a final overcoming of all the struggles of the past and the achievement of the People's Republic in its humanitarian mission.[49] The figures of trackers, in this context, are deprived of the meaning that they as a group had produced for themselves (as it is explained above in this chapter). In the Three Gorges Square in Chongqing, boat haulers are represented by the national and local government and signify what these institutions wanted them to: namely, the worker's victorious struggle under the guidance of the Chinese Communist Party materialized in the form of the Three Gorges Dam.

To provide another example of such an appropriation of the boatmen in the new cityscape of Chongqing, we should look at the most important edifice in this largely refurbished and gentrified city—Chongqing's Three Gorges Museum (Chongqing Sanxia Bowuguan 重慶三峽博物館). On September 9, 2000, the General Office of the State Council of the People's Republic of China initiated construction of this new key edifice, marking it as one of the main cultural institutions of the PRC.[50] At the cost of 650 million RMB the project was meant to synthesize Three Gorges, Chongqing, and Eastern Sichuan history as an element of the greater Chinese heritage, with a firm emphasis on the national experience and its contributions to world history.[51] The museum was opened on May 21, 2000.[52] The temporal span covered by the exhibitions ranges from early human history to World War II. It also contained a permanent exposition on Yangzi history and geography ("Majestic Three Gorges" [Zhuangli Sanxia 壯麗三峽]). A story of the boatmen with vivid sculptures and multimedia occupied a major part of this exhibition. Similar to the square, the exhibition celebrated the inclusion of boatmen history in the general narrative of Chongqing, Eastern Sichuan, and Yangzi histories. It pictured boatmen struggling along the river before the background of the beautiful landscape of the Gorges (figs. 9.4 and 9.5).[53] Their emaciated, naked, and twisted bodies and somehow apelike faces clearly symbolize the cruelty of pre-Liberation river labor. The group of trackers is also surrounded by a number of historical artifacts giving an impression of the labor conducted on the rivers, and especially of the brutal toil implied by such work. The exhibition and a sculpture group do not, however, historicize boatmen labor. Instead, they turn it into a perennial occupation of the people residing in the region.[54] Interestingly, the museum does not integrate the trackers into the exhibition of the city's past, which is on display in a separate exhibition titled "Chongqing: Streets of the City" (Chongqing: Chengshi zhi lu 重慶: 城市之路). As we have seen in the songs, the boatmen pictured and exploited urban space (both physical and social) equally to that of the rivers. For the constructors of the exhibition, on the other hand, the world of the city (with plans, streets, shops, etc.) and that of boatmen with their river labor were entirely alien to each other. Boatmen and their river were part of a different culture, and in fact, part of nature, now regulated and modernized by the Three Gorges Dam project (explained at the end of the Magnificent Yangzi exhibition), which the museum, at least in name, celebrated. Conversely, it pronounced a similar relationship between river, city, and boatmen as the one forged on the Three Gorges Square, but simply with less pathetic and populist overtones.

FIGURE 9.4. “Majestic Three Gorges” permanent exhibition in the Three Gorges Museum, Chongqing. Photograph by the author.

We have seen that through highly stylized forms, boatmen and river became part of the official urban narrative; at the same time, little effort was spent on adopting historical boatmen narratives and their culture in the vision of Chongqing or any other Sichuan city. Local government’s appropriation of the boatmen culture signaled the final reversal of the river-city relationship, where the river, vanquished by the technology and the power of the People’s Republic, became recast as its obedient servant. As such a perspective has been projected into the future, it also allowed for reinterpreting the past. The junkmen could be represented as original warriors against a malevolent nature and the PRC government as an actor who finally finalized the works of Great Yu, a legendary hero who stopped the mythical deluge and controlled the rivers of China. This association put the Communist Party government in a position of an ultimate savior of its people who were no more to labor as beasts. Concurrently, it turned a deaf ear on the cultural tradition produced by Sichuan boatmen and the manner in which they defined the city-river relationship.

The Sichuan boatmen’s professional culture and the story of its appropriation by the CCP in the late 1990s and the 2000s shows a process running

FIGURE 9.5. Representation of the dehumanizing labor of the Sichuan boatmen. Part of the "Majestic Three Gorges" exhibition in the Three Gorges Museum, Chongqing. Photograph by the author.

counter to the one discussed in the other chapters of this volume as "rivers lost, rivers regained." The People's Republic of China, both on the national and regional levels, is contemporarily (as of 2015) engaged in sweeping modernization projects, which through application of heavy engineering to the environment attempts to improve the country's productivity, efficiency, communication, wealth, and health. To my mind, such an approach to environment, and most particularly to the waterways, unreflectively copies previous models of economic development, which both in China and the West brought more often than not very limited positive results. The persistence of the CCP-run China in conducting such projects into the twenty-first century reveals an unceasing conviction of the Party elites about the beneficial nature of such a model of development and, by the way, of the antagonistic relationship between the rivers and the industrial, urban, and "modern" human society. This chapter therefore does not speak of the "rivers lost, rivers regained," but rather of the "rivers possessed, rivers lost."

CHAPTER 10

THE GANGES AS AN URBAN SINK

Urban Waste and River Flow in Colonial India in the Nineteenth Century

Awadhendra Sharan

> Our knowledge on this question [mixing of sewage with river waters] is based on general epidemiological considerations and on observations made by many European scientists . . . and by Hankin on the self-purifying powers of the Jumna, as regards India. It must, however, be noted that rivers vary very markedly in their bacterial and chemical contents and in their powers of self-purification. This depends on the soil and the nature of the country and watershed they drain, on their volume, and on the relation to this volume of polluted water (sewage) they receive, on their rate of flow, the nature of their bed and on the climate.
>
> —L. D. SPENCER, OFFICIATING SANITARY COMMISSIONER WITH THE GOVERNMENT OF INDIA

Rivers in India, especially the Ganges, have been much in the news in recent times on account of their pollution and the difficulties associated with their possible rejuvenation, marking yet another episode in the complex relationship between India's ancient, sacred river and the many cities that dot them.[1] The Ganges offers life to millions,[2] and its sacred nature continues to draw the faithful to its worship on a routine basis.[3] Alongside this, however, there are also concerns about the large quantities of untreated waste that are discharged into the river each day, and the impact the building of canals, dams, and other human artifacts has on the river's flow. Both of these features of the river—its sacred nature in the eyes of the faithful and the continual transformation of its properties in the view of engineers and environmentalists—have been characteristic for over two centuries now. Its history, I suggest, therefore has much to offer to our current efforts toward maintaining river flows and preventing water pollution.

As the introduction to this volume and the quote above suggest, for much of this modern period—the nineteenth and twentieth centuries—rivers and cities have been thought of in comparative terms. This comparative understanding offers several interesting narrative possibilities regarding European and colonial projects of urban modernity, resource use, and river rejuvenation. As David Arnold has argued, although at one level the sanitary and environmental issues facing Indian cities in the nineteenth century were no different from those that were encountered in the cities of Europe and America, "the understanding of India's cities was overlaid with a colonial perceptual grid that served to distance it from Western precedents. There was repeated insistence that these were 'tropical' as well as 'oriental' cities, oppressed by climatic, social and sanitary difficulties that were not encountered (or not encountered to the same degree) in the temperate zones."[4] The idea of colonial difference, in other words, was stamped at the very moment of origin of sanitary modernity in the colony.[5] Indeed, difference and imitation, when considered together, implied a distinctive colonial form of modernity, one in which people, nature, materials, techniques, and fiscal resources in India were all arranged differently than in Europe.

Similarities and differences between the metropolitan project and the colonial one are also evident in the interweaving of two different languages and imaginations, anchored in science and engineering on the one hand and faith and ritual on the other. Cities and rivers have coevolved and shaped each other in multiple ways. Their attachment is manifold, linking trade, travel, and worship, but also waste, pollution, and disease. Representations of rivers are thus overlaps of several differently situated projects and outlooks, and in the Indian context one of the more significant overlaps has been between the religious and the natural. Like the meandering branches of the river, these were, and are, sometimes drawn together, and at other times pull apart. Nearly a century ago, Gandhi wrote that in the modern rush, the chief use we had for rivers was to empty gutters into them and to navigate cargo vessels upon them. By contrast, there was little time or inclination "to stroll down these rivers, and in silent meditation listen to the message they murmur to us."[6] In a more contemporary account, the anthropologist Kelley Alley points to the distinction made by the people of Banaras between different conceptual couplings such as *pavitrata/apavitrata* (pure/impure) and *saaf/gandagi* (clean/dirty) and what official terms such as *pollution* or its Hindi equivalent *pradushan* may include or elide.[7]

This essay revisits these overlaps, between the metropolitan and the colonial and between the scientific and the religious, through a focus on the issue of pollution and river flow, as the many uses of the Ganges were sought to be controlled, modified, and rearranged in the nineteenth century. In the process, I seek to move beyond the study of the relationship between a river and "its city" to a conceptualization of a river in flow as it gets entwined with developments in the many cities that dot its banks, specifically the cities of Haridwar, Calcutta, Banaras, and Kanpur. I begin with a brief consideration of the idea of a sacred river, as exemplified in the annual bathing ritual in Haridwar. Then I trace the history of the river in the plains of north India and the emergence of the *modern* problematic of pollution around the turn of the nineteenth century. This is followed by an examination of the debates on the dangers posed by such pollution, through an understanding of the special nature of a tropical river. In the final section, I return to Haridwar to look at the issue of sacred flows, as distinct from contemporary notions of ecological or environmental flows of rivers in India. The conclusion reflects on the ambiguous nature of colonial modernity and the limits placed on the fuller realization of the modern sanitary project.

Colonial Settings and Indian Rivers: The Ganges at Haridwar

The British in India saw themselves as a departure from their predecessors, the Mughals, due to a specific governmentality of establishing practices that were not exclusively about securing territory but also about improving the conditions of life (or what we might today call the bioprospects) of the native subjects. The old native rulers, they believed, regarded the people simply as producers of wealth from whom the means of paying for the expenses—or extravagances—of the court were to be extracted. By contrast, their idea of government was one that sought to protect the people and promote their well-being; in their regime, they believed, people would no longer be the property of the government but instead the government itself existed to promote their prosperity and well-being.[8] People alone did not matter in this quest for "improvement"—land, forests, and rivers were all to be reshaped to create new regimes of property and rework relations with nature.[9] The rivers and streams of England, Ireland, Scotland, and Wales, observed the Collector of South Canara (a district in southern India) in 1868, were all the private property of individuals or companies. The rivers of India, by contrast, belonged to the state—offering, in his view, greater facilities "for uniform treatment, and direct advantages both to the State and the public."[10] At first there were interventions by way of the making of canals, to provide water for irrigation that would ensure greater productivity and generate more revenue. This was followed by sanitary interventions, concerned with water supply and sewerage works.[11] Much of this intervention was guided by a calibration of "natural difference"—regarding the size of rivers, the nature of flows, etc. A simultaneous goal, however, was to imitate that which had already been developed in the advanced countries of Europe: "a diffusion of the knowledge of the latest European improvements in the supply of water and drainage of towns among the existing engineering staff," the best course for which would be "to frame rules that could enable Public Works officers [in India] to visit at public expense engineering works in the United Kingdom."[12] Improvements undertaken by British municipal authorities at home, officials believed, built upon "old traditions." "The real technical work," in India, by contrast, had to be initiated "as something altogether new."[13]

The pandits (priests) of Haridwar, the site where the Ganges exits from the hills to enter the plains of India, and which was considered an especially holy place for that reason, were not entirely sanguine about these prospects. They expected a speedy termination of the river's sanctity, fearing that nothing would be able to withstand the British power. "Let but the Ganges cease," one brahman is reported to have said, "and nothing would remain to the Hindoos but to embrace the Christian faith."[14] Another believed the imminent demise of the river to be the "determinations of God." The river had but sixty-six more years to continue, and the gathering of sand was a mark that her end was nigh. The "purposes of God" were not to be averted![15] Perhaps the pandits exaggerated their fears. India did not become Christian, nor did the Ganges lose its sacredness.[16] Indeed, the British were quite reluctant to intervene in matters of faith, especially following the armed revolt of 1857, which had been partially a consequence of hurt religious sentiments.[17] From now on, India was to be governed directly by the British Parliament and the queen was keen to assure her Indian subjects that the religion of the rulers would be kept separate from the affairs of the people: "Firmly relying ourselves on the truth of Christianity, and acknowledging with gratitude the solace of religion, we disclaim alike the right and desire to impose our convictions on any of our subjects. We declare it to be our royal will and pleasure that none be in anywise favoured, none molested or disquieted, by reason of their religious faith or observances, but that all alike shall enjoy the equal and impartial protection of the law."[18] Indians would henceforth be left "free" to conduct their religious affairs in their own manner, free to treat religion as an "inner" domain where they could express their sovereignty, while recognizing the superiority of the British in an outer, material domain.[19] The great bathing festivals (Kumbh Mela) organized on the Ganges would seem to logically feature in this inner domain.[20] However, this was not quite the case, for these festivals, as Kama Maclean points out, were located on the cusp of the inner-outer binary.[21] While the British were reluctant to interfere directly with the activities undertaken by the pilgrims, they were extremely cautious about sanitation (fearing disease) and policing (fearing rebellious activities of a crowd they understood little). Occasionally attempts were made to prohibit the gatherings, especially in the wake of large-scale outbreaks of diseases such as cholera.[22] However, on each such occasion a counterargument was proposed: The large body of the people who flocked to Haridwar could be induced to stay away only by direct proclamation, and by a considerable show of force, and officials believed that "it would be highly inexpedient for many reasons to prevent the assemblage, a gathering which, while it is attended with great risk and danger, is, and had always been, a religious function of great importance to almost every section of the Hindu community."[23]

The sanitation of the fair and prevention of disease, wrote a magistrate in 1891, were by far the most important parts of his duty in connection with the assemblage. No effort could thus be spared to keep the town clean, to limit the number of persons who could stay in a particular building, to prevent nuisances, and to disperse the crowds in an orderly manner. Care would also be taken to ensure that no sullage water remained in the city and that the night soil was properly trenched, leaving little waste to flow directly into the Ganges.[24] The risk, indeed, was of a different kind—of people drinking water from the same ponds in which thousands had bathed earlier, and consequently falling prey to disease. There had been a suggestion on the occasion of the Kumbh in 1879 that a strong current of water should be directed through the sacred pool of Har Ki Pauri, so that the water might be constantly renewed.[25] The magistrate observed: "When it is remembered what numbers bathe at the same time during the fair, and that each of these must drink some of the water, it is unnecessary to say how charged the water must be with human emanations, and how dangerous these under certain circumstances might be if not instantly removed by a strong current."[26] It was agreed therefore to request the authorities of the Ganges Canal to make arrangements to send a constant stream of freshwater into the pool, which would quickly remove the impurities. Similar arrangements were also proposed for other sacred tanks. The cost was great, wrote the magistrate, but were the project not undertaken the outbreak of cholera was certain. In 1892 the Mahavaruni fair was canceled, the only festival ever canceled on sanitary grounds. Quickly following this, in January 1893, the government formed a committee to advise them on how to prevent cholera at festivals in the future. Following the committee's recommendations, the Bhim Ghoda tank was supplied with water by an underground channel in 1903 and within another five years the bottom of the Har Ki Pauri was paved, directing a current to provide a constant flow of water, and an iron railing was erected to provide support for bathers.[27] But things were not always this smooth between the authorities engaged in "improvement" and local religious personages whose control over the management of affairs at Haridwar was sharply declining. There were a series of local conflicts in the late 1890s, but one that emerged most prominently in the early decades of the next century concerned the possible impediment to the free flow of the Ganges on account of proposed engineering works, a conflict that I shall return to in the concluding section of this essay.[28]

The Ganges, Contaminated and Polluted

Contamination of the river water was, evidently, a relatively minor issue in the hills and the sanitation of Haridwar possibly well managed. It was another matter in the plains, where millions more resided on a permanent basis and where the British were engaged in providing a piped drinking water supply and managing wastewater through flushed underground sewerage networks in the major cities. Among these were Calcutta, the second city of the empire; Banaras, as holy in the Hindu imagination as the city of Haridwar; and Kanpur, an important military station and industrial town, the making of whose waterworks in the late nineteenth century became occasions for outlining the possibilities and limits of colonial urban modernity.[29]

Beginning sometime in the 1860s, cities in India began to discuss and implement a scheme for a piped supply of water, and relatedly, of drainage. The question was both of relieving the city of its waste (including wastewater) and finding an eventual sink for it. Sometimes authorities worried about only the former—whether the drains were adequately flushed—at other times they expressed concern about the likely health impacts of the wastewater on those who directly relied on the river water for provision of their daily supplies. Calcutta, at the very tail end of the river but the most important of the empire's cities, was the first to consider this matter of urban drainage and river pollution. There was a systematic and willful pollution of the river Hooghly abreast of the town and above the shipping of the entire port, wrote David B. Smith in his *Report on the Drainage and Conservancy of Calcutta* (1869). Day by day, he observed, the water that thousands of Calcutta's inhabitants drank and that was consumed by every sailor in the port was defiled, "by orders of the Municipality, with I know not how many tons of night-soil."[30] The mouths of the main sewers of the city opened up to the banks of the river, some near the main bathing *ghats* (platforms) in the city: "as if the sacred stream was not already poisoned enough by the daily Augean deposit of night-soil, the bathers are furnished a special fountain for themselves!" Worse still, the night-soil was hurled into the Nimtollah Ghat, where "the natives drink it with the greatest appetite, bathe in it every day, to clean both their bodies and souls, and carry it to an immense distance as the greatest imaginable treasure." The comparison was with England, where "lazy river fouling barbarism" was no longer permitted after the passing of the Thames Navigation Act. A similar act, Smith believed, ought to be introduced with regard to the Hooghly too, for to argue otherwise would be to encourage "the thin edge of a vicious principle."[31] The practice of throwing night soil into the river was eventually stopped in the following decade, but the question of river pollution due to the actions of other cities returned to haunt Calcutta soon. Equally importantly, the idea that the law that prevailed in England must also be enacted in India was subject to much dispute.

A modern waterworks and sewerage scheme for Banaras, upstream of Calcutta, had been first proposed by the Municipal Board in 1875 but remained largely on paper as the engineer seconded to it soon died.[32] Meanwhile, local community organizations joined in. In 1881 a petition was sent to the viceroy by two residents of Banaras bemoaning the unsanitary condition of the city and the discharge of sewage into the Ganges, rendering it "unwholesome" for both drinking and worship. They pleaded therefore that the government build a large subsoil drain to receive all of the filth of the city and discharge it into the river at a designated spot downstream. The idea was rejected, the municipal board objecting to the sewage flowing directly into the Ganges and unwilling to bear the large expenditure. A few years later, in 1886, the local king took the initiative and the Benares Ganga Prasadini Sabha, an association of local gentry formed at his initiative, decided to raise the necessary funds to make arrangements for redirecting the sewage of Banaras, keeping the bathing *ghats* free from pollution.[33] The matter was soon rendered official with the presentation of an address by the municipal authorities to Prince Albert Victor of Wales in January 1890, where they reversed their original stance:

"Our water will be drawn from the holy river above the city," the address read, going on to add that "we also intend to commence a system of under-ground sewers which will be commanded for flushing purposes by the water pipes, and the sewage will be put into the river somewhere below the city." The aims of the project were thus multiple from the very inception—on the one hand to protect the health of the residents of the city and on the other to secure a religious good by ensuring that the sacred waters of the Ganga at Kashi (the traditional name for Banaras) remained unpolluted.[34]

From the time that the British had first begun building new centralized waterworks, rivers and tanks as sources of water across the country had stood condemned across the country. This was not bad water but contaminated water, and contaminated as a result of "native habits." The view of "dirty Indian habits" commanded large adherence among official and unofficial British circles, as for instance in Florence Nightingale's observation regarding the "objectionable native habit of bathing and washing in the tanks and using the same water for drinking and erecting privies close to wells."[35] Many Indians joined them in deprecating these practices. Above all, Gandhi was to return to this theme often, writing of the holy city of Haridwar, "Whilst I realized the grandeur of the holy Ganga and the holier Himalayas, I saw little to inspire me in what man was doing in this holy place. To my great grief I discovered insanitation both moral and physical. . . . Thoughtless ignorant men and women use for natural functions the sacred banks of the rivers where they are supposed to sit in quiet contemplation and find God. They violate religion, science and laws of sanitation."[36] However—and this was critical—while native habits could well suggest long-standing contamination of Indian rivers, these alone could not account for *modern* pollution. That would be caused by the new sewerage works that accompanied the modern waterworks from the last decades of the nineteenth century.[37]

As had been the case in Europe and America, where population growth and new technologies such as the water closet were beginning to pose distinctively new urban problems (see introduction), so too in India were habits beginning to be considered alongside the potentially harmful impacts of new technologies. The view was put forward most powerfully by G. Hutchenson, the sanitary commissioner of the northwestern provinces.[38] Hitherto, he wrote on the occasion of the making of the Banaras sewerage works, the pollution of the rivers of upper India by the flow of crude sewage had been more or less unknown, the discharged amount being too small to have any major effect. The pollution question thus arose only "when it became known that an attempt would be made to dispose of the excreta of such a large town as Benares by the wet method and by discharging the sewage under a general system and without any purification into the Ganges."[39] In this he was supported by Surgeon General W. R. Rice, who opined that "to run into a more or less stagnant river millions of gallons of liquid sewage daily, through well-aligned drains," was by no means "the same thing as allowing it to filter slowly through the soil on the banks."[40] The medical view, however, was not entirely unanimous. Surgeon-Major J. Richardson, serving as inspector general of civil hospitals in the northwestern provinces, wrote by contrast that too much was being made of the danger to the riparian population posed by the discharge of sewage into the water of the Ganges. Nor was sufficient information available on how many people relied on the river as a direct source of drinking water, nor were there any data on how many among them fell ill to a greater degree than the population that used water from the wells. Until such time as enough information was available, he would have no hesitation in affirming that there was a "tendency to exaggerate the dangers" arising from the discharge of sewage into the Ganges at Banaras.[41]

Communities downstream were not so easily assured. Within ten days of this proposal the sanitary commissioner of Bengal wrote that if the proposed scheme of underground sewers discharging into the Ganges were permitted to be carried out, it would seriously affect the health of the people living in the towns on the banks of the river below Banaras, Calcutta being the most prominent.[42] The lieutenant governor, the highest civilian official of Bengal, wrote in support that there was little doubt that if the sewerage of a large city such as Banaras, containing a population of nearly two hundred thousand, were allowed to flow into the Ganges, it must pollute the river to an extent that would seriously endanger the health of the populous districts and towns on the banks of the river for many miles.[43] The Public Health Society of Calcutta added to the chorus, writing that the practice of ejecting crude sewage into such a major river as the Ganges was open to the greatest possible objection. The superiority of a water carriage system, they wrote, was not in doubt, but in contrast to Europe and the United States, where the practice had been first introduced, the volume of water in Indian rivers was not conducive to direct transfer of untreated waste.[44] Commissioners inquiring into means of preventing water pollution in England, they further pointed out, had amply demonstrated that the effects produced by passing various waste products into rivers depended largely on the character of these products. Many such products—dyes and certain kinds of chemical waste—were speedily disseminated and diluted to a degree that deprived them of their deleterious qualities. But crude sewage consisting chiefly of excremental matter was "a waste product of a different character."[45]

The government of the northwestern provinces could hardly be expected to concur. There was nothing new regarding such discharge of the drainage of large towns into the rivers Ganges and Jumna, they responded.[46] And if the concerns of the Bengal government had to be taken seriously, the imperial government must frame the general principles with respect to waterborne sewerage systems without special regard to Banaras or any other particular town.[47] In the debate that ensued one issue figured most prominently—the difference between the rivers of India and those of England, linked to which were two other subsidiary issues: the nature of the waste that flowed into the river and the risk posed by such flows of waste streams. A long note was penned by Surgeon General Rice, who doubted very much that the sewage could be rendered innocuous through the self-purificatory powers of Indian rivers.[48] It was certainly opposed to the laws of England that prohibited the sewage of towns to flow into rivers in their neighborhoods unless it had been rendered innocuous by chemical or other means.[49] There were two other concerns to be borne in mind, he added: first, in India, many villagers were prone to taking drinking water directly from the river and not from wells as was the case in England. Second, the nature of rivers was also different, for unlike England, where rivers were rapid-running with a very small range between full current and low water, in India the rivers were known to spread out into wide, slow-running streams and in dry weather break up into many backwater and stagnant pools that were practically settling tanks for sewage. Under the circumstances, little reliance could be placed on the possibility of the oxidation of sewage in "rapidly running voluminous streams." On the other hand, there was little denying the great benefits that would ensue from the building of a modern sewage system, leading Rice to suggest an inquiry into:

a) The procedure regarding the disposal of town sewage, as authorized by the law of England, with reference to its discharge into inland rivers;
b) The circumstances under which conditions existing in this connection in India differed from those in England, and the extent to which the question of sewage disposal was affected by this difference;
c) And, if discharging into the rivers was not advisable, the alternative proposals of treating it with chemical disinfectants, and of utilizing it in manuring sewage farms.

Others protested that this would be too casual an inquiry, and at any rate, considering that there would be several proposals for drainage schemes that involved the flow of sewage into the rivers, only a general principle in this regard would suffice. Indeed, precisely because the rivers of India were so different—their beds being swept clean each monsoon—they countered, the laws of England could not be applied except after a great deal of deliberation and expert opinion. Some others opposed the idea of a general principle altogether, arguing instead for a commonsensical scheme adapted in each case to the circumstances of the city concerned. It would be impossible, they believed, to introduce into India English legislation in regard to the pollution of the rivers unless a lot more was known about the way in which Indian rivers dealt with sewage and other causes of pollution. Eventually, the government of India expressed itself in favor of this view. Based on the available information, they reasoned, it was not a good idea to lay down any general principles as to the discharge of the sewage of riparian towns into Indian rivers. They suggested instead that the Sanitary Board of the North-Western Provinces make suitable inquiries and ascertain "by actual testing and experiment whether the effect upon the water of the rivers is such as to render it dangerous to public health."[50]

In the debate that ensued, Sanitary Commissioner Hutchenson elaborated on his earlier concerns, writing at length against the idea of conveying sewage into the river.[51] The amount to which a river may undergo pollution without injury or danger to health was an unknown quantity, he argued, and while the question of the proper removal of sewage of towns and cities should be considered in light of the expertise of scientists, there were difficulties with respect to all the known sciences—chemical, bacteriological, physiological, and sanitary. It was therefore no easy matter to practically demonstrate the relation between disease and the pollution of rivers due to sewage. And yet he remained convinced that no sanitarian worth his salt would argue that the absence of palpable evidence implied the absence of pollution. Nor would they accept that the incapability of demonstrating the presence of microorganisms (on account of their diffusion) implied the absence of risk and danger. Under the circumstances, the thing to do was to apply the same logic as had been used in England, where the Parliament had made the pollution of rivers actionable. If the nuisance caused by the pollution of rivers such as the Thames and Clyde was being treated at huge costs, so must care be taken in India regarding rivers such as the Ganges. To cite the large volume of the water in the Ganges as an excuse for extending an "already condemned common and unscientific method" of sewage disposal was entirely undesirable, especially as the river water both was considered sacred and served as the major source of supply for drinking purposes. The aim of the government, therefore, ought to be to avoid "impurity and hazard" and to establish "purity and safety."[52] Contesting these views, as we saw above, was Surgeon-Major Richardson, who remained convinced that while there could be some validity to abstract claims of harms to riparian populations there could be no definitive argument made until such time as actual evidence was forthcoming. It was futile, therefore, to refer to dangers, unless a practical alternative, technically sound and fiscally feasible, could be proposed. By the turn of the century, with Hutchenson transferred and Surgeon Lieutenant Colonel S. J. Thompson taking over as the new sanitary commissioner in place, the tide tuned firmly in favor of this view, this time in the context of the drainage and sewerage of the city of Kanpur, upstream of Banaras.

As initially conceived in the 1870s, the scheme in Kanpur had been to provide drainage only for the removal of excess rainfall and refuse water for domestic purposes with the use of two canals, leaving the conservancy of the city to be dealt with through the dry conservancy method.[53] Even with respect to this limited scheme, the results were mixed, leading to another being proposed in 1891. There were two key features to the new scheme: (a) the sewers would carry off the rainfall of the city, as well as its sullage and sewage; and (b) the system would terminate in a long intercepting sewer running through the cantonment area and discharging into the river at the downward extremity of the cantonment limits.[54] Though technically sound, the scheme was expensive, and the municipal board conveyed in no uncertain manner that unless a considerable sum were promised by the government of India, the whole project would require reconsideration. This the imperial government was loath to do, remaining firm in its opinion that local improvement must be achieved with local resources only.[55] It instead asked for a reframing of the scheme, and to consider whether the sullage alone was best disposed in the river or onto a farm. The views of new sanitary commissioner, as pointed out above, favored discharge into the river. There was no large habitation for some distance from Kanpur, he pointed out, and in any case the persons residing in the riparian villages were more likely to obtain their water from wells than directly from the river itself. Moreover, recent investigations and inquiry had led to much greater value being attached than formerly to the germicidal powers of sunlight and free oxygenation. And if this was so, in his view, the conclusions of the River Pollution Commission in England had only limited application "to the circumstances of large Indian rivers, teeming with active organic life which feeds on impurities and flowing under an eastern sun."[56] Indeed, if sewage was as dangerous as his predecessor had made it out to be, there should have been frequent outbreaks of cholera in places downstream. However, cholera, when it happened, had occurred upcountry and against the flow of the stream. The risks, in other words, were far less than in England, though it would still be desirable to conduct a series of chemical and bacteriological tests to determine more conclusively the self-purification power of Indian rivers and to ascertain how quickly the water recovered its original purity.[57]

The government of the northwestern provinces was naturally inclined to go along with this opinion, but the question was whether the government of India could "now without risk and without stultifying itself contemplate the discharge of crude sewage although diluted into the Ganges."[58] The question was both of principle and of public health.[59] Would it be expedient for government to formally concede the principle involved of passing sewage of however "light" a character into a river of the special character of the Ganges? No *additional* pollution was in question, and the main point then was the expediency of the formal concession to the devolution of this sullage. If sanction was not accorded, the sullage in a less diluted state would still get access to the river at the most dangerous part (above the city and the cantonment). However, it was the normative principle that was more vexing, and to arrive at that it was imperative to consider another related issue: Did a danger to public health arise from permitting the entry of this foul sullage to the river? From a sanitary point of view, it was of no consequence that this procedure was already followed, if there was an element of danger. And regarding this, the sanitary commissioner to the government of India felt that there was certainly a case to be made. At the same time, it was important to retain commonsensical distinctions. It was a well-established sanitary axiom, he opined, that sewage should not be discharged into a river, but that rule "must be interpreted with discretion and due regard to the circumstances in each case." Similarly, while it would be impossible to give a scientific guarantee of harmlessness, or, short of this, a decided opinion, without much further information regarding the actual nature and quantity of the sullage and rubbish, volume of the river, its rate of flow, etc., together with chemical and bacteriological observations at different seasons and at different points of the river's course, still it was possible to accept, on the basis of common sense, and considering that it was not sewage but sullage that was being debated, that the natural agencies of purification could be relied on to avert harm. All that the local government was proposing for the moment was for drainage works that would only transfer the sullage and the refuse that at present flowed into the river where they were seriously offensive to an outlet where they would be virtually harmless. This the government of India could approve.[60]

The hedging and the qualifications evident in the sanitary commissioner's conclusions were not accidental, for they point to the extreme reluctance on the part of the colonial government to make significant outlays on the project. The consequence was that on all related issues—undertaking the series of experiments recommended by its own officials, providing the necessary expertise, constructing sewage farms—there was a constant deferral, leaving both the waste of the cities and the fate of the Ganges largely unaddressed in the nineteenth century, and indeed into much of the twentieth century.[61] When the question of river flows eventually emerged as a public issue, then, it was not the question of the flow of the river as it bore upon the question of pollution in and around Calcutta, Banaras, or Kanpur, but the extent of the unimpeded flow of the Ganges at Haridwar that was necessary to preserve its sacred character.[62] As described by James Lochtefeld, beginning sometime in the mid-nineteenth century the British had begun to technologically "manage" the Ganges for myriad purposes, aimed at increasing revenues through better irrigation and facilitating transport. A major project in this regard was the making of the Upper Ganges Canal, which widened and deepened the small branch of the river that ran past the town of Haridwar to serve as the canal's supply channel.[63] In the early years of the twentieth century, the government decided to turn this into a more permanent structure, confident that just as the people had appreciated the supplying of pools with water through underground channels, or the making of the headworks of the Lower Ganges Canal at Narora, where the Ganges had been completely controlled since 1878, there would be little hard feeling this time around too. This was not to be. Several voices of protest were raised beginning in June 1914, under the leadership of the Hindu nationalist leader Pandit Madan Mohan Malviya, arguing that this would impede the free flow of the Ganges and thus make it lose its sacred character. Opinions were wide-ranging—the Maharaja of Kasimbazar wanted the Ganges to flow freely from the Himalayas to the sea, whereas others had no objection to the proposed permanent works, since farmers needed the water. Most opinions lay between these two extremes—recognizing the need for some type of construction, yet reluctant to allow for complete control over the flow of the Ganges. Eventually, a compromise solution was arrived at, the government reasoning that in the context of an ongoing war, it would be imprudent to antagonize "Hindu" opinion.[64] And as far as native opinion was concerned, though it was largely representative of its elite sections, there was little doubt that asserting control over the Ganges and over the religious festivals associated with it had become a powerful symbolic issue with which to confront the colonial state.

Rivers in colonial India, I have argued, evoked two distinct senses of purity and pollution, one drawing upon sacred texts and ritual performances and the other on the science of sanitation. Official interventions, for the most part, were based on new forms of expertise and the new science of sanitation, which the British believed was unavailable in traditional cultures. Based on this, what developed in nineteenth- and early twentieth-century colonial India was mimicry of the metropolitan culture, with important differences. The Indian city, as both different from and facing similar problems as cities of the West, underwrote much of the colonial intervention in this field. Indian rivers too, as a natural body similar to rivers elsewhere, but with properties that distinguished them from rivers in the temperate zone, were exceptionalized. The character of the Indian people, so remarkably different from the British in their habits and dispositions, further accentuated this sense of difference. Together, these material and subjective differences led officials to believe that only a modern scheme of things, undertaken by British engineers borrowing from the latest European techniques, could put things right. And yet, such modernity itself would be problematic, and on two counts. First, a river such as the Ganges may well have been contaminated through native habits, but pollution, as we understand it today, was not a product of the traditional arrangement of things but of the centralized water and sewerage works that were inaugurated by the colonial regime itself. Second, such new works had also always to meet the test of "practicality," this being ensured on some occasions by redrawing schemes, on others by making a distinction between sullage and sewage, and at almost all times by citing the need for greater evidence of harm before more resources could be devoted to sanitary matters. This ensured a continual deferral of the schemes under review, such that the Indian city could never parallel the metropolitan urban.

Indians, by contrast, both partook of this modern science and drew upon Hindu notions of purity. In their book, *Sanatan Dharma: An Advanced Textbook of Hindu Religion and Ethics*, published in 1903–1904 by the Theosophical Society, Annie Besant and Bhagwan Das, suggested that the rules of purity also included "a duty of the good citizen to see that the rivers in the neighborhood of cities are not poisoned, nor filth allowed to accumulate to the injury of public health."[65] Much the same can be deduced from Gandhi's statements cited above. Veneration for the Ganges, the traditional leadership of the king, and Western (rather than Hindu) notions of cleanliness and pollution, Arnold writes, were conjoined.[66] Others have described this as an ambiguous sanitary modernity, where the dependence on colonial bureaucracy for creating infrastructure could coexist with appeals to traditional and religious values.[67] This essay demonstrates too the significant intersections of differently oriented thoughts and practices in the making of urban waterworks in the colonial period. However, rather than read these as ambiguous responses, I prefer to focus on their overlapping nature, each modifying the other to suit local contexts. Further, this essay suggests that accounts of cities and their rivers benefit by considering local developments and wider geographies simultaneously. Thus on the one hand, while the Ganges was (and is) considered holy all through its journey from the mountains to the sea, it was especially revered at Haridwar and Banaras, which in turn had important bearing on the nature of improvement that could be effected in these cities. On the other hand, I have also demonstrated that what happened in Banaras or Haridwar also had serious implications for people living elsewhere, especially communities downstream. Considered in this manner, localized urban histories of waste and sanitation may be intertwined with histories of dispersed effects, offering a narrative framework that resonates in significant ways in our own times.

CHAPTER 11

POLLUTED THAMES, DECLINING CITY

London as an Ecosystem in Charles Dickens's *Our Mutual Friend*

Agnes Kneitz

ENVIRONMENTAL PROBLEMS AS A SYMBOL OF SOCIETAL CHANGE

Dickens knew London like the back of his hand, enabling him to deliver not only a geographically precise depiction, but also a vivid description of its sanitary conditions.[1] Extreme population growth over the first half of the nineteenth century had multiplied the city's problems.[2] Besides severe air pollution,[3] London's greatest challenge turned out to be removal of waste, both organic and inorganic, from its main lifelines—its waterways—especially the river Thames. Consequently, the city's metabolism was disturbed by its extreme tractive force of both humans and resources overwhelming the prior equilibrium, while at the same time it was unable to efficiently dispose of everything collected.[4]

Regularly wandering the streets, Dickens had become aware of the offal's transforming effects on the city scape, directing his interest to London's waste and waste removal for more than a decade before he settled on the topic for *Our Mutual Friend*.[5] Most prominently, this manifests in an essay entitled "Dust; or, Ugliness Redeemed," published in his journal *Household Words*, a periodical designed "to do the Sanitary cause good service."[6] Besides his publishing efforts, he also (politically) engaged in solving the social and urban challenges of his time, as in fighting the 1834 Poor Law.[7] On the basis of his personal history, Dickens presents an informed awareness of the interconnections between physical health, pollution, and social status. Given his active engagement in public health politics, he was certainly aware of the contributions of Florence Nightingale, who was equally convinced that "the connection between the health and dwelling of the population is the most important one that exists."

At the same time, he was also familiar with Charles Darwin's *On the Origin of Species*, and infused his novel with related conclusions even though he did not necessarily agree with all of Darwin's major arguments.[8] Dickens was convinced that environmentally unjust realities would evoke dire social consequences, strongly disliking the idea of a hostile, competitive society, supported by Darwin's idea of the survival of the fittest that had been integrated into the second edition of *Origin*. In a letter to medicinal and sanitary reformer Ernst Hart, he wrote, "my knowledge of the general condition of the sick poor in workhouses is not of yesterday, nor are my efforts, in my vocation, to call merciful attention to it. Few anomalies in the land are so horrible to me as the unchecked existence of many shameful sick wards for paupers."[9] Together with Hart and others, Dickens had been involved in the London sanitary movement, also as a founding member of the Metropolitan Improvement Society (MIS), whose first general meeting was held in 1842. Compared to other initiatives, the MIS stood out by addressing issues of smoke abatement, sanitation, and water supply, as Dickens noted.[10] This favoring of a comprehensive solution serves as another account of the author's understanding of general underlying ecological ties. But he did not only theoretically reflect on these subjects, to make his hometown a more livable place. Alongside Edwin Chadwick, he campaigned to bring London into the provisions of the Public Health Act of 1848,[11] which integrated many of the discussion points of the manifold civic societies and associations. Being a contested political topic, however, societal and (urban) environmental change did not remain in the political sphere, but became increasingly communicated in contemporaneous media. One popular—and at the time probably the most famous—account is the flourishing "milieu" of the Victorian novel, the genre in which *Our Mutual Friend* is situated.

Ecosystem Dysfunctions in the Novel

Building on Dickens's interest in urban and societal reform, the novel's background story is pivotal for the narrative: A greedy misanthrope, having made a fortune in waste trading, bequeaths it to his son John Harmon, but under the precondition that he marries an unknown young woman called Bella Wilfer. Until this event, the fortune should be managed by Harmon's loyal servant, Mr. Boffin. Unfortunately, John falls victim to an attempted murder, and gets thrown nearly lifeless in the Thames. Shortly after that—and here the novel begins—a corpse is pulled out of the waters, falsely identified as John Harmon. Under the new identity of John Rokesmith, the survivor signs in as secretary for the trustees of his prospect estate; a feint leading to various societal complications. But the Boffins' household also takes in Bella as an act of welfare toward the former prospective bride. John falls in love with her, but, interested in a rich marriage, she rejects him. After a while Rokesmith reveals himself to the Boffins as the presumed-dead John Harmon, and together they plan a play to demonstrate to Bella quite plainly the vanity of her aspiration for monetary wealth. To its end Bella and John are thrown out by the Boffins, which finally leads to their marriage.

Common literary interpretations of the novel mainly point out a critique of the upper class, its greed and presumptuousness. But Dickens saliently features the entanglements between social decline and the pollution of the river Thames. The "dust-heaps" of the Harmon family have usually been interpreted as material representation of the moral bankruptcy of a society committed to capital.[12] Connected to its components that constitute them,[13] this reading has been further advanced to an equation of waste and money,[14] not without obvious Freudian interpretation toward the pleasure that humans draw from either.[15] Patricia Gilbert, however, for the first time dispatches it from mere capitalist critique and places *Our Mutual Friend* in its immediate context of sanitary reforms that had developed almost naturally as part of the metropolis's rapid expansion.[16]

An ecological interpretation of Dickens's novel based on environmental aspects, however, is grounded on the influence of pollution on the English metropolis's different social ranks, mirrored in the novel. While the satirically painted upper class gains profits from this misery, they leave the lower class suffering both economically and in their physical health, as presented through characters such as Betty Hidgen.[17] Upward mobility provides the only way out of menacing livelihoods in the novel, which focuses especially on ways of improving London's living environments, most prominently displayed through Lizzy and Charly Gaffer (Hexam). From an ecocritical perspective, it can be argued that Dickens saw London as an ecosystem, with the Thames as its lifeline, and whose balance is heavily disturbed by social inequality, especially due to unjust and unequal exposure to environmental pollution.[18]

In simple terms, an ecosystem can be described as a collective entity of plants and animals interacting with each another and the abiotic environment in a particular given place.[19] But as Donald Worster points out in his history of ecological ideas, this "concept kept blurring the lines between living and non-living components, so it mixed mechanistic and organic metaphors to a confusing degree."[20] This initially science-based definition allows for integrating both humans and urban environments in the sense of defending internal and external nature of both body and the city, attentive to both humans' and ecosystems' health.[21] The idea of ecology originating during the same era, formalized by Ernst Haeckel's 1866 description as "the investigation of the total relations of the animal both to its inorganic and organic environment; including above all, its friendly and inimical relations with those animals and plants with which it comes directly or indirectly into contact," must be kept in mind when approaching Victorian literature from an environmental angle.[22] In analyzing the relationship between historical actors and their specific ecosystems in order to identify proto-ecological perspectives, an integrative approach from historical and (historical) literary studies can be made particularly fruitful.[23]

Literary criticism assigns a major role in *OMF*'s narrative to the matters of waste and filth, both in close connection with corpses and the characters' experiences of death, respectively. For this interpretation Dickens's high value for life itself is important, as expressed throughout his oeuvre, for example, by foregrounding the importance of selflessness, compassion, and sympathetic love not only for personal development but as a basis for social reform. Over the course of the novel he highlights this value by contrasting life and death.[24] In a circular flow, the Thames provides and takes life at the same time, John Harmon/Rokesmith being only one example. Loss of identity—then almost equal to death—is also depicted repeatedly, also circular. Dickens seemingly pictured the Thames as a vital part of an urban metabolism. Consequently its pollution and dysfunction can be interpreted as its own loss of identity, as it is being stripped of its vital function for the city: the mystical river is lost. Thus, the mid-nineteenth-century embankment of the river is an attempt to retrieve its identity (analogous to the protagonist's crisis of identity) and indicates a means to regain the Thames for London society in both literal and metaphorical terms, improving both its ecological health and its overall water system. Damming the river signifies human intervention in the ecosystem to control it, prohibiting the river's "interventions" in the human living environment, i.e., through the spread of diseases, by soaking the riverbanks, or overflowing into water pipes. Meanwhile, simultaneously constructed leisure spaces meant to improve public health on another level.[25]

FIGURE 11.1 Illustration in *Our Mutual Friend*, "The Garden on the Roof," engraving by Marcus Stone, 1895 edition. Courtesy of the Library of Congress.

Analyzing environmental interconnectedness is not an unfamiliar approach for the small field of Victorian ecocriticism, as ecosystematic thinking can be perceived as almost natural to Victorian literature, comparable to later Continental naturalist literature with its affinity to contemporary science. Based on her readings of Darwin, Lyell, and Mayhew, Regenia Gagnier argues that in comparison to the rationalized disciplines emerging over the previous century, which ascribed causality to either nature or culture exclusively, the predisciplinary Victorians were closer to a more ecosystematic way of thinking.[26] In developing ecocritical readings, the concept of ecosystems can thus help avoid biased views oscillating between a romanticized vision of untouched nature and literary anthropocentrism.[27]

A central theme of Dickens's novel is waste, often referred to as "dust." Waste has always been abundant and omnipresent in the London metropolis, determining cityscape and river since medieval times, with the latter being the central supply organ of both water and goods.[28] Waste removal and its marketing offered a means of capital, enabling both upward and downward mobility, or stagnation, at the same time. In the novel, Dickens presents the ecosystem of the city—and by extension its society—disturbed by the waste, the ugly, the polluted. Thereby many things fall by the wayside. It is debris, falling by the wayside, more precisely social debris—exemplified in the novel by pauper characters, but also personal identity loss or crisis.

From the literary description in the novel two kinds of waste can be distinguished: wet waste, like the river's debris or (human) excrement, and dry waste, like refuse in the streets. The former is "bad" waste, as it disables the flow of the ecosystem in both river(s) and streets, and its inherent negativity is ascribed to the humans dealing with and exposed to it; e.g., the body-strippers, scamming their financial outcome by illegally digging in strange belongings (*OMF*, 25). The latter is of higher value, as it can be used "honestly," as in founding a waste-removal service, leading to social wealth, while representing a service for society (*OMF*, 13). Despite conflicting implications, the two kinds of waste are dependent on each other. This sense of interconnectedness and circularity resonates with Tina Choi's reading of thermodynamic aesthetics in *OMF*, accentuating the "novel's multiple encircled and circulatory qualities, the intricate circuits linking causes and effects, old and new, and the transformations and continuities traceable throughout the whole. Within this formal structure, resolution, like the rainfall, must be drawn from that bounded system itself, from the circulation of characters and other elements already present."[29] Grounded in the bourgeois moral discourse, the two kinds of waste stand as exemplary for classes of paupers: usable waste and unusable waste. Their existence signifies a dysfunction within society, unable to clean itself, and equated to the waste it produces. Already in *David Copperfield*, Dickens had been describing lower-class character Martha as human waste, seemingly "part of the refuse [the Thames] had cast out, and left to corruption and decay, the girl we had followed strayed down to the river's brink, and stood in the midst of this night-picture, lonely and still, looking at the water. . . . There was that in her wild manner which gave me no assurance but that she would sink before my eyes." While usable waste could be disposed of, even profitably, reinstalling the city's cleanness and ensuring a productive (eco-systematic) flow, the unusable waste could not, and residues remained. Some paupers are assigned the potential of being (re)integrated into society, even contributing to its operability. Without such potential paupers would remain disturbing to society, like "dozing women-drunkards," with their "companionship of the trodden vegetable refuse, which is so like their own dress that perhaps they take the Market for a great wardrobe," and who as being "such damaged-orange countenance, such squashed pulp of humanity, are open to the day nowhere else" (*OMF*, 729). The assessment of value in both cases, however, is conducted by the ruling social class.

One distinctive example for this dysfunction is the Gaffer (Hexam) family, who make a living by fishing for corpses and selling their pockets' contents.[30] By occupation alone, the Gaffers' social status is low, their poverty is bitter, and reflected in their livelihood. Accustomed to the "ugly" business and dwelling at the muddy riverbank (*OMF*, 70–71), its inhabitants are described as "a knot of those amphibious human creatures who appear to have some mysterious power of extracting a subsistence out of tidal water by looking at it" (*OMF*, 74). From the father's perspective, there is no escape from the mud. His daughter Lizzy, on the other hand, becomes aware of an escape from the miserable living situation—through education. Thus, she ensures that her little brother Charly leaves the family to attend school even against their father's will. As Old Gaffer does not hold a high opinion of either school or societal education, the family is divided (*OMF*, 30).

With the potential to become a valuable and respectable member of society, Lizzy's brother represents "dry waste." Her father, on the other hand, rejecting (self-)education, represents "wet waste." His conscious decision becomes even a moral conduct leading to perdition, and finally to death (*OMF*, 173), almost as a punishment for withholding others from personal progress. Charly senses this negative traction, radiated by his family: "It'll be a very hard thing, Liz, if, when I am trying my best to get up in the world, you pull me back" (*OMF*, 278, similar 711–12). Lizzy Gaffer herself represents a transitional figure in the story.[31] In the beginning she can be categorized as wet waste, all the more so as she rejects help from well-meaning acquaintances out of filial piety toward her father (*OMF*, 69–70). Her father's negative influence also hints at obstacles from society's similar traditional family structures hindering integration. Rebelling by "freeing" her brother initiates a transformation process in Lizzy, who is slowly becoming "dry waste." She recognizes and opens herself to other concepts of life, averting an existence as hindering social debris.[32] Lizzy's upward mobility is finalized with her marriage vow to Eugene Wrayburn at his sickbed: "Live for me, Eugene; live to see how hard I will try to improve myself, and never to discredit you" (*OMF*, 753).

More examples of societal dysfunction are depicted in the characters of John Rokesmith, Jenny Wren, Silas Wegg, or Rogue Riderhood. Middle-class Rokesmith disturbs the societal order with falsehood and receives payment in form of personal complications, including a persistent and difficult reversible change of identity. Jenny Wren, on the other hand, remains social debris due to her handicap, seeming dysfunctional despite her successful sewing shop and openness to education (*OMF*, 233ff). Silas Wegg also sits at the bottom of society, partly integrated through reading services at the Boffins' house, but, handicapped by a missing leg, his status remains vague (*OMF*, 77–78). This stands in relation with the Boffins adopting a place in society that is not designed for them. Although not by their own doing, their prosperity is based on dishonesty; hence, their place on the sunny side of society cannot be sustained in the long term. Reality and appearances do not match—partly indicated and problematized by lack of education—making the couple vulnerable (*OMF*, 655–56). Attempts to using the newly gained fortune to help the socially disadvantaged turned out to be problematic, as part of the motivation is based on self-portrayal (*OMF*, 108), although it shows their adaptation to common practice in London society at the time.

In this context of lower-class struggle for improvement, hints of environmental (in)justice can be recognized: livelihood and influence by polluted (living) environments are dependent on educational status as well as an openness and ability to adapt, but also on the possibility of defying these risks.[33] This can be motivated by personal change and/or by leaving the social and spatial sphere, initiated from outside. The social debris is disadvantaged and hinders itself, yet it is a factor of disturbance for society's cleanliness and inner harmony. An attempted breakout of the inherited status results in a dysfunctioning system; thus, social mobility is prevented rather than supported.[34] Disruptions can figure within the microsystem of a family, as shown by the Gaffers, or within the macrosystem of London society, as shown by John Harmon or the Boffins. After Harmon's fall into the Thames and the loss of his place in the societal order, various obstacles present challenges to him, until his place can be regained, his true identity reclaimed, and the dysfunction overcome.

Transgressing Ecosystem Boundaries

In the nineteenth century, identity loss came rather close to being dead in a monstrous city like London. In *OMF* the absence of a fixed identity, or its loss, is closely connected with immersion in the river, identifying the mutual permeation of city and river. Individual loss of identity is mirrored by the permeation of the city boundaries through the river's sickening waters. The city itself hereby also loses its value as a secure entity. As mentioned above, Dickens's novel is based on and relies on the unembanked state of the river, its pollution of the surrounding land and the uncertainty of its boundaries, consequences of the nature of the Thames as a tidal river. More than occasionally overflowing its basin, the tidal rhythm washed offal of all kinds and sizes on the shores, while pushing diluted nuisances into the muddy banks and through insufficient sewage pipes farther into the water system of the city—a grievance genuinely addressed since the middle of the century by a municipal committee dedicated to dredging and embanking the river.

Dickens's principal characters are likewise struggling to contain and define their own porousness in relation to the Thames. Social detritus can be compared to human waste, and the Thames, as both the organizing metaphor and the landmark of the novel, seems at key moments to incorporate the city along its banks. Rotherhithe, for example, is "where [the] accumulated scum of humanity seemed to be washed from higher grounds, like so much moral sewage, and to be pausing until its own weight forced it over the bank and sunk it in the river . . . among vessels that seemed to have got ashore, and houses that seemed to have got afloat" (*OMF*, 436). When Riderhood goes to misidentify Mr. Gaffer as Harmon's murderer, he leads Eugene Wrayburn and Mortimer Lightwood to the water, walking through hail and leaving "marks in the fast-melting slush that were mere shapeless holes; one might have fancied, following, that the very fashion of humanity had departed from his feet. . . . It seemed as if the streets were absorbed by the sky, and the night were all in the air" (*OMF*, 157). Dickens's portrayal of the inseparability of land and water highlights the interdependence of the two realms rather than seeing them as a clear-cut binary, and the ecological understanding that informs his view of society. On the river itself, "everything so vaunted the spoiling influences of water—discoloured copper, rotten wood, honey-combed stone, green dank deposit—that the after-consequences of being crushed, sucked under, and drawn down, looked as ugly to the imagination as the main event" (*OMF*, 172). Worse is that not only metaphorically speaking, everything that goes into the river comes back out. And it gets incorporated into the city's living bodies and their livelihoods, such that Eugene feels that he is being transformed into a "half-drowned" criminal after he has swallowed what feels to him like a gallon of the "wash of the river" (*OMF*, 156).

But damaged as it is, the river even represents redemption, a symbolism, as laid out in Peter Ackroyd's biography of the Thames, that has infused the literature on it as strongly as its treacherousness, constructing a mystical and timeless image. Especially toward the end of the Victorian era, this image serves as a mirror for anxiety and insecurity facing an unknown and seemingly uncontrollable modernity. This is to be extracted most artistically from the writings of Joseph Conrad, who knew and valued the Thames as much as Dickens did, yet also cherished the dark and gloomy image of the river that the city had obtained through rapid urbanization and industrialization since the late eighteenth century. In *OMF* John Harmon loses his identity during his waterside adventure, but is also able to fashion a new and more secure self after his slow rebirth. Being thrown into the Thames paradoxically revives him; nevertheless, the narrator describes him as a "living-dead man" (*OMF*, 373). Later his new wife Bella also loses her identity as Mrs. Rokesmith, in order that she and her husband may be reborn as their true selves, the Harmons. Only through the loss, through destruction, does renewal become possible, as John Parham deduces from Dickens's review of *The Poetry of Science*, for the latter's reconstructionist social vision—impossible without technological intervention.[35] Despite the river being lost in pollution, it still contains its natural life-giving potential. As the human creatures living with the river are losing and regaining themselves through the very same force, this generally offers hope for the river being regained in its entirety eventually, which seems deeply rooted in the public memory of the "Sacred River."

Value of Life and Death

In order to understand the full range of the novel, another level of interlinkages must be added to common interpretation, as understood by Dickens: the destruction of Nature will result in destruction of humans themselves.

Dickens set a supreme value on human life, and *OMF* depends on this feeling. The proximity and mutual dependency of life and death, both literally and symbolically, make it often difficult to distinguish the value of the two. Death, as pauper Jenny Wren describes it, is a good place (*OMF*, 247). But the value of death depends in part on the dominant values of life, and the moments of near death, as experienced by the Rokesmiths, make this very clear. Referring to Dickens's contemporary and friend John Ruskin, Catherine Gallagher has discussed the connections between value and life—"wealth" in the latter's work, and opposed it to his other term, "illth"—that which does not avail itself of life may even cause death and therefore has no value.[36]

In both Ruskin's and Dickens's writing the emphasis on death follows certain economic theories (most importantly Malthusianism) "that claim to valorize life most fully," as Gallagher states. "For those," she goes on, "are the tendencies stressing the commodity's need to have a life-giving potential that makes up for its life-draining origins."[37] This economic narrative, which traces the transformation of death into life, and of waste into vitality, holds an epistemological transformation as well. Life and death, decay and regeneration are mutually dependent and virtually the same. The striving self must abase itself in order to be vivified, to gain wealth or to become wealth.[38]

Societal pollution equals environmental pollution of the city as in waste or degrading river(s). The waste problem in the streets was caused by dry waste and could be rectified while made productive. Since the Metropolitan Police Act of 1839, London's waste disposal had become increasingly more regulated, attempting to minimize "nuisances" for residents and passersby. As certain hours of the day were designated for waste removal of different kinds, this would eventually allow a sophisticated and lucrative system of private enterprises to grow, to which the Harmons' dust heaps are to be added. Nevertheless, the reuse and recycling of human and nonhuman waste needed to be conducted in a (socially) acceptable and sustainable manner, as the metaphorical example of the Boffins shows. The problem of wet waste, on the other hand, especially when focusing on the pollution of the Thames, led to a more sustainable problem: the disruption of the river's self-cleaning ability.

While a certain amount of waste and debris can be diluted by a healthy riverine ecosystem, the river in both reality and its Dickensian depiction was heavily polluted, with debris exceeding levels for a natural solution, turning the body of water into a sewer. What Dickens's use of the metaphor is pointing at is the observation that solutions to social problems could not be achieved through prevailing Victorian societal structures but only by reorganizing and restructuring society as a whole. London saw its social debris increasing at an alarming rate, such that traditional means of relief and social security systems fell short of reinstalling a balanced state and ensuring a "normal" and natural flow within the metropolis. Traditionally, this would be defined by all members of the community remaining within their designated ranks and places, with only those individuals holding potential for development ascribed from above rising in their levels of a functioning society. Although even under ideal conditions some sediment of social debris would remain, it could theoretically be "washed away" by a healthy flow within the ecosystem.

With a basic triumph of moral and social values over material pursuit, contrary to the Darwinist visions incorporated in the novel, Dickens imagines a more desirable state for society.[39] But even with this positive attitude the ecosystem's inhabitants struggle in "an intricate pattern of mutual relationships conducted in a chaotic environment by individuals seeking their own advantage."[40] With the uncontrollable growth of the metropolis in the first half of the nineteenth century, its output of waste had increased, in conjunction with the growth of a "problematic" poverty society and risks of pauperization, leading to a collapse, with the system incapable of "cleaning" itself. Furthermore, the filth of the slums had started penetrating all quarters of the city—which Dickens personifies through Silas Wegg, for example, who integrates and lodges himself at the Boffins (*OMF*, 187ff). Finally, this results in a categorical composition of the matter constituting the dry waste in the streets, which in turn influences the ways of dealing with it—for example, the possibility of reuse and recycling.

Following this thought, the transgression of the whole society can be compared to the river Thames overflowing its banks, soaking the shores, as was possible only before the construction of embankments, especially in the city center. Scientifically well informed, Dickens describes the processes within a river's ecosystem, with society exemplifying the processes within the river, whose self-cleaning ability is disrupted, much like the traditional infrastructure of waste disposal is overstrained. Being exposed to waste with all its negative implications and unhealthy features, alongside the absence of possibilities to change one's status by one's own means, is highly problematic from an environmental justice point of view, but determined by Victorian social structures.

Yet waste does not only symbolize the social focal points within society. On a level of picturing a societal portrait it becomes apparent how street pollution through waste and filth as well as the pollution of the Thames influences the living environment and living situation of the characters significantly. In *OMF*, two basic kinds of disadvantaged livelihoods can be distinguished: the settlement along the river Thames and the poor quarter. Both descriptions are modeled on examples presented to the public in Edwin Chadwick's sanitary report.[41] Along the riverbanks, the Gaffers dwell in an ecological problem area (*OMF*, 70–71), defined by mud, affected by the tides, incorporating sanitary defaults caused by river pollution, plus a distinct cold due to the proximity to a body of water (*OMF*, 157). Different but not less problematic are the slums themselves, such as the neighborhood of Jenny Wren (*OMF*, 504ff). Waste, both dry and wet, dominates the street view, and in combination with poor housing conditions produces severe health risks. The author refers to it as the "Dismal Swamp" (*OMF*, 323ff).

Ecosystem Structures

A historical perspective on the novel, along with its contextualization into burning topics of the time, allows the deduction that Dickens's choices of metaphor and narrative were consciously targeted toward environmental questions. He was well aware of the interdependency of health, pollution, and social status, as numerous examples of his letters and speeches suggest, and in his oeuvre, the pollution and disruption of the ecosystem affects some inhabitants more negatively than others.[42] Very fitting in this respect are the critical voices responding to the novel who claimed that the author would only "lament" about the Poor Law, painting too dark a picture of the metropolis's current state. On the other hand, Dickens's work is ascribed with a positive influence on his contemporaries' social awareness.[43]

One main interpretation of Charles Dickens's last completed novel is the symbolic connection of pollution of the river Thames and the societal degradation of the city. A closer analysis of his depiction of social decline during the mid-Victorian era, however, suggests an analytical layer to probe deeper. Dickens's critical view on pollution can thus be read as the author's attempt to highlight the ecological downside of flourishing "progress" and to take a reflexive view on (Second) Modernity as would be proposed by Ulrich Beck more than a century later.[44] Dickens understood the connection between the influence of urban growth and pollution of the metropolis on its human inhabitants. But the Thames itself, this flowing lifeline of an ever-growing urban monster, was presented as likewise producing ecological and societal risks.

From a certain point in history traditional (infra)structures were no longer able to control these risks, thus transforming the river into an uncontrolla-

FIGURE 11.2 King's Cross, London: the Great Dust-Heap, next to Battle Bridge and the Smallpox Hospital. Watercolor painting by E. H. Dixon, 1837. Courtesy of the Wellcome Library, London.

ble risk itself (and for itself). In *OMF*, it can be argued, the novelist describes this critically dangerous nexus: He points out how human agency gradually destroys the environment, which consequently results in the destruction of humans themselves. London society is lost, once the river is lost. How far the process of degradation had already progressed can, for example, be read from the near-extinction of the once-flourishing native salmon population, this affecting not only the aquatic ecosystem but also local dwellers.[45]

Spending most of his life in London, Dickens suffered from insomnia and developed a habit of taking long walks through its then distorted-looking nighttime landscape.[46] Not surprisingly, his novels are quite specific about locations, producing accurate geographical descriptions.[47] In applying this observational practice, he somewhat followed the footsteps of urban researchers, like Henry Mayhew, with whom he was in close contact.[48] In one of his nighttime adventures the novelist also describes the state of the Thames, and focuses on the embankment and the eroding of boundaries: "But the river had an awful look, the buildings on the banks were muffled in black shrouds, and the reflected lights seemed to originate deep in the water, as if the spectres of suicides were holding them to show where they went down. The wild moon and clouds were as restless as an evil conscience in a tumbled bed, and the very shadow of the immensity of London seemed to lie oppressively upon the river."[49] With his unique power of observation, his descriptions also allow the readers to sensually experience the sights, sounds, and smells of the old city, which were filled with manure, filth, unwashed bodies, and waste of all kinds. In his famous biography Peter Ackroyd notes, "if a late twentieth-century person were suddenly to find himself in a tavern or house of the period, he would be literally sick—sick with the smells, sick with the food, sick with the atmosphere around him."[50] This condition would eventually lead to the introduction of large double windows in which nicely smelling plants would be kept.[51]

Through his vivid mode of description, Dickens was able to portray conditions in London, which he and his contemporaries perceived as both quotidian and monstrous: human filth is physically present in the very heart of the city. Excrement, blood, decaying corpses—dirty or decaying bodies are always present in the novel. The body will eventually become waste, but in some cases it is already waste. The natural cycle of decay and organic waste matter was—at least for some of Dickens's contemporaries—still strongly connected to spiritual questions of birth and rebirth, which are also addressed in John Harmon's experience of both dying in and being revived by the river.[52] The author seems to integrate Christian implications in order to balance a world of self-seeking advantage, especially given the Malthusian causation underlying Charles Darwin's evolutionary theory.[53]

The city, on the other hand, stands for the material manifestation of the state of the social ecological body: sick, decaying, dead in parts or at least near death. Following William A. Cohen, this grotesqueness is in a way the structuring mode of the novel,[54] and further, makes it seem somehow dreamlike and unrealistic in its portrayal of space. Leaving aside geographical and spatial representation, filth and dirt and pollution are employed as metaphors for the sick social body, but also reflect a new assessment of environmental risks. Dickens borrows here from sanitary discourse, integrating a thematic distinction by the contrast between wet and dry waste.

The Thames had always ingested what the city's open sewers were discharging into it. Like residents in other urban centers in the industrializing world, Londoners used to draw their water from the river, regardless of whether sewers emptied their wastewater close to these stretches. Although Londoners were in the process of abolishing the medieval structure of their capital city, early Victorian scientists retained the coeval medical belief according to which infectious diseases were caused by miasma or foul air. Within the framework of urban sanitation, this idea was even extended to the degree that such inferior air quality could potentially affect the atmosphere of the whole city; this further indicates that the theoretical struggle against invisible health risks had already begun. Most of the problematic medical conditions, however—like the "hydra" cholera—were not caused by bad air but by bacterially contaminated drinking water. But due to existing popular medical beliefs it took until the cholera epidemic of 1854 to prove this, and to locate the origins of the disease in insufficient sanitary systems as a source for public health hazards.[55]

Nonetheless, a changing attitude toward sanitation was closely tied to the morals of an emerging middle-class feeling threatened by filth and foul air, those above all possibly causing various kinds of immoral behavior.[56] Yet the riddle of how cholera spread was not solved by studying bacteria but by close observation of the affected "ecosystem," and spatially analyzing pedestrian patterns of where residents fetched their drinking water. In 1854 the epidemic was ended by removing a well, which had been poisoned by infected filth and human waste from a nearby damaged sewage pipe. Despite the positive outcome, this event visibly contributed to the ongoing realization that not all risks could be easily controlled. Outdated social and infrastructural systems had already started to become an integral part of the problem, not its solution.

Increasing pollution was commonly perceived a necessary risk for people to accept at the onset of another stage of modernity. But this risk was also seen in the light of its impact on nature and human life. It can be assumed that Charles Dickens was not the only one who believed that environmental injustice might engender dire social consequences.[57] It almost seems as if he perceived the city as an ecosystem threatened by its pollution. "Ultimately, Dickens feared that the combination of polluted air and dreadful sanitation would bring into being an 'Unnatural humanity,'"[58] as John Parham interprets the author's view as articulated in *Dombey and Son* (1848). Dickens began his work on *OMF* when the construction of the Victoria Embankment finally began, and efforts were put into a better sewage and drainage system for the East End, where slums were mushrooming.[59]

The centrality of his use of the urban metabolism, and the metabolism of an ecosystem, is set at the very beginning of the novel.[60] Here, while dragging the first corpse out of the water, old Gaffer responds to Lizzy's aversion to the Thames by insisting that the river is indeed "meat and drink" to her (*OMF*, 3). As it is "crushing," "sucking under," "drawing down," the "mutual friend" is acknowledged as an active entity in the ecosystem of the city—even despite its actual biological death. The river's agency becomes akin to human agency, as it actively transforms things that find their way into it. Naturally, through the same process in which it takes on bits and pieces into its waters, it changes itself through the absorption of whatever is intruding. As long as it still runs through the boundaries of the city, the river gives itself back to the city, but due to its pollution the underlying idea of the river as a fertile nourisher becomes corrupted. But transformation of death into life and waste into vitality is no linear process, but a circular one. Lives are lost, regained, then lost again. But in London outside as well as within Dickens's novel, the ecosystem of the city is corrupted as its humans actively destroy it, and the Thames retaliates and reacts through diseases, identity theft, or death, taking away human value. So in order to survive—biologically—the inhabitants need to rescue their river from the deathbed, in the long term helping themselves to regain "wealth" in the Ruskian sense.

The "Ecosystem" Taken Further

This essay wants to suggest that Dickens perceived the city of London and its human and natural actors as one big ecosystem, seemingly employing London as a metaphor for all urban development in *OMF.* Within it, the human actors are also invested with a particular level of understanding. Despite their many trips to the sea or the countryside the principal characters must still go back to this place and face the worst of the river to claim their true identities.[61] That is, it can be read, they must overcome the filth for which they are in part themselves responsible.

Much as the human protagonists need clearly defined boundaries of their selves in order to live and survive, the river does as well, or so at least Dickens seems to suggest. For the Thames, the "project of embankment, is providing definite boundaries between land and water, herewith containing the pollution therein."[62] Reinstalling these violated boundaries—violated by human agency—is a means to vivify the river, to save it and open a better perspective for its future. Of course this embankment of the river and its agency is also an attempt to control the risk that the Thames had become through its pollution—through human agency. Demonstrating man's supreme power over nature, enabled by technical progress, underlies this wish for further developing hydrological engineering.

In continuation of the Dickensian symbolic, the embankment under construction is more than a sanitary measure; it is a means of environmental and social policy, in order to keep the river, and its social filth, at bay. Still, in an ecological sense it could only combat symptoms. The city becomes cleaner and sanitized as the Thames is prohibited from transgressing its boundaries. Its obvious (un)natural cause, however, the river's human-induced pollution, remains untreated. In fact, an ecological improvement could not yet be initiated, as proven by continuing epidemics and the embankments' collapse in the 1928 flood. Also not in a social sense, as all attempts of reforming the society did not tackle the problems at their roots. Not attempting to implement sustainable social reforms or a thorough policy of pollution prevention, no long-term solution could be instituted, as the city's expansion would soon outgrow its improved infrastructure. Moreover, all arrangements only partly functioned during Dickens's lifetime, as even the construction of more sophisticated sewage systems merely relocated the problem of the "Great Stink" toward the Thames estuary.[63]

The author uses his novel as a means of communication to point at the impacts of urbanization and progress on society and city ecology, despite his partly comical, and in comparison rather jovial, social critique. Already formulated by Howard Fulweiler, an unanswered question of the novel raises another interesting subject relevant for the environment and the ecosystem of London: What is valuable enough to be bequeathed? And how can the legacy of the past develop itself in the future?[64] In this particular framework of waste and material circulation, modern questions of sustainability are touched upon.

Dickens regarded natural science as a transformative force, and a basis for social change. In this discourse he saw himself as a translator between science and the public, and shared his well-researched knowledge in publications like *Household Words* and *All Year Round*. But despite all his "sympathy with scientific issues in general and evolution in particular," the author could not "accept as the scientific explanation of nature an ultimate appeal to the irrationality of chance, joined in marriage to the mechanical economic theories of late-eighteenth- and early-nineteenth-century capitalism."[65] Within the context of capitalist critique, London's lifeline had gotten an additional level of meaning not only through Dickens and Joseph Conrad, but from William Morris, whose *News from Nowhere* (1890) depicts a journey through a socialist utopia—on the Thames.

At the end of the century the traditional symbolic charge also stands for and reflects the end and change of an era. In this sense those images stand in a tradition of reevaluating nature in the context of social and environmental change by deconstructing traditional meaning while ascribing them to urban and industrial settings, as determined by Berbeli Wanning for German realism.[66] Although Morris or Kenneth Grahame's *Wind in the Willows* (1908) already evoke less gloomy images than the Victorian social critics, a century-old sense of mystery remains and still characterizes the public imagination of the Thames. During the twentieth century, however, its depiction in belletristic literature is hardly as symbolically charged as during the time of the famous British novel. Yet the timeless myth of the river Thames remains timeless. But with social and media change its monuments are to be found elsewhere, as in songs of modern bards like The Clash or The Kinks, who in their 1967 *Waterloo Sunset* serenade the "Dirty old river," where "Terry and Julie cross over the river where they feel safe and sound / And they don't need no friends / As long as they gaze on Waterloo Sunset, they are in paradise."[67]

Chapter 12

Living on the River over the Year

The Significance of the Neva to Imperial Saint Petersburg

Alexei Kraikovski and Julia Lajus

This chapter explores the seasonal cycle of changing interactions between residents of Saint Petersburg, the new capital of the Russian Empire founded in 1703 by Peter the Great, and the Neva River, which flows through the city. The constitutive role of the river in the life of Saint Petersburg was observed early on. Mid-eighteenth-century naturalist and poet Mikhail Lomonosov described the founding of Saint Petersburg as a big surprise to the Neva River, which he personified as a potential partner for the young capital.[1] During the past three centuries, inhabitants of Saint Petersburg have lived side by side with this partner who could be very generous in one instance but mortally dangerous the next. German naturalist Johann Gottlieb Georgi, who came to Russia in 1770, noted that Saint Petersburg was full of disadvantages due to wet ground and an overabundance of water. Construction was costly and difficult. Wooden buildings quickly decayed, and deep wells and good cellars were lacking. Roadways deteriorated quickly and streets always seemed dirty. Disruptive and even dangerous floods occasionally threatened the city. At the same time, the location of the city offered many advantages. Situated on the delta of this very large river, Saint Petersburg was connected by water to almost all of Russia. The Neva's estuary provided a spacious harbor for seagoing ships, and all sorts of commodities could be easily transported. Moreover, the great river and its branches delivered fresh, healthy water to all parts of the city. Georgi concluded that these benefits fully compensated for the disadvantages and proved that Peter the Great was wise to found his city in this place.[2]

The story of the interactions between the city and the river can be told from the perspective of two different timescales. The first scale is annual. This includes routine cycles of ceremonies and operations repeated year after year that framed customary and everyday life for city dwellers—from the tsar himself to paupers. The second scale comprises the three-hundred-year history of the city on the Neva's banks and shows more permanent changes in city-river interactions, as well as in human perceptions of the river and river life. This chapter primarily considers the annual scale. We look at routine and long-term traditions connected to the Neva, examining how these traditions were born and changed over time. Our sources include narrative materials: memoirs, excerpts from Russian literature, and Saint Petersburg newspapers, which are also good sources of visual materials. We limit ourselves to the imperial period of Russian history, leaving the Soviet era for future research.

This work continues our previous studies of spatial relations between Saint Petersburg and the Neva.[3] However, here we focus on the role of the Neva River in everyday urban life, rather than its role in connecting the city with the wider world inside and outside Russia, the subject of our previous article. This topic is certainly not new for the history of Saint Petersburg,[4] because it is not possible to write such a history without discussing the interactions of the city with the river. Nevertheless, these narratives are usually secondary to other objectives and rarely become central. A remarkable exception is a recent PhD dissertation by Randall Dills on the Neva River and its role in the culture and environment of Saint Petersburg, although it covers only the nineteenth century.[5] We fully agree with the author's statement that "most stories of St. Petersburg . . . stop at the edge, on the docks, quays, and harbors, instead of seeing the integral relationship of those spaces with the water that flowed nearby."[6] In one chapter Dills also uses a seasonal perspective for describing this integral relationship between the river and one particular location in the city. We developed this approach independently and, as much as we can in this chapter, we try to apply it systematically throughout the text by expanding the types of activities under consideration. In particular, we examine fisheries in greater detail—an important urban activity that is often neglected by historians.

Spring: The Neva Becomes Fluid Again

We chose spring as the beginning of the annual cycle, the time when ice melts and life on the riverbanks becomes vivid once more. Residents of Saint Petersburg considered the revival of the Neva to be the most important sign of early

FIGURE 12.1. The barges near the pier of the Metalworking Factory, May 1915. The depicted part of the river upstream from the delta had been cleared of ice very recently. Despite the fact that the picture was taken in early May (late April according to the modern calendar), snow is still visible on the roofs and on the pier. However, the navigation had already begun: in the picture we see the vessels for transportation of commodities and passengers. Noticeably, the passenger steamship belongs to the factory itself, which can be deduced from the inscription visible on the nose of the boat. Photograph by Karl Bulla.

spring. For instance, Gavrila Derzhavin, a late eighteenth-century poet, started his description of spring with the fact that "the Neva runs to the Baltic and the vessels are launched."[7] This first part of the annual cycle included several major elements: the clearing of ice from the Neva, opening it to navigation; the building of floating bridges; the appearance of smelt entering the river from the Gulf of Finland on their spawning run; and the Russian Orthodox feast of Mid-Pentecost.

Clearing of the Neva

The breaking up of the ice was among the most significant events in the annual cycle of urban life. As a rule, the day when ice flow began was carefully recorded: the first known instance in the history of Saint Petersburg is April 3, 1704.[8] Later, three shots from a cannon atop the Peter and Paul Fortress signaled that ice had just broken on the Neva and water flow had begun.[9] In his poem "The Bronze Horseman," Alexander Pushkin mentions this cannonade among other remarkable moments of Saint Petersburg life connected to the "smoke and thunder" of its major citadel.[10] This also marked the time when some venturesome citizens made bets during the winter on when the ice would start moving.[11]

The annual revival of the Neva separated major parts of the city from each other. Moving ice could easily destroy floating bridges, which were frozen during the winter and used as roads, so they were usually disassembled in advance. Boat transportation was also impossible because of ice blocks moving downstream from Ladoga Lake. Thus, the first permanent bridge built in 1850 between the admiralty side of the city and Vasilyevsky Island remarkably improved city life.[12] However, the custom of crossing the river on foot even during the dangerous spring thaw persisted among Saint Petersburg citizens. European travelers described this as a strange local tradition that sometimes had tragic consequences.[13] Even in the early twentieth century, when the Neva's banks were connected by permanent bridges at several points and crossing the river during ice melt was prohibited, tragic incidents on moving ice still frequently occurred.[14]

Opening of Navigation

Water transportation always played a crucial role in the life of Saint Petersburg, so it is not surprising that the opening of navigation in the spring was among the most important moments in the annual life cycle of the capital. Yet some vessels, such as the tugboats responsible for relocating floating piers on their summer moorings, did not wait for the official opening of navigation but appeared on the Neva immediately after the main volume of ice had passed through the city.[15]

As a rule, navigation opened officially on the first Sunday after most of the ice was gone. The commandant of the Peter and Paul Fortress crossed the Neva to obtain permission from the emperor to start shipping. The fortress guns were discharged and a special flag displayed to signal to steamboats waiting near the piers that shipping was once more permitted.[16] As new technologies developed and replaced older ones, popular attitudes toward this highlight of the yearly cycle changed. In the early twentieth century, after diesel-powered steamboats appeared, journalists connected the opening of navigation with increased oil pollution in Neva waters.[17]

Even the official opening of navigation did not mean that the Neva was fully free from ice, and shipping remained rather dangerous. Nevertheless, steamboats tried to get under way as early as possible. The most memorable tragedy took place on April 7, 1905, when moving ice crushed the old steamboat *Archangelsk* and thirty-nine people perished.[18] The story was often repeated later on, but with inflated numbers of people lost (up to a hundred) and different locations for the accident.[19]

Building the Floating Bridges

Floating bridges were used for transportation across the Neva from 1727, when the first such bridge was constructed, to the late nineteenth and early twentieth centuries, when the present network of permanent bridges was completed. The earliest bridge was built on ordinary barges, but after it was destroyed by a storm in 1734 special pontoons were constructed so that the bridge could withstand harsh weathers.[20] Late eighteenth-century observers stressed the advantages of floating bridges for the natural conditions of Saint Petersburg. Technology of the time was not sufficient to build permanent bridges across a river as large and changeable as the Neva. Floating bridges could easily survive seasonal changes in water level and had good load capacity. Broken pontoons could easily be replaced and when the ice appeared, the bridge could be disassembled in a couple of hours and reconstructed again in four to six days.[21] Late nineteenth-century memoirists described the construction of floating bridges as a complicated operation that involved several tugboats and took about two days to complete.[22] The bridges themselves were called "ugly" and considered incompatible with new permanent constructions.[23]

A notable detail is that the permanent bridges arched much higher over the water than the floating ones, and their construction to some extent severed the link between the Neva and Saint Petersburg's populace. Mid-nineteenth-century writer Nikolai Gogol's novel *The Nose* describes a person on a floating bridge as he stares into the water, watching the strong current and the fish returning to spawn at the end of March—this was a popular everyday amusement for Saint Petersburg inhabitants.[24] Such a pastime was impossible on a permanent bridge with pedestrians more than ten meters above the water.

Spring Fisheries and the Spawning of Smelt

Smelt fishing in the rivers emptying into the Gulf of Finland was reported in documents as early as the late fifteenth century.[25] Eighteenth-century writers mentioned smelt among other fish in the Neva River, observing that to enjoy the particular taste of this fish one had to eat it hot and crispy, a preference that persisted into the early twentieth century.[26] To Saint Petersburg inhabitants, smelt was considered to be a very special food. For instance, in Gogol's comedy *The Government Inspector*, the mayor of a provincial town knows only one fact about Saint Petersburg: in that city, people eat two unbelievably tasty sorts of fish—smelt and whitefish.[27]

Saint Petersburg fishermen caught smelt using nets and traps, sometimes starting to fish before ice flow. In early April 1910, ice flow combined with the absence of a sea wind interrupted smelt fishing. Journalists noted this trouble, for suburban fishermen had hoped to have a lot of smelt, which were "so popular among the dwellers of St. Petersburg."[28] Smelt was not the only spring fish in Saint Petersburg, however. Artist Mikhail Grigoriev mentions in his memoirs that the symbol of spring for Saint Petersburg fishermen was the lamprey. During its spawning period, dilapidated fishing boats checking lamprey traps could be seen even near the Winter Palace.[29]

Spring fisheries were always considered very dangerous. Fishermen had to spend a lot of time on unstable ice and quite often were lost at sea. For instance, on March 22, 1903, twenty-five to thirty fishermen were swept out to the open sea on moving ice. Icebreakers, including the famous *Ermak*, the most effective of that time, were sent to rescue the trapped men, but their efforts were useless. Only five fishermen were saved, thanks to the keeper of Tolbukhin Lighthouse, which was situated on an inlet of the Neva.[30] In 1910, however, a similar story ended happily—all fifty fishermen swept out to sea on ice shoals were rescued.[31]

Mid-Pentecost

Mid-Pentecost is celebrated on the twenty-fifth day after Easter, usually between the end of April and early June in the new calendar. By this time the ice had finally drifted away. The sacred procession took place around the walls of the Peter and Paul Fortress, and many citizens crossed the river from the admiralty side to see it. Eventually, crossing the Neva became a sort of ceremony in itself. Describing this "pious and poetic trip" on the Mid-Pentecost of April 17, 1828, Prince Petr Vyazemsky wrote to his wife that the weather was bad, but there were still many people in attendance. Furthermore, he observed, in nice weather "the Neva is dotted with vessels transporting people" during this feast.[32] A watercolor by I. A. Ivanov depicts the Mid-Pentecost in 1815: the procession on the fortress walls and a flotilla of boats full of festively dressed people crossing the Neva. The river, wide, picturesque, and shining in the bright sunlight plays a central role in this festive atmosphere.[33]

SUMMER

Summer was the peak of river transportation, when the Neva and its tributaries were full of life. The importance of the rivers and canals as transportation routes for Saint Petersburg was recognized from the beginning, and emphasized by early eighteenth-century observers.[34] A visible peculiarity of Saint Petersburg's summer that distinguished it from most other Russian cities, river transportation has attracted a lot of attention from researchers, so there is no need to explore this issue here.[35] Instead, our chapter examines summer recreational activities that were closely connected with the Neva, as well as salmon spawning, an important summer component of the seasonal cycle.

Summer Recreational Season

Summer season meant that a lot of people left Saint Petersburg. Early twentieth-century artist Mikhail Dobuzhinski recalled that during his childhood the city became empty in the summer. People of higher social strata moved to their country dachas or abroad, and the city filled up with commoners building or repairing houses and roadways.[36] The Neva played a major role in the recreational activities of both the educated classes (who left the city) and the ordinary people (who summered in Saint Petersburg).

The Neva basin included exclusive districts where the nobility and elite of the imperial capital could enjoy leisurely summers. Upstream from the city, nobles' manors occupied both banks of the river as early as the eighteenth century. Some of these mansions have been preserved, but the general character of that region was transformed in the late nineteenth century when the area was turned into an industrial zone. Tavricheski Palace, the residence of Prince Grigorii Potemkin, closest confidant of Catherine the Great, is a good example of a summer estate. Although separated from the river and scarcely visible from its bank today, originally the architect Ivan Starov included the Neva as a fundamental part of his composition.[37] The same is true for the manor of Count Bezborodko, designed by Giacomo Quarenghi.[38]

The northern part of the Neva delta, called simply "the islands," was another recreational area for the nobility. This area boasted big summer estates, including imperial residences on Kamennyi and Elagin Island. In the nineteenth century these mansions were replaced by much smaller, though still luxurious, summer homes for wealthy people. By the early twentieth century, however, the islands had opened up to all echelons of educated Saint Peters-

FIGURE 12.2. The view of the Neva enbankment in the Mid-Pentecost, 1815, I. A. Ivanov. The twenty-fifth day after Easter could be considered as a boundary between the seasons. The painter depicted the bright religious ceremony on the wall of the Peter and Paul Fortress and numerous citizens crossing the river in their picturesque colored boats.

burg society, with numerous restaurants and entertainment establishments. Fyodor Dostoevsky more than once mentioned amusements on the islands.[39] The "white nights," the long-lasting daylight that Saint Petersburg enjoys in June and early July due to its northern latitude (60th parallel), added a romantic sparkle to summer on the water.

For the common people who didn't leave the city in summer, the Neva itself became a significant source of relaxation. Floating bathhouses were situated at different points on the Neva. Visitors could choose to swim inside or outside the basin in the open water.[40] Journalists also mention professional "divers" at the bathhouses, who would dive for money that bathhouse visitors threw into the water. This strange amusement was especially popular with merchants.[41] The lack of bathhouses in the city meant that residents had to refresh themselves in the open water, and the abundance of drunk swimmers meant that many bathers drowned (up to five hundred per season in the Neva).[42] A cemetery was reserved on the northern coast of the Gulf of Finland for those people who could not be identified, or whose relatives could not be found.[43] Pleasure rafts and boats hosting music and song were also a key part of folksy fun in the summertime.[44] Trips were normally organized on the outskirts of the city in working districts, upstream from the imperial center.

Summer Fishing and the Salmon Run

For professional fishermen, summer was a busy time due to the overlapping spawning seasons of anadromous fish species entering the Neva and going upstream to spawn. The Saint Petersburg regional survey of 1877 listed 283 fishermen operating in the Neva River.[45] In the early 1880s, 229 local and 120 seasonal fishermen were mentioned.[46] When whitefish and salmon entered the Neva in mid-June, the most profitable part of the fishing season had begun.[47] Summer for citizens of Saint Petersburg was when fresh fish was available at city markets.[48] Local fishermen delivered live hauls in boats; the fish could be sold at open markets or in specialized fish shops where they were kept alive in tanks of water. The most important local-market fish were whitefish from the Neva and the Volkhov Rivers, the Neva lamprey, eels from the Gulf of Finland, trout from small rivers in suburban areas, and salmon from Lake Ladoga.[49] Citizens also used to buy fish at floating fish cages—pictures of such a fish cage on the Fontanka River (a Neva River tributary) are well known[50]—and directly from fishing stations at different sites on the Neva, predominantly downstream from the city. Late July was considered the peak of the salmon spawning run. Then, numerous visitors came to these places, informed by newspapers about the magnitude and quality of the catches.[51] Peddlers played a special role in the fish trade. With tubs on their heads, they could easily be seen in most corners of the imperial capital, even near the imperial Palace Embankment.[52]

Saint Petersburg's populace considered fishing on the Neva to be an essential summer pastime and a great form of recreation. Amateur fishermen can easily be seen in visual materials from different periods and different parts of the city, including elite areas. For instance, a fashionably dressed gentleman with a rod is shown on the bank of the Fontanka River near the Summer Garden in the mid-1700s in an engraving by M. Mahaev;[53] another can be seen in the northern part of the Neva delta on Petrovskii Island in an 1815 painting by S. Shchedrin.[54] Angling was also quite popular among the common people—a peasant fishing with a rod near the Peter and Paul Fortress is clearly apparent in a painting by B. Patersen completed in 1799.[55] Late nineteenth-century memoirists also stressed that this activity was popular among all sorts of urban residents, including workers.[56] Even members of the imperial family were among the amateur fishermen, though they normally fished at summer residences or in the Gulf of Finland.[57]

Another summer entertainment was visiting fishing stations to watch professional fishermen at work. Such idle observers can be seen in the landscape of the Kamennyi Island estate completed by Silvestr Shchedrin in 1803.[58] Memoirs mention this activity as a popular way to finish a sleepless white night cheerfully spent at a restaurant, along with a portion of *ukha* (fish soup) made from freshly caught fish.[59] Spectators could also gamble by purchasing, for a fixed price, the future harvest of a net. A three-ruble investment (in the early twentieth century) could yield a lucky gambler a catch of fish worth perhaps thirty rubles. If unlucky, the gambler got nothing.[60]

In the 1870s, by our estimation, the annual catch of salmon in the Neva River ranged between 3,500 and 7,000 fish.[61] By the 1930s, the abundance of valuable migrating fish, such as Atlantic salmon and whitefish, declined. This was mostly because of intensive fishing not just in the Neva River but in the Baltic Sea basin in general. Following this, fisheries gradually lost their economic significance. In addition, it is likely that fewer suitable fishing locations remained within the city, but also that industry could provide more attractive occupations than fisheries.

Autumn

Autumn in Saint Petersburg was a time when its citizens returned from their summer dachas and normal levels of activity began again. The most important aspects of the Neva's autumn cycle were its periodically unstable water levels and the early freeze that heralded the temporary removal of the floating bridges and the start of ice fishing for lamprey and burbot.

Period of Water Instability

Describing life on Vasilyevsky Island in his novel *The Islanders*, nineteenth-century writer Nikolai Leskov begins his story with crossing the Neva in summer in fine, sunny weather. However, unexpected wind from Lake Ladoga whips up the waves, which begin to flounce madly and to toss the hero's yawl so violently that he is barely able to stay in the boat.[62] Such rapid weather changes were much more pronounced in autumn.

Hydrologists note that the variability of the Neva's water level is exceptional.[63] The main factors causing wide annual fluctuations are the water level in Lake Ladoga, ice jams, and hydrological processes in Neva inlet. Downstream, where Saint Petersburg is situated, the influence of the Baltic Sea is decisive and wind plays the most important role. Cyclonic winds that often occur near Saint Petersburg in the autumn cause the water level to rise and fall significantly.[64] Interestingly, sources rarely mention sudden decreases in water level even though decreases of fifty to seventy centimeters below average

FIGURE 12.3. Theater square, Flood, November 12, 1903. The water level in the Neva River and canals this day was high enough to cover the streets and even to damage the pavement. However, as we can see in the picture, the residents continue life almost normally. The autumn floods were an ordinary part of the urban life until they became really catastrophic. Photograph by Karl Bulla.

could expose water intake pipes, cut off water supplies, and close transportation routes.[65] But rapid decreases were never as catastrophic as rapid increases, with severe floods being the extreme example of this phenomenon.

The history of Saint Petersburg floods has been studied in numerous works, and is beyond the scope of a chapter more or less about normal, everyday life.[66] Sources list the consequences of bad autumn weather on the banks of the Neva, which significantly upset daily activities in Saint Petersburg. Most concern water transportation. Instability of the water level, combined with strong winds, impeded the loading and unloading of barges.[67] This also made crossing the Neva in small boats particularly dangerous. Late nineteenth-century memoirists recalled that Neva ferrymen and Estonian and Finnish sailors were considered experts, among city-dwellers, in water transportation, since they readily departed from their piers in all weather, even during autumn storms.[68] This branch of the urban economy was controlled by the communities of experienced boatmen who came to Saint Petersburg from the upper Volga region from the vicinity of Tver (up to several thousand people in the second half of the nineteenth century). They were employed on both long-distance and local water routes, and thus the Neva also provided income to the peasant families living very far from Saint Petersburg.[69]

Highly variable water levels affected parts of the city differently due to their elevation. Even when the water didn't reach maximum flood levels, what was not dangerous for inhabitants of the central parts of the capital could be catastrophic for those who lived by the sea. Cannon shots signaling rising water are mentioned more than once in fictional literature and in newspapers as a backdrop for anxious and sleepless nights.[70]

Freezing of the Neva

Since the eighteenth century, observers have noted that ice normally covered the Neva quickly. According to Georgi, in some years freezing took place in one day.[71] First, small chunks of ice moved downstream from Lake Ladoga. They became bigger and bigger until, finally, ice covered the whole river. As a rule, this took place in late October or early November. Journalists considered it necessary to record this occurrence in the newspaper to inform the public in Saint Petersburg and other cities.[72] Ice that appeared too early was dangerous for shipping and economically damaging for the sea port of Saint Petersburg.[73]

At the appearance of ice, the floating bridges had to be deconstructed.[74] This tradition started rather early, when the Isaacs Floating Bridge was the only one in existence in the city. It was initially taken down between autumn and the following spring, but in 1779 Catherine the Great ordered that it be built again after the Neva froze solid. To accommodate her wishes, a channel was cut in the ice every year for this purpose.[75] This practice was also used later for other floating bridges.

Autumn fisheries were also connected to the major environmental changes that accompanied freezing. Salmon fishing stopped in the early autumn; salmon caught during the first half of October were considered extremely late.[76] The freezing of the Neva and its tributaries heralded the beginning of ice fishing. Lamprey and burbot were the main target of fishermen in November, who used the strong ice to establish traps and kiddles in areas where fish were prolific, such as the Okhta Bridge.[77]

Winter

Winter in Saint Petersburg was usually described as severe and frosty. An early eighteenth-century observer noted that a typical German winter was summer compared to a typical Saint Petersburg winter.[78] For Pushkin, among the most important elements of Saint Petersburg winter were the motionless air,

FIGURE 12.4. Electric transportation line "The Senate–Rumiantsev Square," seen from the University embankment, 1913. In the 1910s a light electric tram crossed the ice-covered Neva with the passengers on board right in the city center (the State Admiralty is clearly seen in the background). In the picture one can see sophisticated seasonal infrastructure including the electric lines, rails, and the wooden bridge for pedestrians. Photograph by a member of Karl Bulla workshop.

the frost, and the sledges skimming the Neva's surface.[79] Georgi noted that in the late eighteenth century winter was considered the healthiest season due to the cold and relatively dry air.[80] Citizens believed that the wet, cold summer was not a particularly healthy season for Saint Petersburg residents and that winter, with the frozen Neva, skating rinks, opera, and gas lamps, was much better.[81] The ice-covered Neva was an essential part of everyday life in Saint Petersburg in winter.

According to evidence provided by Georgi, Neva ice was generally between 61 and 92 centimeters thick, about 70 centimeters thick on average.[82] However, even in the middle of winter when it should have been covered by thick, hard ice, the Neva could be quite dangerous. For instance, during the night of December 19, 1902 (early January in the new calendar), a storm from the southwest caused a flood that was quite destructive to the city.[83]

In the following sections, we examine the role of the frozen Neva in city transportation, winter recreational activities, and the ceremonial life of the imperial capital. However, we will start with the example of perhaps the most direct interaction between the city-dwellers and the ice. Cutting ice blocks for storage houses was a key activity of Saint Petersburg's winter economy. This hard and dangerous work provided salaries for numerous poor peasants who wintered in Saint Petersburg while waiting for the summer—the peak of seasonal work in the capital in the sphere of construction, repairing, and navigation. One icehouse could bring them up to fifteen rubles, but this work involved several people and horses.[84] Plenty of good, clean ice for keeping food and drinks fresh in hot summer months was considered a crucial advantage of Saint Petersburg's environment.[85] The technology of ice production has been studied in detail and thoroughly described in historical sources and in secondary literature,[86] but one notable advantage should be emphasized: ice produced in winter was later sold in summer, thus providing year-round revenue. For instance, in late May 1912, during extremely hot weather, ice cream vendors needed lots of ice to make ice cream, which was in great demand. They paid up to 1.5 rubles per trolley of ice to the icehouse owners.[87] The ice had of course been produced during the past winter on the Neva. Therefore, the freezing Neva was directly involved in this significant and profitable industry.

Winter Transportation

As soon as ice covered the Neva, shipping stopped. If ship owners couldn't pull vessels out of the water, they tried to find a safe place for the ship to remain during the winter. During Saint Petersburg's early years, ship owners used the canal to the north from Zayachii Island, between the Peter and Paul Fortress and the Kronwerk.[88] In the early twentieth century, safe places for ships during winter were located in relatively slow streams, not far from manufacturing plants and factories providing winter repairs—for instance, near Vasilyevsky Island upstream from the Malaya Neva.[89]

Using ice for transportation was considered the greatest advantage of Saint Petersburg's winter as early as the eighteenth century.[90] F. H. Weber, a well-informed German observer, noted in the 1710s that citizens began to safely cross the river on strong ice immediately after it froze.[91] Not only could the Neva be crossed on ice, but transportation along the river flourished even on the Neva inlet, between Saint Petersburg and Kronshtadt. In winter during the eighteenth and nineteenth centuries, the city administration created a sophisticated infrastructure on this ice road between Saint Petersburg and Kronshtadt, including road signs and a temporary pub midway.[92] At the beginning of the twentieth century the ice road between Kronshtadt and Oranienbaum was even equipped with electric lamps.[93] In general, the seasonal infrastructure for ice transportation became more complicated and technologically advanced over time. In the early twentieth century, in addition to bridges, winter roads were specially constructed both along and across the river.[94] Several seasonal electric tramlines operated on the ice between 1895 and 1911,[95] and skaters used ice tracks to pull small sledges carrying one or two passengers (a sort of rickshaw).[96] The transportation here was much less controlled in comparison to the streets.

Winter Recreation

Obviously, the role of the Neva in winter recreational activities was central even in the earliest days of Saint Petersburg. Peter the Great himself established a traditional winter pastime inextricably linked to the Neva. Passionate about sailing, the tsar took to skating with a sail, and then sailing an iceboat. This became his favorite form of winter recreation in the new capital.[97] Later, this pastime was taken up by others in Saint Petersburg, and in the early twentieth century observers described numerous iceboats and individuals skating with sails on the ice of the Gulf of Finland.[98] In the early twentieth century, attempts at technological improvement aimed to make this amusement useful. In February 1910 an engineer named Vladimir Lebedev constructed an iceboat with a propeller and tested it on the ice between Saint Petersburg and Kronshtadt. The new machine went from forty to eighty kilometers per hour,[99] but it never became practical, perhaps because of the First World War and the Revolution, which interrupted many projects in the city.

Inside the city, ice formed an integral part of recreational areas together with the open spaces on the shores. Numerous amusements were situated there, such as skating rinks, ice hills for riding (a necessary part of the Russian *Maslenitsa* [Shrovetide] in Saint Petersburg), seasonal hippodromes for horse races, and areas where Finnish and Saami people organized recreational sleigh rides every winter.[100] Visual sources locate most of these amusements on the wide part of the Neva between the Peter and Paul Fortress and the Winter Palace, and downstream from this place between Vasilyevsky Island and the English Embankment, in the most spectacular part of the city. In their paintings in 1792 and 1813, respectively, D. A. Atkinson and M. F. Damam-Dematrais depict the ice mountains on the Neva near the English Embankment, but in an engraving by N. Serrakapriola made in 1817 we can see the ice mountains near the Peter and Paul Fortress.[101] In 1814, A. O. Orlovskii depicted races on the Neva near the Spit of Vasilyevsky Island (the cape on the eastern end of the island, across the Neva from the Winter Palace and Peter and Paul Fortress).

In the background of the watercolor we can see the Rostral Columns and the Cathedral of Prince Vladimir—recognizable landmarks in the city center.[102] In an early nineteenth-century engraving by E. Korneev, denizens of Saint Petersburg (including the fashionably dressed and obviously rich) are shown skating on the Neva with the magnificent Academy of Fine Arts building in the background.[103]

Pancake Week (Maslenitsa) was and still is the most important winter holiday. This week before Lent marked the end of winter and, as shown above, a significant part of its traditional celebrations took place on Neva ice. There was an enormous demand for fish at this time, including hauls from winter fisheries in the Gulf of Finland. The winter frost was considered an advantage that allowed the catch to be frozen immediately and then delivered to the city on sledges. The frozen fish could be partially revived in water tanks, making live fish available to consumers.[104] Newspapers in the early twentieth century described large deliveries of all sorts of fish before fasting periods, not just during Shrovetide, as well as seasonal drops in the market prices of fish.[105]

Winter Ceremonies

Ceremonies that took place during winter can be roughly divided into routine, annual, and extraordinary. Christmas and Epiphany were the principal annual religious celebrations that took place in winter. Christmas was mainly considered a family holiday and information about big public ceremonies outside churches was scarce; however, during the time of Peter the Great, both Christmas and New Year festivities took place on the water square near the Peter and Paul Fortress.[106] The Christian holiday of Epiphany was strongly connected to the water, especially to the Neva if we are talking about the Orthodox residents of Saint Petersburg. Lavish imperial Epiphany ceremonies are described more than once in archival sources[107] and in published literature.[108] Numerous visual materials from the eighteenth to the twentieth century show that ceremonial decorations were more or less the same throughout this period and included a wooden pavilion with columns and a small dome.[109] At the same time the ceremony itself, including the military parade, was shortened and partially moved inside the palace to avoid the frostbite and colds common to severe Saint Petersburg winters.[110]

Among extraordinary winter ceremonies connected to the Neva, imperial funerals had the greatest import. In the eighteenth century, if a member of the ruling family died in winter, the funeral procession crossed the Neva from the palace to the fortress and the ice square became a parade ground for military display and salute.[111] This large open space also had another function: it could become a stage for enormous firework displays, including hundreds of rockets and fire fountains. Obviously, this was the safest place in the city for such a display.

Volatility and Persistence in the History of the Neva

Historical sources clearly show that the Neva River occupied a significant place in the routine annual cycle of urban life in Saint Petersburg. Many diverse links connected city-dwellers and the Neva, some of them retaining their importance for hundreds of years. These links shaped the lives of the people of Saint Petersburg, who came to use the river in all possible ways, be it liquid or solid, cold or warm. The Neva was used comprehensively: as a stream of fresh, high-quality drinking water; as a transportation route; as habitat for commercially important fish; as an inexhaustible source of high-quality ice; and as an open space, the largest square of the capital city.

People constantly needed to develop new skills and technologies to take advantage of the Neva's potential, but the rewards were, for the most part, worth their efforts. New technologies could strengthen ties between society and the river (like the introduction of steamboats and tugboats), or decimate or destroy them (like the construction of permanent bridges, improvements in ground transportation, and water pollution). In either case, the history of Saint Petersburg cannot be correctly understood without the Neva—constantly changing and simultaneously remaining the same over centuries.

Further study of the technological and social changes of the twentieth century might reveal the evolution of these connections up to the present time, when interrelations between the Neva and the annual cycle of urban life have changed significantly. Seasonality in these relations is now much less manifest and the river's influence on everyday life in the city has diminished. For instance, across the Neva and upstream from the center of the city, a large cable-stayed bridge was constructed ten years ago, which has opened the opportunity to cross the river at any time of the day or night. Construction of the sophisticated complex of protection dams in the Neva Bay has put an end to Saint Petersburg's history of floods. The winters have become much warmer and citizens can no longer rely on the Neva ice in the cold season. Walking on the ice is not allowed; for most of the winter it remains artificially broken. Last but not least, water transportation for local passengers has become a thing of the past; today, boats serve only tourists—for a price. However, the Neva remains central in terms of the spatial perception of the city, its landscape, and also its environment (citizens still drink water from the river, which is now cleaned by modern technology). Seasonal commercial smelt fisheries still operate within the city area of the river every spring, and amateur fishers remain a visible part of the city culture.

PART IV.

RIVERS REGAINED

CHAPTER 13

"A RIDICULOUS FAILURE OF GOVERNMENT"

The Chicago River in the Age of Ecology

Harold L. Platt

The half-century battle to reclaim the Chicago River from an open sewer to a recreational waterway clean enough for swimming began as a crusade to save Lake Michigan. In 1966, four years after the publication of Rachel Carson's *Silent Spring*, the city joined her campaign to rid the environment of dangerous chemicals. The staid *Chicago Tribune* sounded the alarm, launching a series of front-page articles under a graphic logo pleading with readers to "Save Our Lake." It was "sick" from digesting too much pollution and in imminent danger of suffering the same "death" as Lake Erie.

The newspaper built on Carson's explanation of the ecology of pesticides to help its readers understand how phosphates in their laundry detergent were poisoning the environment. Pumped into the lake as partially treated sewage, the chemicals were overfeeding microorganisms. They, in turn, were rising to the lake's surface as massive algae blooms, depleting its water of oxygen, killing its fish, and making its smelly shorelines off-limits for recreational use.[1]

Over the next two years, Lake Michigan suffered one unnatural catastrophe after another, hammering home Carson's and the *Tribune*'s warnings about dire unintended consequences whenever the balance of nature was upset. First came the revelation that the very government authorities entrusted to enforce antipollution laws—the US Army Corps of Engineers and the Metropolitan Sanitary District (MSD)—were among their worst violators. Both were using the lake as a free dumping ground for the highly toxic sediment they were dredging from the Chicago River.[2] Next came the invasion of the alewives, six billion dead fish washed up on the lakefront beaches. A combination of pollution, overfishing, and invasive species had wiped out all of the native predators of the four-to-six-inch herring-like alewives. This nonnative species was left virtually alone to feast on the algae-enriched waters of the world's sixth-largest body of freshwater.[3] In 1966, they had stayed offshore, but had clogged the city's water supply intake wells. The following summer, however, a massive die-off and easterly winds left the city's beaches buried under thick layers of rotting fish. "They keep coming in," exclaimed the man in charge of cleaning up the stinky mess. "In some places they are a foot deep," he stated, "look out over the lake and there they are as far as the eye can see." Many of his workers gave up and walked off the job, leaving the beached alewives to generate a biblical plague of bugs, including swarms of mosquitos that attacked the elite residents of lakefront apartment towers. To complete the catastrophe that kept at least 2.5 million beachgoers away that summer, the depletion of the lake's alewives allowed the algae to flourish, wash ashore, and bury the beaches under piles of decaying "seaweed."[4]

By the fall of 1967, the campaign to save the lake had reached a critical mass of support that turned it into a full-fledged environmental movement. Responding to reader requests, the *Tribune* printed 4,300 bumper stickers and 6,600 copies of a forty-eight-page booklet on the subject. Fearing the ghost of Lake Erie, citizen groups mobilized to force elected officials at every level of government to give the conservation of Lake Michigan serious attention for the first time since the Progressive Era. In September, Secretary of the Interior Stuart Udall convened a conference of policymakers and engineers from around the Great Lakes to draw up plans to save them from pollution. Out of this meeting would come a barrage of regulations, litigation, and legislation leading to the passage in 1972 of the comprehensive Water Pollution Control Act, more commonly known as the Clean Water Act.[5]

In sharp contrast, changing perceptions of the Chicago River as anything other than an industrial corridor filled with wastewater would take a much greater stretch of imagination and efforts spanning a fifty-year period. Its unique "envirotechnical" conditions and deeply entrenched attitudes among engineers and the local political elite made the process of building a constituency of stakeholders in favor of reclaiming the city's river a much more complicated process than saving its lake. In 1970, Mayor Richard J. Daley (1955–1975) would confess that he had a dream of being able to fish on the river someday. For the next thirty years, his vision would become the single

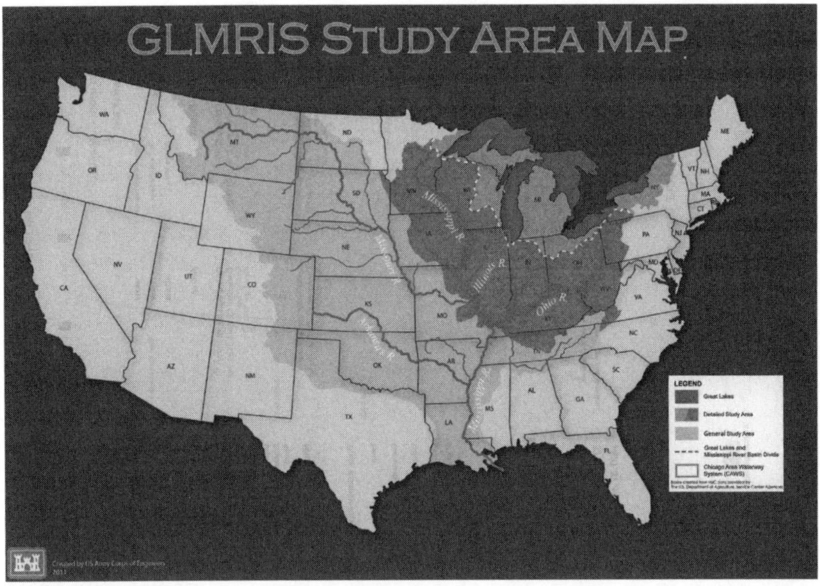

Figure 13.1. The US Great Lakes–Mississippi River Basins. Source: US Army Corps of Engineers.

most repeated expression of bitter disappointment in the failure of City Hall to reclaim the river for this and other recreational uses.[6]

Chicago's location made flood and pollution control of its waters an interrelated problem because its flat land rests on an impermeable layer of clay. The predominant topography of the metropolitan area is prairie wetland, with patches of woods rising in between innumerable sloughs and shallow streams. Fed only by rainwater, its rivers are normally sluggish except during thunderstorms, when the waters overflow the low banks to spread across wide floodplains. Although the Chicago River was a major economic resource in the growth of the city during the nineteenth century, it took an equally large toll in damaged property and lives, as well as environmental degradation.

The transformation of this already artificial river into a "sanitary and ship channel" became Chicago's greatest myth of heroic engineering, despite its serious flaws as an envirotechnical system. It had been built in response to the great flood of 1885, which had inundated the central business district and set off a panic of fear of a plague that was sure to follow. While the canal kept some pollution out of the lake, it drained billions of gallons a day from it through several locks to dilute the city's chemical and organic effluents and

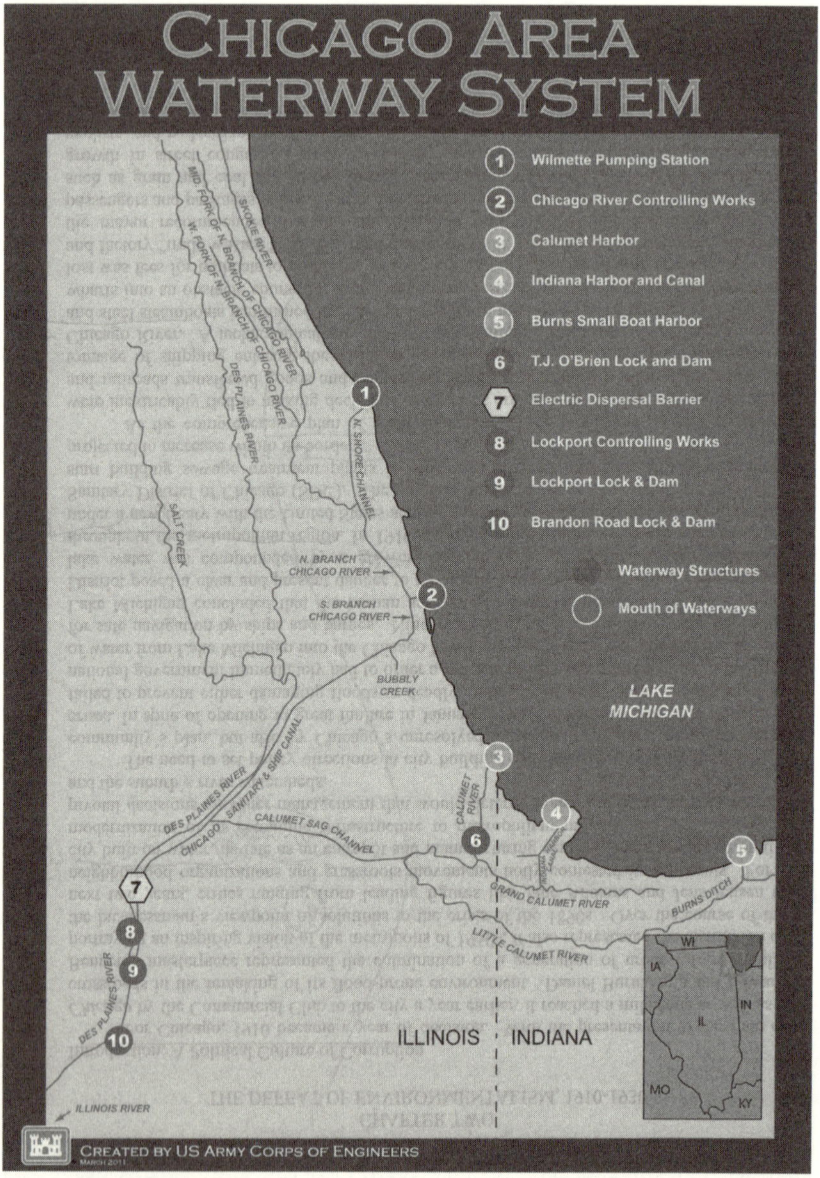

FIGURE 13.2. The Chicago Area Waterway System. Source: US Army Corps of Engineers.

render them harmless as they merged into the Illinois and Mississippi Rivers. Yet the river still became so overloaded with industrial wastes and partially treated sewage that it needed constant dredging in order to keep its shipping lanes open to barge traffic.[7]

Unlike all other streams entering the Great Lakes, the flow of the Chicago River was reversed in 1900 in order to stop the city's wastewater from contaminating its drinking supply and causing deadly epidemics. By reversing the river's flow, Chicago became the one point of contact between the continents' two largest watershed basins: the Great Lakes and the Mississippi River. The normal, reversed flow was "downstream" from Lake Michigan into the river, including a daily average of two billion gallons (7.6 billion liters) of partially treated sewage. At the same time, invasive species were moving "upstream" from as far away as the Gulf of Mexico.

In addition, the MSD had to open the locks during severe thunderstorms, vomiting tons of raw sewage into the lake. During these predictable seasonal downpours, the MSD's underground infrastructure of 5,500 miles of combined sewer and storm pipes not only emptied their contents into the river from over six hundred emergency overflow drains; the overloaded pipes also backflowed into basements in the city and the almost two hundred suburbs lying within the sanitary district's boundaries. Costing millions of dollars in damages to mostly uninsured homeowners, it put at least 1.5 million people at constant risk of flooding.[8]

Nevertheless, the city's cult of the "technological sublime" drew a sharp dividing mental line between the riverfront as its back alley for getting rid of its liquid wastes and the lakefront as its front yard for recreational use. Chicago had kept this myth of big technology alive in the contemporary era by building the world's largest sewage treatment and water filtration plants. More important here, the engineers of the MSD put forward what would eventually become the biggest urban public works project in the nation's history, the Tunnel and Reservoir Plan (TARP).

Reacting to sharp public criticism following particularly damaging thunderstorms for two years in a row in 1957–1958, the MSD proposed to build an "underflow" system of deep tunnels to store the floodwaters until they could be pumped back up and treated. They promised that the multibillion-dollar TARP would solve the overflow problems of flood control and lake pollution. Yet its planners never questioned the status of the river as an industrial corridor, nor its use as the "ultimate sink" for its own partially treated effluents, the daily equivalent of a million people's raw sewage.[9]

Local traditions of lakefront beautification stretched back into the past as much as the heroic engineering of the artificial waterway, to the days of the World's Fair of 1893, Daniel Burnham, and his influential Plan of Chicago. The *Chicago Tribune* tapped into this Progressive Era spirit of the conservation of nature and the city beautiful in its "Save Our Lake" crusade. In a similar way, Mayor Richard M. Daley (1989–2011) would draw upon the Burnham legacy to build public support and private funding for his grand plan for a makeover of the city's front yard, Millennium Park. At the same time, however, he would still be referring to the Chicago River as the "second lakefront" as late as 2007. And he was still giving it subordinate priority in the city budget.[10]

Behind the mayor's back, so to speak, the mobilization of stakeholders in seeing the river transformed into a recreational waterway was not so quietly growing in numbers, diversity, and organizational strength. They dreamed of the Chicago River as an outdoor playground and tourist attraction of "nature in the unnatural world." The story of the materialization of their visions of an urban wilderness parallels the situation of the Rust Belt city of Cleveland, Ohio, its burning Cuyahoga River in 1969, and its "dead" Lake Erie, as studied by David and Richard Stradling. Stradling and Stradling illuminate a self-reinforcing process of structural and attitudinal changes that caused a complete metamorphosis of the meaning, and even the iconic image of the "river on fire."[11] As in Cleveland, the deindustrialization of Chicago's riverfront and the related loss of jobs brought about a corresponding loss of stakeholders in the river as an economic resource. At the same time, it was gaining new constituencies of users among outdoor and fishing enthusiasts, city beautification groups, and real estate developers. As the river showed slow, albeit steady, improvements in environmental quality, its permanent base of stakeholders in its full restoration for recreational use grew in proportion. It would take fifty years before Chicagoans could fully reimagine the river and the lake as a single urban ecology.

This reciprocal process of river reclamation and stakeholder mobilization can be divided into a number of convenient stages of development. Each incremental improvement in the physical condition of the river fed psychological expectations for further gains and enlisted new recruits to the cause of saving it from pollution. In this way, the relationship between the city's people and its hydraulic environment underwent change.

As we have seen, the period from 1962 to 1972 was dominated by a popular movement for the conservation of the Great Lakes, culminating in the historic Clean Water Act. Riding this wave of popular support, the MSD was able to gain approval for its grandiose plan of flood control of the river disguised as the final solution to the problem of pollution control of the lake. From 1973 to 1986 the MSD built the first phase of the Deep Tunnel, while simultaneously adding chlorine to its treatment process in order to prevent its stormwater discharges from causing an epidemic. The chemical additive killed not only the dangerous microbes in its effluents but virtually all forms of life in the Chicago River.

In spite of the river becoming a dead zone, the formation of the Friends of the River (FOTR) marked the beginning of its rebirth from a sewage channel to the "wildest area within the city limits." Like the *Tribune*'s earlier "Save Our Lake" crusade, the August 1979 issue of *Chicago Magazine* acted as a catalyst in the coalescence of a reform movement. Under the headline "Our Friendless Chicago River," it brought together a wide range of stakeholders in the waterway's reclamation, who had previously been working independently of each other. Written by city planner Robert Cassidy, the lead article made a similar appeal as the "Save Our Lake" campaign to "reclaim our lost resource." In spite of the river being "neglected, fouled, and abused," he asserted, it "is the second greatest gift that nature has bestowed on this city." He highlighted the benefits of riverfront improvements in other American cities such as Denver, San Antonio, and Saint Louis. The new lobby's acceptance of the MSD's definition of it as a "working river" reflected the initial weakness of its cause on behalf of the urban environment. The FOTR's diverse coalition of special interest groups had to assume a cautious political posture at first. It attempted to include everyone and alienate no one. At the same time, the magazine's call to arms included a master plan for riverfront redevelopment that established both an urban vision and a political goal for the future.[12]

With the long-anticipated opening of the Deep Tunnel in sight, perceptions of the river were set to undergo a major change for the better in the public imagination. Poised at the beginning of a sustained building boom, increasing numbers of developers joined architect Bertrand Goldberg in looking at the river as a marketable natural amenity. The two-thousand-strong FOTR represented another growing lobby in favor of its improvement. In the long battle to reclaim the Chicago River, the early green activists and visionary prophets had made a difference. Their example had established a solid foundation of stakeholder support for the emergence of a full-fledged movement of environmental reform. From garden patches and river cleanups, it would gain momentum and move up the federal system to find defenders against its worst enemy, City Hall.[13]

Save Our River: The Rebirth of "Wild Nature," 1985–1997

From 1985 to 1997, the river entered a third era of rapid commercial and residential redevelopment. Private investors created an urban renaissance in an expanding central business district (CBD). During the nineties, the downtown area would gain twenty thousand residents as condominium apartment towers joined the new office skyscrapers lining the riverbanks. In addition, the MSD, renamed the Metropolitan Water Reclamation District (MWRD), stopped adding chlorinate to its effluents because the opening of the Deep Tunnel meant it could hold one billion gallons (3.8 billion liters) of stormwater overflow. Portrayed as a cost-saving dividend of TARP, the river was the primary beneficiary. It was rewarded with the return of sport fish, nesting birds, native plants, and other wildlife.

The sea change in public perceptions of the Chicago River owes the most to its dramatic ecological metamorphosis after May 1985, when the MWRD started operating the Deep Tunnel. Adding thirty-one miles (50 km) to the forty-seven-mile (76 km) underflow system reduced daily overflows of raw sewage into Chicago's waterways by about 85 percent. As the water became less contaminated with bacteria and no longer poisoned with chlorine, oxygen levels began rising high enough to sustain aquatic life. And with the revival of plants and fish in the river, other forms of wildlife began returning to what had once been natural habitats along its banks.

The rebirth of nature in the city, in turn, recast the river in a more positive light and reinforced the ranks of reform activists, who had a stake in reclaiming it for recreational use. Drawn to it was a base constituency of sport fishers, birdwatchers, boaters, and other outdoor enthusiasts. It also began winning new groups of stakeholders in the city center, including office workers on their lunch breaks, apartment dwellers on their balconies, and out-of-towners on their tour boats looking up at the renaissance of architecture in the CBD.[14]

Nonetheless, the local sanitary authority in charge of pollution control of the Chicago River continued to contaminate it with the equivalent amount of raw sewage produced by a city with 150,000 inhabitants. The MSD/MWRD refused to bear the cost of employing existing technologies at its treatment plants, which could remove 98–99 percent of the impurities. Instead, they were operating at a 90-percent level of treatment, constantly pouring dangerous microbes into the river and blanketing its bottom with toxic sludge.

More troubling for the local agency was the failure of its TARP to stop ruining basements and closing beaches with bacteria-laden floodwaters. In fact, the problem of storm runoff overwhelming its Deep Tunnel only got worse during the next two decades of explosive expansion of the metropolitan area. As the MWRD's ability to dig more tunnels and reservoirs fell further and further behind the pace of sprawl-induced flooding, its big-technology strategy of water management would come back into the political arena of civic discourse.[15]

During this period, the remarkable resurgence of the river's biotic community boosted the mobilization of support for further enhancing its natural habitat. Of course, this reciprocal process of interaction between physical appearances and psychological attitudes was also driven by human agency. Impersonal structural shifts in economic and demographic patterns of settlement caused both a "multicentering" of jobs and homes throughout the metropolitan region and a refocusing of urban culture and leisure activities in the cosmopolitan downtown. Individual choices made by investors, developers, tenants, and buyers also played a role in reclaiming the river for recreational use.

As this constituency of stakeholders grew rapidly in the nineties, it would come increasingly into conflict with the MWRD over its current policies and future plans. The state and national governments were also becoming more fully engaged by enforcing their pollution control laws. With increasing frequency, their environmental protection agencies (EPAs) would clash with City Hall over local water management strategies. A quarter century after the passage of the Clean Water Act, the Chicago River would finally begin to receive its protective battery of defenses against pollution.[16]

Crisis-driven decision making during another episode of severe flooding in 1986 proved to be pivotal in directing the course of public policy for the next quarter century. During the last two weeks of September, 14–16 inches (36–41 cm) of rain fell in northern Illinois, providing the preconditions for flooding by saturating the ground and raising river levels. When cloudbursts struck on October 1 and 4, the entire metropolitan area was inundated, putting four thousand homes underwater and causing $35–40 million in damages. After the Chicago River rose five feet (1.7 m) above normal, the MSD opened the gates three times, discharging polluted stormwater into the lake.[17]

The president declared Chicago a disaster area, the governor sent in the National Guard to relieve exhausted first responders, and the MWRD doubled down on TARP. Proposing more of the same to solve the shortcomings of its big-technology approach to water management, it now called for 110 miles (117 km) of tunnels and three giant reservoirs at an additional cost of $2.7 billion. For eight years preceding the flood-induced crisis, however, its plan faced rejection after rejection by the federal government's accountants.

Afterward, the politicians overrode the bureaucrats to begin construction of the "flood control" phase of TARP. By the end of that rain-soaked week in October, a deal had been worked out in Washington for a new funding formula for the first project of the eight-year plan. In exchange for sweetening the state and local governments' share of the costs from 20 to 35 percent, Congress authorized $55 million to start building reservoirs to hold another billion gallons (3.8 billion liters) of Deep Tunnel overflow.[18]

Like the early grassroots activists, the few voices of protest against big-technology solutions to environmental problems during this period went largely unheeded. A month after the latest "worst flood disaster ever," the *Chicago Tribune*'s environment writer interviewed several experts on alternative flood control strategies. Significantly, none was from the MWRD. David Hunter, an Army Corps engineer, drew a historical dividing line marked by the publication of *Silent Spring*. The traditional "structural" approach was to refashion nature into an envirotechnical machine of flows. Since Carson's book, however, he had witnessed the alternative, "nonstructural" approach of designing with nature gain support within every level of government. For example, Donald Hey, the head of an environmental planning firm, was helping suburban Gurnee return a three-mile (five-km) stretch of the Des Plaines River back into "a marshy wetlands—the kind of swampy stream that settlers encountered in the 1860s." In another suburban initiative, the mayor of Skokie was installing "rain blockers" on the street drains, which slowed the rate of runoff entering the sewers. The streets themselves served as temporary storage reservoirs. Without having to add public infrastructure, he was taking a further, ecofriendly step to prevent overflows of the sewer system. He was ordering homeowners to disconnect their downspouts from it and to let their lawns soak up the stormwater.[19]

But for the vast majority of Chicagoans, the amazing rebirth of the river following City Hall's heralded opening of the main stem of the Deep Tunnel was a convincing reaffirmation of belief in the myth of big technology. Foremost among the converts to the cause of saving the river with TARP were the big investors in downtown commercial and residential real estate. Since the early days of the first Daley administration, they had set up an insider organization, the Chicago Central Area Committee (CCAC), to promote their plans for the future. In the city's long tradition of private plans and public policies, the new mayor had anointed it the supreme planning authority. Moreover, he created a city department to rubber-stamp its vision of high modernism with an official seal of legitimacy.

Following the guidelines of the CCAC's original 1958 master plan of urban renewal, Mayor Daley had largely fulfilled its goals before his death in the mid-seventies. The business elite had envisioned an all-inclusive makeover of the CBD, ringed by the construction of a high-security cordon sanitaire of luxury apartment towers. In 1974, the topping off of the world's tallest skyscraper, the Sears (now Willis) Tower, marked the symbolic completion of downtown's transformation into a world-class business center.[20]

After a five-year recession lull, Bertrand Goldberg's River City announced the beginning of a new wave of construction frenzy in the city center. His project fit into the master plan by helping to fill in remaining gaps in the ring of "defensible space" around the CBD. Its southwestern and northwestern flanks on both sides of the river were lined with obsolete relics of an industrial past. These corridors of light and water became especially attractive sites for redevelopment. By the late eighties, at least thirteen major projects including ten thousand residential units were coming off the drawing boards and taking material form. They ranged from the erection of the mega-scale "Cityfront Center" at the mouth of the Main Branch to the conversion of the Montgomery Ward warehouses into a mini-neighborhood on the North Branch. And on the South Branch below River City, developers were putting the final touches on a blueprint for a mixed-use "Chinatown Square" of commercial offices, retail shops, premium townhouses, and senior citizen apartments.[21]

By then, the building boom had turned the riverfront into a speculative land rush. What a CCAC planner only a few years earlier had termed City Hall's policy of neglect as "laissez-faire" was now considered to be unacceptable chaos. With support from the FOTR, the elite businessmen's group drafted a riverfront protection ordinance. The incoming administration of Mayor Richard M. Daley duly ushered it through the city council. "By following these guidelines," he trumpeted in March 1990, "we will transform our river into the centerpiece for an expanded downtown . . . a beautiful tourist attraction and recreational development." Yet following local traditions of talk-but-don't-act, almost a decade would pass before he budgeted any city money to help achieve this goal. In the meantime, the new regulations still gave private planners a relatively free hand to design the required public access spaces.[22]

The main beneficiaries of their work were the inhabitants of the brand-new offices, stores, homes, entertainment venues, and light manufacturing facilities. These urban pioneers represented a unique wave-in-reverse of white flight from the city. On the contrary, an exodus from the boredom of the suburbs was under way among both older empty-nesters and younger, white-collar singles and couples without children. Their attraction to the river covered a broad range of attitudes toward the environment. For some cliff dwellers, a beautiful river view served as a status symbol, a three-dimensional wallpaper for their glamorous lives. For other denizens of these tower blocks, the riverfront walkways and open plazas offered an oasis of healing, an escape outdoors from the surreal interiors of their workaday lives.

Others sought to find a hybrid, urban nature in the river's industrial archaeology of rust. In the spring of 1985, for example, an MWRD trustee and founding member of the FOTR, Joanne Alter, stood at the northern tip of Goose Island, on parkland she helped create. "Look at this view," she declared, pointing toward the North Avenue Turning Basin. "Isn't it spectacular? It's amazing how scrap metal can be so interesting." In coming years, in fact, the avant-garde would portray these abandoned industrial sites as pop art landscapes of urban doom.[23]

And finally, there were growing numbers of recreational users who looked at the recovering river as a generator of wildlife in the city. Five years after its rebirth, the US EPA upgraded its status from "toxic" to "polluted." Biologists found plant life on the bottom of about a quarter of the North Branch, where none had been found a decade earlier. And under the surface of the water, where formerly only sludgeworms, leeches, and carp could survive, now there were at least seventeen species of fish, including bass, perch, minnows, and bullheads. Waterfowl also returned and began nesting along the river's banks, adding to the diversity of its ecosystem. A year later, in the spring of 1991, beavers would be spotted not only in the Des Plaines Wetlands Demonstration Project but also on the edge of the CBD, swimming in the South Branch.[24]

During the remainder of the decade, in turn, this self-reinforcing process of biotic regeneration had a cumulative effect in changing more and more people's attitudes toward the urban environment. Their perceptions of the Chicago River underwent a decisive shift from an industrial corridor to a recreational parkway. Unfortunately, this kind of attitudinal change of mind cannot be measured with precision.

Though anecdotal, one of the many stories of grassroots activism helps expose how this reciprocal process of environmental and attitudinal metamorphosis worked to build a political movement to save the river. At the FOTR's fourth annual "Great Chicago River Rescue Day" on June 2, 1996, its organizer Joyce Peralta noticed that "every year you have to work harder to find trash." Five hundred "river rats" still hauled away ten tons (91,000 kg) of trash on that Sunday, "peel[ing] away layers like an arch[a]eologist." Yet as the river corridor improved in appearance, people no longer thought of it as the place to get rid of their (illegal) waste. "People aren't dumping in the clean spots," Peralta observed. "The more we clean our areas, the less people will use them as a garbage can."[25]

Twenty years after the passage of the Clean Water Act, the bar was even rising on the minimum standards of toxic industrial effluents, which the national EPA was willing to tolerate in the Chicago River. By then, it had already forced many of the polluting factories in the metropolitan area to stop dumping their liquid wastes into the sanitary district's waterways. Instead, the companies were required to pretreat their effluents before they entered the sewer system. The US EPA had been cracking down hard on the heavy industries along the Calumet River.

This industrial landscape had already undergone the most striking ecological transformation of the six watersheds in the metropolitan region. In the mid-eighties, the MWRD began promoting a recycling of the Calumet–Sag Channel subdistrict with a market plan of land sales. Its planners envisioned a "River Edge Renaissance" of private development on 1,200 acres (486 ha) of its land, with about a third set aside for parks and recreation. Over the next decade, the cumulative results of deindustrialization, TARP, and enforcement transformed this depopulated environment into a more natural than suburban landscape. By 1993, a scientist could see "sponges, crayfish, gnats, dragonflies, earthworms and turtles" in the Grand Calumet River. Chinook salmon and steelhead trout could also be found there.[26]

In effect, the MWRD became the last polluter standing as the federal EPA became relentless in tracking down and prosecuting industrial violators of the Clean Water Act. In 1997 alone it won court orders against five big companies on the Grand Calumet River. These companies had to begin paying $5.5 million in fines that went into a special cleanup fund for its environmental remediation. After the triumph of the "Save Our Lake" crusade in the early 1970s, in contrast, the Illinois EPA had quickly snapped back into playing their traditional role of political noninterference in the local government of Chicago. Saving its rivers depended on Washington, a spokesperson for the Citizens for a Better Environment confirmed, because "the Illinois EPA is blindly committed to a voluntary pollution prevention program which is obviously not working."[27]

Chicago put Illinois at the top of the list of states in the amount of toxic chemicals dumped into its waters by local sanitary authorities. In 1992, for instance, they poured a record-high total of 72.2 million pounds (32,750,000 kg) of dangerous compounds into them, with the vast majority contributed by the MWRD. In contrast, the amount added to Illinois waterways by private industry was on a steady downward trend. It dropped from the previous year by 11 percent to a record low of 5.7 million pounds (2,585,000 kg). As these numbers underscore, the MWRD stood out as the biggest polluter of toxic chemicals in the Great Lakes and Mississippi River basins. It also drew the attention of the national EPA because it fell further and further behind other sanitary districts in upgrading their treatment technologies to disinfect organic wastes.

Except for City Hall, a self-reinforcing process of environmental restoration and stakeholder mobilization had produced a sea change in public perceptions of the Chicago River. As the Clean Water Act approached its twenty-fifth anniversary, the planning department kept insisting that "it's primarily a working river." It denied the petitions of the swelling community of downtown residents to rezone the remaining patches of industry on the riverfront for redevelopment. Wielding virtually dictatorial power over local government, Mayor Daley had successfully sustained a united front of defiance against the law for a decade. With a Democratic administration in Washington, he had enjoyed an exemption from bringing the MWRD into compliance with its standards of water quality.[28]

Save Our Waters: The Dawning of Urban Ecology, 1998–2012

But even this all-powerful city boss was not strong enough to hold back the environmental movement when it came to saving the Great Lakes. City Hall's refusal to disinfect the one billion gallons (3.8 billion liters) of partially treated sewage it poured into the waterways every day became the flashpoint of conflict in this phase of the battle to reclaim the Chicago River. The MWRD's failure to stop discharges of polluted floodwaters into Lake Michigan again came under the spotlight of public scrutiny. With each beach closing due to TARP overflows, political pressure mounted for enforcement of the law from both the top-down of the federal government and the bottom-up of grassroots activism. In sharp contrast to the earlier "Save Our Lake" crusade, however, the focus of attention as Public Enemy Number One became a local government rather than a private industry.

A generation of changes in environmental conditions, aquatic sciences, and popular perceptions of the Chicago River had accumulated to the point of shattering the city's myth of big technology. In its place, a political consensus began to emerge out of new attitudes toward Chicago's environment that reunited the river and the lake as a single, interdependent ecosystem. The rebirth of the Chicago River after 1985 turned out to have been a mixed blessing. On one hand, the rebounding albeit rapidly evolving biotic communities on both sides of the locks had led policymakers and recreational users alike to reenvision them in more holistic terms. On the other, invasive species could now cross back and forth across what had previously been a toxic dead zone between the Great Lakes and the Mississippi River basins.[29]

The period after 1998 became characterized by the resistance of the Daley administration against bringing Chicago into compliance with the standards set in the Clean Water Act. As pressure mounted from both the grassroots activists and the EPAs, the mayor stubbornly held on to the old division between the lakefront as a showcase of the city beautiful and the riverfront as a plumbing fixture hidden out of sight and out of mind. Refusing to bear the costs of fully disinfecting its sewage, Chicago became the only big city in the nation holding out against upgrading its waterways to recreational use.[30]

Moreover, the MWRD became the single worst polluter of Lake Michigan. Suburban sprawl continued its relentless paving over of the land, which caused ever greater, faster, and more frequent runoffs during thunderstorms. The $3.5-billion-and-counting, still incomplete tunnel and reservoir system filled up, the overflowing river threatened to flood the CBD, and the MWRD was forced to open the locks, discharging millions of gallons of pollution into the lake. The beaches were closed to swimming, leaving their frustrated sun worshippers to bake on the sand without relief.

Caught in this crossfire of opposition from the grass roots and the federal government, Mayor Daley was still able to forestall a capitulation until his retirement from office. The master politician to the end, he made a narrow escape from being held accountable for what Henry Henderson of the National Resources Defense Fund labeled a "ridiculous failure of government."[31] Nevertheless, the environmental reform movement had been able to penetrate the smokescreen of "junk science" thrown up by his loyalists in charge of the MWRD. In the November 2010 elections, green candidates for its board of commissioners had pledged to reverse the loyalists' "good enough" approach toward the waterways. The election returns ensured that a new majority would vote in favor of the EPA's plans, when its final orders of compliance were issued the following May. By then, a new mayor with long-standing credentials as a eco-guardian of the Great Lakes, Rahm Emanuel, had moved into City Hall.[32]

Despite his willing submission to the EPAs in 2011, invasive species on both sides of the barrier between the river and the lake have further complicated the interrelated problems of flood and pollution control in the Chicago area. Consider the Asian carp, a "monster fish" that both eats and out-eats all of its rivals. Set loose in the Mississippi River, it threatens to enter the Great Lakes and destroy its thriving sport fishing industry. At the same time, the explosive growth of zebra mussels in Lake Michigan and the Chicago River is also upsetting the balance of nature in unpredictable ways, like the alewives fifty years ago. These unintended, dire consequences of human interaction with the environment have finally forced policymakers to start planning in holistic ways that treat the two bodies of water as a single, interdependent ecosystem.

The battle for the reclamation of the Chicago River is over. Inspired by Rachel Carson's book, the reform movement for its protection from pollution won. Over a half-century period, a reciprocal process of nature restoration and stakeholder mobilization changed public attitudes toward the urban environment. Perceptions of the city's waters shifted from a bifocal vision of front-yard beauty and back-alley eyesore to a holistic, ecosystem point of view. Mental maps of downtown moved the river from the edge to the middle of this expanding zone of cosmopolitanism. Coming full circle, the rebirth of wildlife in the Chicago River has inspired environmental reformers to set aside patches of well-manicured beaches on the lakefront to evolve back into their natural habitats.

According to the consent agreement, Chicago will make its rivers clean enough for swimming. The MWRD will install the required equipment at its sewage treatment plants to bring the Chicago River into compliance with the Clean Water Act. In return, the state and federal governments agreed to help it pay for more digging until TARP reaches a storage capacity of nearly eighteen billion gallons (68 billion liters) over the next twenty years. The reformers also used the federal courts to force the state and the local governments to implement strategies of flood control that include nonstructural, green technologies.[33]

Executing this U-turn in public policy, Mayor Emanuel also reversed the course of City Hall in planning the future development of the waterways. He has taken a leadership role in fulfilling Goldberg's vision of the Chicago River as a front street or parkway of nature in the city. On the heels of City Hall's surrender to the EPA in September 2011, he announced plans to give the river greater priority, beginning with $16 million to build four boathouses, complete with bicycle, canoe, and kayak rentals. "The truth is," he reflected, "because of our lake, we've only thought of it as our front yard. I now want to turn our attention to our backyard and have that same vision [for our riverfront] neighborhoods. . . . I want that river to be for all of us what the lake has been." Drawing on "starchitects" like Jeanne Gang, Mayor Emanuel has turned City Hall into a full partner with the private sector in charting the course of riverfront development.[34]

Perceptions of the river as an urban wilderness are having their own feedback effects on changing attitudes toward the environment. In the spirit of the city beautiful movement of the Progressive Era, virtually all of its beaches and harbors had been turned into highly ordered recreational and leisure spaces. The dramatic rebirth of the river's native flora and fauna created a powerful impression of the successful regeneration of nature. It inspired environmentalists to envision the lakefront too as a wild place. "Chicago's future," the *Sun-Times* stated, "is that of a post-industrial environmentally friendly 'green' city."[35]

Their demonstration patches on the lakefront are duplicating the renaissance of the river's biotic community. Whether on Northerly Island in the downtown area or the Pratt Avenue Beach on the far North Side, these nature preserves have become equally amazing places of dune grasses, wildflowers, and waterfowl not seen, in many cases, for over a century. The return of these plants and animals to their shoreline habitats, in turn, is drawing new groups of stakeholders in the environmental protection of Chicago's waters.

Victory in the long battle to reclaim the river for recreational use has triggered the start of a new one to save the ecosystems of these two great freshwater basins from destroying each other. In 1998 Congress began a national counterattack against the Asian carp and a rogues' gallery of other alien invaders. In 2010, it put the Army Corps in charge of this mission when it created the Great Lakes and Mississippi River Interbasin Study project. Besides building a fourth-generation electric fence, the Corps has been monitoring life below the surface with a full battery of tests.[36]

While the reams of data its task force has been collecting for scientists to study are first-rate, its engineers' continuing pursuit of the myth of big technology is less reassuring. Will their recommendations give ecosystem approaches top priority? "To our knowledge," a Corps brochure on the electric fence boasts, "our barriers are the largest of their kind in the world and the only one on a highly-trafficked, commercially-navigable waterway." It takes special pride, moreover, in its structural approach of building an ever bigger and more powerful fail-safe machine. It still maintains a mind-set that does not envision the waterways' use for recreational purposes, only "treated wastewater, storm water management, and navigation." In 2011, Congress moved up the deadline for a final report by two years, to 2013. Its impatience expressed the nation's anxiety about our ability to protect our waters from the unintended consequences of efforts to turn them into an envirotechnical system of total control.[37]

CHAPTER 14

SHARED WATERS, SHARED CONCEPTIONS?

Two Cities on the River Rhine on the Long and Winding Road to Urban Sustainability

Michael Toyka-Seid

When on February 27, 1945, about 80 percent of the inner city of Mainz, an Episcopal and political center in southwest Germany, was destroyed by Allied bombing, some inhabitants apprehended nothing less than the "end of Mainz."[1] For almost two millennia Mainz had been an industrious and flourishing city, marked by its close relationship to the river Rhine and proud of the famous prospect of the city towering over the river and its banks. The firestorm left the city and its commercial harbor in ruins and shambles, the connection between city and river seemingly broken irrevocably.

When almost to the day sixty years later American president George W. Bush visited the rebuilt city, local politicians, media, and tourism businesses sensed the chance to present a rejuvenated and revitalized city to the outside world, the "chocolate side" of Mainz.[2] But what was meant to become a major marketing event for one of Germany's most historic sites turned out to be a ghostly spectacle, with streets closed for traffic, evacuated buildings, and places void of people. The president and the accompanying media entourage saw no sign of the renovated Rhine prospect of the city, nor could they enjoy the pleasure of a boat trip on Europe's most praised river along the city's riverbanks—marine vessels were on guard blocking all shipping on the Rhine River. A chance to present to the world the rebuilt city center and the progress made over the previous decades with regard to the embellishment of the city's waterscape was finally lost when the president crossed the Rhine to visit

American soldiers in neighboring Wiesbaden, which congratulated itself for having been left out of the presidential tour.

Not only on this occasion did expectations and perceptions attached to city-river relations collide with political realities and everyday routines. The emotional and historical relationship between cities and rivers in general has to be seen on a longer and multifaceted timescale, however. This chapter thus asks for the place of water in the modern city in the context of two German cities on their way to sustainability, or better: to a more sustainable way of producing and reproducing the city.[3] The water is the Rhine, the most prominent part of the visible waterscape of Mainz and Wiesbaden, two neighboring but still rather distinct cities. The Rhine connects these two cities and at the same time separates them from one another. Each city has its own history in connection to the great waterway, and neither of them would have come into existence without its proximity to the river Rhine. In recent history and perhaps even up to now both cities represent "their" river in different ways and deal with it in many different forms. Of course, a major waterway like the Rhine can hardly get "lost" in a physical sense from public view. But the cultural and economic urban presence of the river and its banks undergoes major changes, depending on general urban developments, changing aesthetic perceptions, and ecological conceptions of urban waters. In this sense the post–World War II era and the breakthrough of the "environmental age"[4] proved to be decisive years for the (re-)appropriation and (re-)gaining of the Rhine by the cities of Mainz and Wiesbaden. To shed some light on this oscillating development is the main aim of this study.

Mainz, Wiesbaden, and the Rhine: (Non-)Relations in History

The development of cities and the organization of the life of city dwellers are hardly imaginable without open water. Running waters like rivers and streams played a decisive role in the foundation of cities around the world and have been essential for their survival and flourishing over the centuries.[5] Rivers facilitated infrastructural and industrial development and had manifold impacts on the life of urban populations. However, with the rise of the hygienic movement and the miasma paradigm from the mid-nineteenth century, rivers, brooks, and other urban waters increasingly "disappeared" from the urban environment. Since the advent of the ideal of the "dry" city of the nineteenth century, rivers were hemmed in by industrial or port usages, and brooks and little streams covered up and integrated into the sewer network.[6]

Only recently, as several chapters in this book demonstrate, has this process started to be reversed: long-forgotten streams have been brought back to light in the search for "nature" in the city, and rivers and their banks have become cultural, housing, and recreational areas again in many cities of Europe, North America, and East and Southeast Asia.[7] The late twentieth century in this sense can be seen as a watershed in the changing relationship between city and river.

This chapter, which relies mainly on source material from local archives and civic collections and on interviews with local decision makers, asks for reasons, motives, and manifestations of this process. It will examine perceptions, usages, and rehabilitation strategies of two medium-sized German cities with regard to their shared water, the Rhine. Whereas both cities nowadays at first glance seem to be quite similar with regard to their physical size and population, geographical location, resource availability, political function, and more, their historical relationship to the Rhine has been quite different, as are still today their everyday usages and perceptions of the river.

Both cities are commonly regarded to be closely connected to the river Rhine. Mainz and Wiesbaden occupy opposite banks of the river and brand themselves in their official self-representation on their homepages, by information leaflets, and other local council publications as "a city on the banks of the river Rhine."[8] Indeed, the growth of the Roman foundation Mogontiacum in medieval times, apart from the role of Mainz as an outpost of early Christianization, relied heavily on the trade privileges accorded to this important anchorage ground in the Rhine valley.[9] Its industrial breakthrough in the nineteenth century after a long period of urban decay and the loss of political importance once again was closely connected to its position on the juncture of the rivers Main and Rhine and to the building of large port facilities on the riverbanks of the city.[10] On the other hand, Wiesbaden, which also claims to be a Roman foundation, became a riparian town only with its early twentieth-century city enlargements around the former independent harbor Schierstein; the historic classicist center of the city is approximately five kilometers away from the river.[11] The largest part of the riverbanks in the suburbs Mainz-Kastel, Mainz-Kostheim, and Amöneburg on the right-hand side of the river was handed over to Wiesbaden only after World War II by the Allied administration—AKK, as the three suburbs are commonly called, still today remains a controversial subject in the relationship between the two cities.[12] For centuries the city's fortune depended on the hot springs in the center of Wiesbaden, running from the slopes of the low mountain range of the Taunus north of the city—water that made Wiesbaden in the late nineteenth century one of the most fashionable spa resorts in Europe.[13] For the self-sufficient and rather wealthy Wiesbaden residents, the proximity of the Rhine with its industrial imprint and smoke nuisances was more of a threat than a desirable neighborhood. At the beginning of the twentieth century it was the declared aim of the local administration to keep Wiesbaden free from smoke and industrial development, so as not to endanger the environmental qualities of the famous spa.

There are more reasons to perceive the Rhine, four hundred meters wide at this point and the busiest waterway of Europe, more as a dividing line than as a common resource for these two neighboring cities. While the two cities are in close proximity to each other and to the riverbanks, their geology and morphology differ significantly. Of utmost importance in the context of this chapter is the relative water scarcity of Mainz's hinterland; this results in a water catchment area with very few direct tributaries to the Rhine, a small system of covered streams and rivers in the inner-city area and a system of meadows, artificial ponds, and springs in the outer city.[14] In part this water scarcity can also be explained by measures to straighten and tame the Rhine River in the early nineteenth century.[15] In contrast, Wiesbaden profits from its proximity to the Taunus low mountain range. A multitude of small rivers and rivulets run from those highlands through Wiesbaden toward the Rhine and contribute to the advantageous climate of the city. The floodplains of these small rivers can cause damage to the city as well, although not in a comparable way to the destructions caused by the occasional flooding of the Rhine. Seen from the political point of view, it has to be mentioned that both cities became capitals of their respective federal states after 1945 within the German Federal Republic; both thus boasted their own state legislation and jurisdiction. In recent years, however, local and regional institutions have lost some of their importance compared to the growing influence of national and European regulation. Multinational boards like the Commission Centrale pour la Navigation du Rhin or the International Commission for the Protection of the Rhine, federal organizations like the Bundesanstalt für Gewässerkunde,[16] and regional bodies like the Landesamt für Umwelt, Wasserwirtschaft und Gewerbeaufsicht Rheinland-Pfalz[17] are major players today with regard to economic and environmental concerns related to this international waterway. On the other hand, infrastructural systems like waste disposal and water supply end on the riverbanks and are locally organized.[18] Thus, there are good reasons to choose Mainz and Wiesbaden for a comparative study on how two neighboring, similar, but still different cities transformed their relationship to the Rhine in the second half of the twentieth century.

Mainz and Wiesbaden—Intrinsic Logic and Socionatural Sites of Urban Development along the Rhine

The considerations of this chapter are closely connected to the research questions of an interdisciplinary research project on cities on their way to sustainable development conducted at Darmstadt University of Technology.[19] This research question was originally inspired by the Darmstadt "Intrinsic Logic of Cities" approach to urban development.[20] In this project on the visible water of Mainz and Wiesbaden I argue that urban sustainability, though in itself not a "historical" concept, will be an interesting research area in the quest for this "internal logic" or—maybe—some other urban "logic." Questions of critical junctures, of turning points, and consciously taken new directions have in recent years enlarged our knowledge about ways in which cities over the years have dealt with their natural and nonnatural environments. How cities perceive and evaluate their problems and chances, how they decide and act are also major ingredients in the analytical theory of the "intrinsic logic of cities."[21] A starting point for this way of looking at cities is obvious divergences, which can be observed when looking at the ways in which cities act and react, sense and communicate. Those differences arguably have their roots in a broad sample of physical, social, and cultural imprints, which can be identified at the local level.

An important aspect for the understanding of this "intrinsic logic" is the (if possible interdisciplinary) comparison of two or more cities. "The individuality of cities grows on historical storytelling and experience, but it [is] also shaped when cities compare themselves to other cities."[22] Thus the "logic" of an individual city has its roots in its specific natural, social, and cultural circumstances, but is also dependent on other cities to be recognized.[23] In urban history comparing cities has of course been a useful concept from the very beginning of this historical subdiscipline, as doing so helps us see the distinctiveness of the individual city, but may also help us detect similarities and patterns that apply to a sample of cities or are relevant for all cities.[24] This historical study of course cannot claim to fulfill all of the necessary prerequisites of the "intrinsic logic" approach mentioned above, but it may hopefully add to a better understanding of historical coinages of modern city-river relations.

That urban environments, though quite distinct from all notions of wilderness and "pristine" nature, deserve a prominent place in the history of the environment has been argued convincingly by Martin Melosi in the early 1990s, claiming a place in environmental history for "the physical features and resources of urban sites (and regions) which influence and are shaped by natural forces, growth, spatial change and development, and human action."[25] Since then the study of urban ecologies and the urbanization of nature has played a substantial role in environmental historiography.[26] Cities are part of the physical world, influenced by natural factors and themselves leaving their ecological footstep in their more "natural" hinterland—a perspective stressed by this chapter on the ways in which two cities try to become sustainable.

The question of how environmental history can avoid a disjunction between human society and the "natural" environment has been on the agenda of environmental historians for quite a while. Actor Network Theory (ANT), with its stress on actants and hybrids of nature and culture, has been one of the many fruitful approaches in this respect.[27] While nature is now seen as part of society (and thus to have a history), human activity is considered to occur amid and as an inherent part of natural, material, and sociotechnical arrangements.

Methodologically, the project sketched above follows the ideas of the concept of socionatural sites, advocated by Verena Winiwarter and Martin Schmid and applied in a couple of research projects centering on the Vienna Danube, conducted by different research groups at the Viennese Institute for Social Ecology.[28] Winiwarter and Schmid define the purpose of environmental history as the search for "the relations between men and environment in the past."[29] All those relations shall be seen as material as well as symbolic interactions. To avoid a disjunction between human society and the "natural" environment, Winiwarter and Schmid propose a concept that relies on the constant intertwining of these spheres. Building on a concept developed by American philosopher Theodor Schatzki,[30] the authors suggest concentrating on the nexuses between arrangements and practices in the relation between men and "nature"—of course almost from the beginning understood as nature in its various man-made metamorphoses. While the historical process is fed by the growing diffusion of the natural sphere through human action, Schatzki and, in his wake, Winiwarter and Schmid advocate a perspective that leaves space for the "natural" in human history as well. By stressing the natural component in their concept of the "socionatural site," Winiwarter and Schmid want to make it clear that environmental history always has to keep the natural as well as the social and cultural aspect of environmental processes in mind.

The metamorphosis of socionatural sites in cities will hopefully foster the understanding of the complexities of urban environmental policies. The corresponding questions for urban perceptions of and intervention in the natural environment illuminate the ways in which the historical heritage and mental stamps of cities become relevant forces in urban sustainability processes. The concept thus will help us to conceive of nature as an influential part of the urban environment and an integral requirement for all urban environmental politics.[31]

The following considerations concentrate, in the categories recommended by Winiwarter and Schmid, on receptions, representations, and work invested in city-river relations in Mainz and Wiesbaden in the second half of the twentieth century. This is not to say that the cities' "natural" endowment does not still influence and differentiate urban development in Mainz and Wiesbaden. The damages caused by floods in 1965 and 1988[32] in the parts of both cities adjacent to the river, the reliance of the cities' water supply on groundwater fed by the Rhine,[33] and the various environmental problems traveling down the river demonstrate how important the natural side of the story still is in the modern world.

Mainz and the Reconstruction of the Riversides

Although the civic "Leitbild" of urban sustainable development is usually closely connected with the breakthrough of environmental awareness and local environmental politics in the early 1970s, followed by the Rio conference of 1992 and the aims proposed for urban development with Agenda 21 and beyond, the necessary starting point for the following discussion is the immediate postwar years. Mainz, heavily destroyed by Allied bombing, deprived of approximately half of its former territory, and thrown back for the first time in its long history onto the left-hand side of the Rhine, rather early aimed at regaining its former river-based importance and the natural beauty of its riverbanks.[34] The boom of the German economy since the middle of the 1950s supported the quick rebuilding of shipping and port facilities on the Rhine. The reclamation of the riverbanks and of the historic Rhine scenery for the badly damaged inner city proved more complicated. In 1958 a competition for the architectural and landscape renewal of the riverbanks was held. Fifty years earlier, the leader of the Christian Democratic Party of the city declared, the Mainz riverbanks had been pivots of modernity; now they had fallen behind with regard to modern ideas of city planning.[35] The outcome of this competition was meager: some artwork embellishments along the small river paths. In the discussions of those days, mirrored in a dense newspaper coverage of the municipal debates, a feeling of deprivation and, still worse, lack of perspective for the city-river relations transpired: Not only had the beauty of the medieval city been destroyed by Allied bombing, the city felt "cut in half" by the loss of its historic counterpart on the other side of the Rhine. It needed the two thousand years' jubilee of Mainz in 1962 and a decided attempt by the city officials to delegate the rebuilding process of the city to the hands of a professional city planner, the renowned Ernst May of Frankfurt, to put the renewal of the inner city and the Rhine prospect more effectively on track in the 1960s.[36]

The ambitions of Mainz to regain and improve its connection to the river in the 1950s and 1960s was accompanied by a rather disinterested stance in Wiesbaden. The city, which had survived the war in comparatively good shape, showed relatively little interest in its new riparian environment in those years. The industrial character of the new AKK suburbs may have been one of the reasons for this neglect, as Wiesbaden had prided itself for most of its history as a city of healthy environments and leisure, not of industrial endeavor and zeal. However, in 1962, notwithstanding the still-fierce struggle with the "enemy" on the other side of the river over the "lost" suburbs, the Mainz administration proposed a cooperation to build a new promenade along the riverbanks. Soon first meetings of representatives of both cities took place and, after some preliminary approaches with occasional meetings of city officials in the mid-1960s, in 1969 an official "Neighborhood Treaty" between the two cities was agreed on.[37] Pushed forward by the planning department of Mainz, plans for a tree-lined promenade on what was by this point the Wiesbaden side of the river Rhine and even a footbridge near the mouth of the Main River to better connect the two cities were brought forward by the city officials. But these far-reaching plans met with skeptical reactions in Wiesbaden. On the right-hand side of the river, city officials were reluctant to refurbish the whole riverbanks according to modern "environmentalist" ideas, as those conceptions clashed with their emerging plans to maintain the industrial character of the banks.[38]

Thus, the idea of establishing a comprehensive plan for a "green" configuration of the riverbanks did not bear fruit in those days. When in the early 1970s both cities established local boards for the protection of the urban environment, other problems such as the purity of the urban water supply or dangers connected to nearby nuclear plants were in the foreground. Questions of urban green and of a more natural design of the cities' waterfronts were put aside for quite a while. They also remained in the remit of the cities' boards for urban green and landscape planning rather than being integrated in the environmental sphere. Instead of creating wholesale scenarios for a more natural arrangement and use of the riverbanks, the Mainz city council further pursued for several more years its attempts to endow the rather uninspiring lawns along the river with modern sculptures and monuments, thus continuing its policy of embellishing the rebuilt inner city with artworks bearing memory to the city's golden past.[39]

The end of this first phase of redesigning the border zone where city and river meet was finally reached in 1972. In this year plans for a new "City Crown" for Wiesbaden, consisting of a high-rise administration building, hotel, and multifloor residential buildings on the side of Mainz-Kastel were presented.[40] Alas, the hopes of the Mainz city fathers to gain at least a more impressive view of the (lost) other side of the river Rhine from their new town hall built almost on the waters of the river never materialized, as the cooperative effort of the two cities faded away quickly. For many years to come Mainz residents would have to look on rundown industrial estates and minor commercial activities on the right-hand side of the river. Not even repeated attempts by local bodies like the Council of Amöneburg suburb to put the question of visible amendments of the riverbanks on the local agenda could activate the rather hesitant Wiesbaden administration. This was also due to a growing interest of the former "Weltkurstadt" in its industrial assets. While questions of small river rehabilitation, greening the city, and laying out new nature protection areas became regular topics on the agenda of the Wiesbaden town council in the environmentally eventful 1970s, the Rhine was seen more and more as the place where economic needs and interests of the city should have priority. Plans brought forward by the representatives of the riverside suburbs Kostheim and Amöneburg to provide a planning scheme for the riverbanks aiming at remodeling the area as a public leisure zone were strongly opposed by the local conservative party, the Christian Democratic Union (CDU), which gained the majority of votes in what was up to then a stronghold of the social democrats. For the CDU it was obvious that leisure and industry had to go hand in hand on the banks of the river Rhine, even at the cost of industrial pollution and aesthetic constraints.[41]

The city of Mainz could only watch from afar the discussions of their counterpart on the other side of the river. The momentum of cooperating for the sake of environmental and planning perspectives, seemingly well under way in the 1960s and early 1970s, faltered, although political discrepancies between conservatives and social democrats on both sides of the river were decreasing in those years. But of course not only the visible ugliness of the opposite river side was painful to the senses of the leaders of Mainz. The industrial pollution of the river, the resulting regular stench, and the damage caused to fish populations through waste from the chemical industries upstream on the Main and Rhine had been a constant subject of local debates since the late nineteenth century.[42] Immediately after the war a special board to care for the cleanliness of the Rhine water had been established, the Commission to Avoid the Pollution of the Rhine. But instead of improving, the situation worsened with the ascent of the Swiss and German chemical industry in the 1950s and 1960s. The lament of the local newspaper *Mainzer Allgemeine* in 1969 about the Rhine being Europe's most famous sewer is just one of very many public references to the river's poor state in those years.[43] In the late 1960s and early 1970s there was widespread agreement that no single city could reasonably take measures against the border-crossing pollution. Only a combined international strategy would be able to tackle the problem.[44]

The rehabilitation of the Rhine is a well-researched example of the early successes of the environmental movement in the 1970s and 1980s. Since the beginning of the industrial revolution, the degradation of the natural environment had been an integral and mostly accepted aspect of the industrial age. Rivers and small streams had been sacrificed for progress—even a major waterway like the Rhine had suffered severely from more than a century of heavy industrial pollution. The rehabilitation of the river in the 1970s thus came to be seen as a demonstration that the destruction of the natural habitat of men was not an irreversible process.[45] Mainz and Wiesbaden—the latter as usual rather reluctantly—declared that the quality of Rhine water near Wiesbaden was not as bad as elsewhere,[46] claiming this to be due to the efforts of the city, whereas actually the measures taken by state, federal, and European regulation had been much more instrumental in this. But the cities also had their own share in the complex process of clearing the waters of the river Rhine and drawing attention to this environmental nuisance. The installation of a pollution-measuring gauge near Konrad Adenauer bridge in Mainz in 1976, undertaken by the Commission Internationale pour la Protection du Rhin and cofinanced by the state governments of Rhineland-Palatinate and Hesse, was a major event in local newspaper reporting and debates in the mid-1970s.[47] It underlined the vulnerability of the cities near the Rhine, but it was also a clear indication that something could be done to improve the living conditions of people living near the river.[48] Certainly, the water quality remained precarious and much depended on the pollution prevention measures of industrial sites upstream.[49] Catastrophes like the 1986 Sandoz accident in Switzerland, when thousands of tons of red-colored waters, polluted by fire-extinguishing chemicals, swept along the Rhine,[50] deeply shook the confidence of the populace of Mainz and Wiesbaden.[51] Even bearing such setbacks in mind, one can safely state that the relationship between city and river was changed by installing this pollution gauge. It marked a turning point; since that point the cities were themselves trying, more or less energetically, to establish a new relationship with "their" river.

The small technical innovation of a regular chemical examination of the pollution of the Rhine waters can be seen as an example of how changes in the socionatural arrangements come about, how they change perceptions and usages, and finally establish new ways of constructing the relationship between men and "nature" in a given environment. Surveying the amount of pollution in river water was not a new technology in the 1970s, but it needed the awareness created by the environmental movement beginning in that time period to be seen as a meaningful and—from the point of view of the local population—comforting innovation. The setting up of the measuring station with its appeal to popular attention thus added to earlier, much more massive interventions into the natural quality of the Rhine. It arguably also deepened the understanding of the river as a complex biophysical entity not "belonging" to one or the other city, but which had to be treated and cared for by many actors in both cities and beyond. In the process of regaining the river in Mainz in the postwar period, the small addition of the measuring station to the urban water environment thus can be seen an important turning point and as a stimulant for the further reappropriation of the river.

Looking back on the development of these years it may seem rather ironic that this environmental success story started the very moment that the city of Mainz underlined its affiliation to the river Rhine in a more visible way. The erection of a new town hall in 1973 with adjacent entertainment hall and luxury hotel at a prominent spot on the riverbanks, connected to the heavily damaged historic city center by way of a rather "unnatural" concrete-based structure, marked a major step in the reclamation of the riverbanks.[52] Thoroughly disputed at the time—and ever since—the ensemble by Danish architects Jacobsen and Weitling, built in the International Style in fashion in those years, was an unmistakable statement: overcoming its postwar losses, Mainz claimed to reinvent its image and position itself as a major city on the banks of the Rhine.

The modernist addition to the scenic panorama of the city, executed in the spirit of nineteenth-century city planner Eduard Kreyßig, almost instantly raised new questions about the "natural" quality of the riverbanks.[53] While planning the new town hall the architects were content to add some little flowerpots and a small footpath along the riverbanks as "natural" ingredients to their stone buildings. But now, growing environmental awareness in the 1970s, accompanied by a longing for "nature" in the growing cities and a desire to appropriate the—by now much more agreeable—riverside for urbanites' living quality, increased the pressure on local politicians to strengthen their efforts in relation to the border zones of city and river. Nature protection societies, fishing clubs, and local initiatives like Green Mainz expressed their wishes and found ample support in local media. The Green Party, which in 1984 had entered the city council, became only three years later part of the up to then rather cooperatively working city government. As part of this, the party pressed for a stronger consideration of environmental and ecological issues and for the opening up of the riverbanks for the public. A rapid change of the situation was difficult, however. While in Wiesbaden, where the Green Party also had become an influential actor in local politics since the late 1970s,[54] large parts of the riverside were then still occupied by industrial plants, Mainz was strangulated by huge port premises, dating from the late nineteenth century, north and south of the inner city. Additionally, most of the original floodplain of the river had been destroyed by river regulation in the nineteenth century.[55] What remained served either as land to safeguard drinking water resources of the cities or, increasingly since the beginnings of the environmental movement in the 1970s, as nature protection areas, and could not easily be made accessible to the public.

Institutional changes such as the setting up of environmental boards as departments of the local administration since the late 1970s and the growing share of green votes at local elections did not produce short-term environmental advances. Only minor improvements like the re-naturing of the Maar meadows in the north of Wiesbaden or the listing of the Laubenheimer marsh in Mainz as a natural monument could be accomplished.[56] At the same time, new threats emerged, such as the expansion of an industrial area in the upstream community of Budenheim, which this small town had decided without further consultation of Mainz. This expansion directly affected the adjacent nature resort of Mombacher marsh on Mainz territory and endangered the protection of species in this resort. Such incidents made clear that sustainable improvement of the riverbanks needed a more comprehensive approach.[57]

Deindustrialization and Reappropriation of Riversides

It needed a combination of internal developments and global influences to bring about the up to now most visible and, perhaps, sustainable changes in the relationship of Wiesbaden and Mainz to their river. On the one side there was the quick and rather unexpected decline and demise of industrial companies and trading activities on the banks of the Rhine. This structural change meant that huge areas like the ones occupied by the Linde Gas Company or the Dyckerhoff concrete firm in Wiesbaden, or the port areas in the northwest and southeast of the city center in Mainz became available for urban development. At the same time the demand from local residents to reclaim the riverbanks for fashionable housing resorts, recreation, and water-based "nature" was becoming more and more vociferous, surely fed by the awareness of similar developments in many other cities around the world. As an important additional aspect to this, the Agenda 21 process, started in Rio in 1992 with its stress on the importance of locally based decision-making processes and of civil society participation, triggered new forms of public involvement in local affairs that could not be brushed aside by the local administrations.

Thus, when the demise of river-based industrial companies opened up opportunities to remodel the former toll port areas in the northwestern part of the city of Mainz, those chances were quickly grasped by urban planners. Early ideas about the reconfiguration of the area were also discussed in local nature conservation initiatives like Green Mainz and quickly found their way into the local newspapers. In 1993 the *Mainzer Allgemeine* reported that the wishes of the local populace with regard to greening the old harbor facilities would be soon fulfilled.[58] But when change did come in both cities in the late 1990s, it took the form of mostly small and rather isolated measures like the building of new upper-class housing near the water in the port of Wiesbaden-Schierstein or the old "Winter Port" of Mainz, cultural achievements like an art museum in Mainz, and the opening of recreational spaces on both sides of the river.[59] Former ideas about using the reclaimed riverbanks to foster the green image of the cities played only a minor role in the regeneration of this coveted urban area. To make things worse from the point of view of sustainable urban development, all these redevelopment projects are located in potential flood zones.[60] Thus, while the aesthetic and leisure expectations connected with the Rhine promenades on both sides of the river actually did come closer to fulfillment over the last two decades, the economic and ecological consequences of those developments must be seen as at least ambiguous.

Notwithstanding all the drawbacks and shiftiness with regard to a balancing of ideas of urban sustainability and city development issues, the development of the riverbanks still today remains a top priority of Mainz city leaders. On the other side, Wiesbaden, with its "multifocal approach," as city planning head Thomas Metz described it in a Darmstadt workshop in 2009, only occasionally turned its planning focus to the riverbanks of its industrial suburbs. There is, however, a considerable public feeling in Wiesbaden that the connection to the Rhine has not yet been satisfactorily approached by the city leaders. Not surprisingly, the Stadtentwicklungsgesellschaft Wiesbaden SEG, a board formed by the city council to promote urban development in the city, staged a public discussion in summer 2011 on the subject of "bringing the river back into the city."[61] A century after the first hesitant moves of Wiesbaden toward the Rhine and more than seventy years after the city gained ample access to the river by Allied decision, the city in the early twenty-first century finally seems to have established an emotional affection toward and also a planning grasp of the Rhine.

Still, communication between the two cities does not run evenly in this period of far-reaching urban redevelopment, undoubtedly a consequence of different perceptions and perspectives. An attempt to apply for a jointly organized river-crossing federal garden festival, a rather high-profile and expensive venture, failed at an early stage. The same fate was experienced by other regional projects aiming for river-crossing nature conservation and environmental protection projects.

However, judging from current developments, there is still a feeling of change in the air across the river Rhine. A first attempt to include all relevant urban actors in approaches to the comprehensive improvement of the city's river, taking care of ecological, economic, and social interests of the urban inhabitants, was made with the Mainz Rheinufer Forum in 2000,[62] followed swiftly by the Hafenforum for the future of the old Toll Port in 2004. Similar approaches to civil-society-based improvements were undertaken in Wiesbaden in the same period. At the moment both cities are involved in a cooperative project to develop the Rhine and its banks in a city- and river-overarching way, the Rheinuferentwicklungskonzept Mainz-Wiesbaden, following an advance by the Wiesbaden city leaders in 2009 to establish a common environmental zone for both cities.[63] Perhaps for the first time in their shared relationship with the Rhine, Mainz and Wiesbaden seem to be actively embracing the notion that the waters of this mighty stream may be connecting as much as separating the two cities. Nevertheless, in the process of (re)appropriating the river over the last half century, the current situation may prove to be just one more snapshot in a long and continuously changing relationship between city and river.

Chapter 15

Revitalization of a Tamed River

The Isar in Munich

Nico Döring and Georg Jochum

As a typical Alpine river the Isar has an ambivalent character. The popular etymological derivation of the name Isar as "die Reißende" (the torrential river) refers to its wild, often destructive character. At the same time, the river is often referred to as the "Green Isar," which points not only to the color of the calcareous stream but also to its romantic appearance and the emotional relationship of the people to their river. A well-known folk song goes: "So long as through our Munich the Isar rushes green, . . . so long the coziness of Munich-town shall flourish."[1] In some images the river is represented as a young nymph.

Since Roman times the river has been used for rafting. With the foundation of the town of Munich in 1158, anthropogenic activities led to the development of a preindustrial sociotechnical system using resources including the water itself as well as fish, stones, and wood found along the river. Water was diverted from the untamed, sometimes dangerous river to small mills located a safe distance away, where settlements and towns arose. In preindustrial times there were only limited means for the technical modification of the river morphology such as the building of bridges, mill canals, and flood protection works.

In the nineteenth century new methods for protecting the land from devastating floods were developed. By employing new technologies like flood protection dams, weirs, and hydropower plants, nearly the whole river system was canalized. Accordingly, it changed its character within a few decades and so did the city of Munich. Thus a second man-made sociotechnical system arose with many economic benefits. Regrettably, questions of sustainability, hydromorphology, and biodiversity were neglected, and the measures taken at that time are now interpreted as sins against nature. Furthermore, the daily economic functions of—and the emotional associations with—the Isar disappeared for many Munich citizens, who became increasingly disconnected from their river.

However, two generations after the river had been straightened it remained close to many citizens' hearts, and a movement sprang up with the aim of revitalizing it. This led in the 1980s to a new approach to evaluating the multiple natural and environmental qualities of the river. After another decade the first projects were carried out: from 1994 to 2002 the restoration of the rural Isar in Mühltal, twenty kilometers south of Munich, and from 1995 to 2011 the urban "IsarPlan," which began downstream before working northward to the city center of Munich. Riverbanks were reconverted to a more natural state and leisure areas were expanded. The two projects became internationally recognized examples of a reorientation toward natural principles in river regulation as opposed to purely technical regulation. This reorientation can be seen as an expression of the rise of a new relationship between Munich and the Isar. The river was regained in a double sense: the "dead" river was revitalized and regained some of its natural dynamic and space, and the people of Munich rediscovered the Isar as a place of leisure and as a meeting place.

In this chapter, we discuss these steps in the anthropogenic transformation and use of the Isar. First we examine the natural shape of the river. We then explore the historical changes from preindustrial to industrial times and up to the present "re-creation," which constitute three fundamentally different "ecosociotechnical systems."[2]

The Natural Shape of the Isar River

The Isar formed at the end of the last glacial period and flows from the northern Alps to the Danube River. Several tributaries in the northern Alps—roughly one hundred kilometers south of Munich—flow together to form the Isar. After a few kilometers it becomes a typical Alpine braided river, with gravel islands and wide gravel banks transporting large amounts of sediments. These banks move with every heavy rainfall and reaggregate into different formations at the end of a flood. Naturally, this character continues until the river reaches Munich and farther to the Danube. The braided flow of different affiliated streams disperses into the lateral and subjacent gravel bed, where it is reenergized and its temperature adapted to the groundwater. It then reemerg-

FIGURE 15.1. View from the high banks of the Isar in Munich. Heinrich Adam (1787–1862), *München vom Isarhochufer aus*, 1832. Source: Münchner Stadtmuseum P133.

es as a clean, cold, groundwater-fed spring, which forms a new stream—separated by temperature—in the water body of the braided river. The naturally transported bed loads amount to over one hundred thousand cubic meters of gravel per year, where the Isar leaves the Alps. Wide gravel banks, uncounted highly dynamic gravel islands, oxbows, and many branched river arms—at almost the same height as the gravel islands—are now defined as the natural hydromorphology and the "natural guiding principle" for restoration measures. The natural dynamic zone of the river had a cross-section of up to one thousand meters. The river flowed within a continuous line of gravel islands and interlocked wide-stretched riverbanks. The natural shoreline length had twenty-five shore kilometers per river kilometer.[3] These areas were nearly free of vegetation. Only "pioneer plants" and "pioneer animals" were able to settle there, being well adapted to the dynamic of water and gravel. After almost two hundred kilometers, gravel loads are pulverized to sand and the river's water

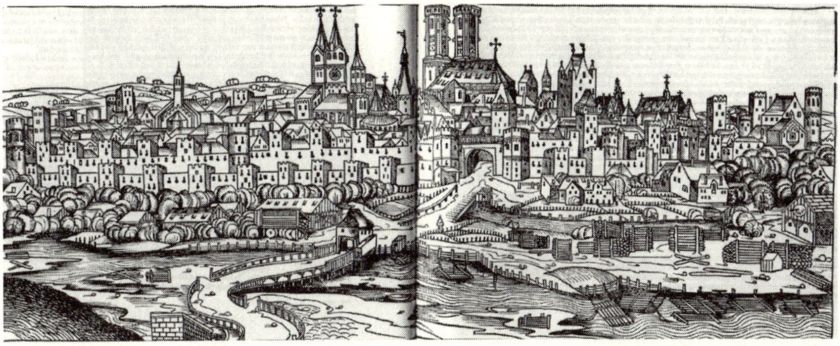

FIGURE 15.2. City view of Munich with the Isar and rafts in the late Middle Ages. Hartmann Schedel, *Die Schedelsche Weltchronik* [*Schedel's World Chronicle*] (1493). Scan from original book. Source: https://commons.wikimedia.org/wiki/File%3ASchedelsche_Weltchronik_d_226.jpg.

flows in wide meanders until "she"—the Isar was a female mythological goddess—empties into the Danube.

MUNICH AND THE PREINDUSTRIAL USE OF THE TORRENTIAL RIVER

The construction of a new bridge over the Isar was the primary reason for the founding of the town of Munich. The city held a bridge monopoly that combined the control and duties of the waterway. Beginning in the Middle Ages, the town developed at a safe distance around the next river terrace to the center. Using the Isar in many ways—for rafting, transporting goods, powering several kinds of mills, fishing, and hunting birds—formed the basis of the economic and social life of Munich. It was normal in daily life for nonprivileged people to go to the Isar to collect wood, limestone, and herbs, and to do their laundry; for children, it was their playground.

From the Middle Ages the town served as a center for the transportation of goods and people by raft. New settlements and commercial activities were located around the trading center near the harbor and the Isar itself. Economically motivated river modifications—local shoreline reinforcements of the riverbed and side canals for mills—were built locally for hydropower, but at that time the torrential river remained fairly untouched. This anthropogenic influence created a preindustrial sociotechnical system with an intensive multifunctional technical, sanitary, and alimentary use of the river.

The oldest picture of Munich in the famous *Schedel's World Chronicle* clearly depicts rafts in the foreground, which serve as an impressive demon-

FIGURE 15.3. Women collecting limestone at the Isar. Joseph Wenglein (1845–1919), *Kalksteinsammlerinnen im Isarbett bei Bad Tölz*, 1883, oil on canvas, 135 × 210 cm, Bayerische Staatsgemäldesammlungen, Neue Pinakothek, Munich.

stration of the significance of the rafting trade for Munich. Evidence of rafting in Bavaria can be found as early as the twelfth century.[4] The rafts transported primarily timber, firewood, and charcoal, which were obtained from the forests of the Isar region south of Munich.[5] Limestone was transported directly to Munich, to be burned there or in lime kilns set up along the river.[6]

From time immemorial charcoal was an important fuel transported to Munich on the Isar. A variety of trades—in particular the metalworking industry—required the resource in large amounts. In 1873 over 1,600 tons were delivered to Munich by rafts, which was then stored on the so-called coal island (Kohleninsel), just outside the city's old boundaries.[7]

But the rafts not only served as a means of transport for timber and building materials from the Isar region to Munich; many carried goods from Italy that had been imported by wagoners to the Alpine town of Mittenwald on the Austrian-Bavarian border. Mittenwald had long been a transfer and transportation center for Italian and non-European goods: figs and wine from Italy, but also pepper, cinnamon, and incense from Asia and Arabia crossed the Alps to reach Munich.[8] Munich did not always serve as the final destination, but often was an intermediate station on the way downstream toward the Danube to Vienna and Budapest. The last thriving period of the rafting trade was in the

Figure 15.4. "The Green Tree" Inn at the Munich raft landing. Joseph Stephan (1709–1786), *Flößerwirtschaft zum grünen Baum*, 1767. Source: https://commons.wikimedia.org/wiki/File:Gruener_baum.jpg.

nineteenth century. During this time, the city was proud of having the largest rafting harbor in Europe. The record was set in 1864, when 11,154 rafts landed in the Bavarian capital.[9] A new trade law with new freedoms of trade was implemented in 1868 in Bavaria and overrode older local regulations.

Some water was guided away from the dangerous Isar in order to drive water-powered mills. The river water was diverted into a regulated, multibranched network of streams and mill canals alongside the uncontrollable, furcated, wide gravel river. The river itself was dramatically torrential during floods, something that bridges and diversion installations were unable to withstand. Mill canals were also used to protect the town, provide drinking water, and transport firewood, food (beer), and handicrafts, in addition to supporting fish stocks and carrying sewage. Mills themselves were used as flour, hammer, water pumping, and, later, saw mills.[10] By the end of the nineteenth century, 142 mills in Munich were powered by river water. At that time, the town was growing and expanding rapidly.

Craftspeople and traders settled near the Isar because they depended on the river and the millstreams diverted from the Isar, and because they supplied the city's economy with timber and charcoal.[11] This was a comprehensive economic use of the river, with further resources coming from bridge and raft taxes. On account of this, the city and its inhabitants strongly oriented their life toward the Isar, including rafting, the bridge, and the mill canals.

River Canalization in an Industrializing Society

The first sociotechnical system—which was characterized by Isar rafting and watermills—was eliminated within a few decades. In the early nineteenth century Bavaria became a kingdom, and Munich, its capital, had fifty thousand inhabitants. The town then grew dramatically. Formerly rural areas around the city—transformed by settlements for poor people and by commercial activities—expanded around the trading centers in alluvial zones. The booming city grew to one hundred thousand inhabitants in the 1850s; after incorporating urbanized villages, Munich expanded to three hundred thousand inhabitants by the 1880s.[12] Continuous problems with flooding and sanitation were one consequence of such expansion. At that time it was the third-biggest town in Germany after Berlin and Hamburg, although it was still limited in terms of industry.

As a result of later industrialization—from around 1880—a new system based upon changing usage of the Isar emerged: the rafting trade soon faced fierce competition from the railway system. The Isar in urban zones was regulated, canalized, and reduced to a small, fast-flowing, and eroding riverbed. A deeper, self-eroding riverbed was intended to be a means of avoiding the frequent devastating floods occurring in the newly settled alluvial areas, and so was reshaped accordingly. Furthermore, the use of the Isar for hydropower plants to produce electric energy changed the city's relationship with the river: whereas the old small water mills had been located along a small grid of millstreams in the vicinity of the Isar, new linear, monostructured canals were built to channel the river water—except in flood periods—to be utilized by several giant hydropower plants.

The 1924 opening of the Walchensee power station—Europe's largest hydroelectric plant at the time—can especially be regarded as an important indication of the declining relevance of old power plants with wooden water wheels, and of the transformation of the Isar into a resource for electric power production. Its construction by the "system builder" Oskar von Miller is a symbol of the sociotechnological metamorphosis of Munich, and of Bavaria as a whole, a change that needs to be discussed in detail.[13]

FIGURE 15.5. Lake Walchensee with the Hydropower Plant. Zeno Diemer (ca. 1925), diorama, Deutsches Museum, Munich. Courtesy of the Deutsches Museum Archiv.

During the 1882 International Electricity Fair in Munich, the young construction engineer Oskar von Miller gained an international reputation by demonstrating—together with the Frenchman Marcel Debréz—the feasibility of long-distance power transmission.[14] The event that established von

Miller's international reputation was his transmission of alternating currents from Lauffen am Neckar to Frankfurt am Main in 1891.[15] Hence, the decisive change in the use of hydropower had been proven: with electricity networks it was possible to separate the site of energy production from that of its consumption. This introduced a fundamental change in all economic structures, and centralized energy services for many people replaced small, local energy supply networks such as mills. Eyes were focused on Munich, and particularly on the electricity corporations settled there—together with Berlin, it became one of the two leading German cities for electric lighting.[16] In 1894 a hydropower plant was put into operation in the south of Munich, with further plants utilizing the Isar's water soon following; the most famous was the 1924 hydroelectric station at the Walchensee, in the Bavarian Alps.

The Walchensee is a natural lake in the Alps near the Isar River, sixty kilometers south of Munich and 801 meters above sea level, situated 200 meters higher than the Kochelsee (600 meters above sea level). Since the construction of the Walchensee power station, the water has flowed through the rock via a tunnel, rushing downward in large pipelines to drive turbines. To achieve maximum economic benefit from the Walchensee, greater quantities of water were required to supplement the natural feeder streams. Water was therefore diverted from the Isar and its feeder rivers by canals and ducts to the Walchensee, flowing back into the Isar fifty kilometers downstream.[17] It is, then, an Isar power station, despite not being located directly by the river. As a result, the Isar and its side branches downstream run dry.[18]

The construction of the Walchensee power plant and the distribution of electricity via the new networks of the Bayernwerk thus marked, as Thomas Hughes has pointed out, "the culmination of activities associated with two phases of the history of electric power systems—the time of intensive development of hydroelectric plants and the subsequent period of the power pool, or grid."[19] Due to their far-reaching technological and social implications, Hughes has labeled these connections a "sociotechnical system."[20]

This new sociotechnical system was to make an essential contribution to the development of Bavaria and especially to the industrial rise of Munich. As a result, a fundamental system change took place: through the generation of electric currents, power supply became separated from the river. These "networks of power"[21] provided by an association of producers enabled the transmission of energy over large distances within a short period of time. In addition, new networks of water supply and new transport systems—railways—developed. Crafts and trades could now diversify regionally, a variety of new enterprises spreading over the industrially developing city of Munich and at distant locations from the Isar. As Hughes points out,

> the northern slope of the Alps . . . had for centuries provided on-site water power. Fast-flowing streams and rivers rushed down the precipitous incline into the Isar. . . . Wooden water wheels housed in mill buildings supplied the power to grind grain, push bellows, raise hammers, saw wood, and perform other tasks. These power plants, however, had to be located on the streams themselves, and their energy had to be used there. . . . After electric power transmission had been proven, von Miller proposed transmitting this power to sites where industry and population were concentrated.[22]

Von Miller himself emphasized the benefits of electrification: "I do hope that the new roads of electricity, which shall distribute our precious natural forces across the whole of Bavaria, will contribute to the fact that . . . industry, trade and agriculture will flourish, and with this the prosperity of the Bavarian people."[23]

There is no doubt that, on the one hand, the usage of the Isar to generate electric power brought advantages for the development of Bavaria, but on the other hand, three significant disadvantages must be pointed out. Firstly, the river lost its natural flow and biodiversity through hydropower plant installments; secondly, it was cut off from the economic activities of the city, and vice versa; and thirdly, citizens were cut off from their emotional contact to, and balance with, the river. Indirectly Munich was still interwoven with the electric power being supplied to all citizens from the Isar plants, which provided 66 percent of the city's electric power in 1926. However, it was no longer a heartfelt pulse in daily life, and so the city turned away from the river.

This was not only a result of the transferability of energy that partly separated industry from the river—additionally, from 1900, large flood protection structures obstructed direct access to the river. As a result of the availability of railways, the rafting trade became worthless, and the bustling harbor disappeared. As early as the turn of the twentieth century, long-distance rafting downstream to the Danube came to an end: The canalized and deepened Isar riverbed, with its new cascades, was a barrier for rafts. On the slowly flowing Danube itself, rafting was increasingly replaced by steam navigation, and in 1905 the last trip from Munich to Vienna via the Isar–Danube route took place.[24]

Beginning with the Mittenwald railway services in 1912—one of the first electric railways powered with energy from the Isar—regular passenger raft traffic ceased altogether. Rafting in the upper Isar region was impossible in

Figure 15.6. The plan for canalizing of the Isar in Munich. The Isar in its natural shape (above) contrasted with the Isar canalized (below), ca. 1905. Source: Archiv des Isartalvereins, Arnulfstraße 60, D- 80335, Munich.

any case, due to the dry riverbed. The rafting trade quite literally ran aground and the last raft was launched in this part of the Isar in 1922.[25] Although the rafters received financial compensation for losing their livelihood, this could not prevent them from looking back to a time before the decline of their old trade with nostalgic feelings: "Beautiful is the green Isar, beautiful is rafting; when the Walchensee project is launched, this will be the end of our trade!" was a common saying on a raft, revealing resistance against the canal project.[26] Today there exist three rafting families offering touristic rafting with Bavarian music, beer, and a traditional atmosphere over a distance of twenty kilometers to Munich.

Flood Protection

An extremely distressing occurrence was the flood of September 1813, when the main stone bridge in Munich, Ludwigsbrücke, collapsed and over 107 people died.[27] In 1829 on the left bank of the Isar the first straight dam was built to protect new districts of the city situated close to the river.[28] The newly regulated river started deepening rapidly. As a desired result, annual calamities and frequent flood catastrophes were under control.[29] At the same time, it transformed from a multibranched river with gravel banks into a stretched, single-branched river constantly deepening by erosion. As an undesired result, the river continued eroding its sole. Bridge foundations and dikes were close to collapse. Uncounted weirs had to be newly established to limit this unending erosion process and to protect buildings and bridges. Nearly the whole river was forced into fixed paths with a narrow 45-meter main channel, adjacent to a 145-meter-wide floodplain and bordering dams. The groundwater regime and the alluvial forests were heavily modified.[30]

The city of Munich benefited from the canalized Isar by developing "useless gravel banks" into "usable urban settlement areas" with new bourgeois quarters, green pleasure parks, leisure areas, rural grassland, and alluvial forests.[31] The banks around the old large landing stage for rafts had been redesigned and bourgeois Isar quays were built.[32] At the Isar new quarters for the privileged arose—today known as Lehel and Maximilianstraße—and on the historic "Kalkinsel" (limestone island), exclusive public spaces were set up. For some years it was en vogue among the upper classes to appear alongside the glamorous new riverside boulevard according to the Parisian ideal. Europe's biggest waltz ballroom and an amusement park aspiring to Vienna's Prater with a giant wheel was installed, inspiring the island's new name, Prater Island. The wide river was tamed in a concrete box section—named the

Boulevard, which is now a protected historical monument—and so was no longer accessible. In 1906, upstream at the former quay of the gravel island used to store coal, wall construction works continued to establish its new use as a site for the Deutsches Museum. This "museum island" was formerly part of a raft port and the coal island.

The Loss of Biodiversity and of Emotional Attachment

Since around a hundred years ago, the Isar watercourse has been forced into a narrow canal over a distance of more than two hundred kilometers. Biodiversity was considered by the system builders to be less important than flood protection and electricity supply. Between World Wars I and II an employment creation measure was enacted that implemented the straightening and stretching of the river's shape, although it was not necessary either for flood protection or for hydropower plants. A chain of forty-four hydropower plants at that time transformed most river energy into electricity. The original lengths of Alpine river shorelines were reduced, on average, from twenty-five kilometers (see also fig. 15.3) to two kilometers.[33] With the change of the riverbed, native animals, plant species, and alluvial forest declined or even disappeared, thus reducing much of the river's biodiversity. Its bed has continued to erode locally by up to eight meters; attempts to reduce this ongoing erosion were made by constructing numerous weirs. As a consequence, fish routes were interrupted and braided side streams, oxbows, and municipal streams were cut off, with the consequence that breeding areas and habitats for all kinds of animals were lost.[34] Several bird species that bred en masse on gravel islands are now extinct along the Isar.[35] Other birds that formerly bred throughout the braided river, like the little ringed plover (*Charadrius dubius*), still exist in very small numbers and did not benefit from the very narrow restoration areas in Munich.[36] The water exchange with the groundwater stream dropped; as a result, many groundwater springs dried up. Experts roughly estimate that the fish population declined to 20 percent of its former state.[37] Reptiles and insects were affected, with some species becoming extinct.[38] The less-visible of the native plants also disappeared altogether.[39]

A perception of the Isar disconnected from nature became common. Gravel loads were valued as a precious construction material in the booming town, and the waterside was used as a sheep run and as a place to deposit construction waste. After World War II, known flood areas turned into slums through the settlement of displaced persons. After the war—during which much of the city center was destroyed—the waste land was reshaped as grassland and fortified against erosion using integrated plastic grids. To summarize, the Isar had become a barricaded area in many ways and fairly inaccessible in comparison to before 1900. The significant loss of biodiversity was a visible sign of the deteriorating river quality. Additionally, increased sewage output and public waste contributed to the alarmingly poor water quality. Because of new dangers at the weirs, and owing to water pollution, swimming in the Isar in Munich had been forbidden since the 1970s. Through the loss of the river's basic life-sustaining qualities and the withdrawal of physical and emotional contact, Munich lost its heartfelt intimate relation to the Isar.

A River Revitalized

For about a hundred years the power plants at the Isar determined the life of the river, and waste disposal from Bad Tölz brought many hygiene problems in conjunction with the slow motion and low water level of the river. In the 1980s the riverbed almost collapsed, and conservation activists invested their power and time into restoring habitats for fishes and birds, and the Isar itself. In 1993, a group of people launched the idea of a "new life for the Isar," resulting in restoration plans and projects, which were finished in 2011. They gained great international appreciation and recognition and are seen as a leading example of the restoration of rivers. Reviving the river dynamics has simultaneously had a significant impact on the redevelopment of Munich and on the leisure activities of its citizens. We will take a close look at the steps of this process, from the development of a vision to the point of its realization as it is now. Further perspectives for the future are in the pipeline.

Of all the former braided rivers in the Northern Limestone Alps, today only a few kilometers remain nearly unchanged in their natural dynamic. These sections are the Pupplinger Au, twenty kilometers south of Munich near the Mühltal with a dimension of about three kilometers, and the five-hundred-meter Flaucher falls within Munich, which were kept untouched. This was the place where one felt the genuine character of the Isar, with branched streams and islands of mixed gravel coming from the white northern Tyrolean limestone, which were continually rearranged after floods without the emergence of green slippery algae. These sections of the Isar have been a frequent attraction for many people, a kind of "sacred zone" for lovers of the river. The few untouched areas remain a valuable site for flora and fauna and provide some orientation as to the guiding principles for recovering Alpine rivers, while serving as a model for interdisciplinary sciences—such as hydro-

FIGURE 15.7. Restoration guiding principle at the Isar River: Pupplinger Au between the Alps and Munich, 2010. Photo by Nico Döring.

morphology, or botany and zoology studying endemic species—through the study of the natural Isar environment.[40]

The variety of measures in the past that displaced the natural pathway of the Isar, in addition to the narrowing of the riverbed, were beneficial—from the technical point of view—only with respect to flood protection, land gain, and energy supply. From a wider perspective, such technical adaptations had negative consequences upon the vital organisms of the river itself, and to all organisms in the areas directly or indirectly served by the Isar. As became clear in the 1980s, the Isar River was nearly a dead body, and so a group of visionary people, NGOs willing to engage in dialogue, and open-minded politicians began a search for solutions that would revive the river.

The Isar Alliance was a voluntary association of multidisciplinary nongovernmental organizations as well as competent individuals (in nature, sports, leisure activities, fishery, hunting, art, health, biology, geology, landscape architecture, and stream engineering) that in 1993 formed a dedicated, synergetic network based on nonhierarchical communication among participants.[41] In the first years the Isar Alliance worked only in cases where there was unanimity and represented, with its different societies, more than five hundred thousand members. This was an essential step in coalition building that demanded the burial of old resentments between group members to find a common and effective way of functioning as a homogeneous group. The first step was forming a network of volunteers, in which decisions were made only when all agreed in unison. The team developed multidisciplinary integrative approaches for the utilization and protection of the Isar.[42] The second step was to develop an ecological and economic vision, and the third step was to introduce it publicly.

Campaigns started when the Isar was recognized by officials and hydropower companies to be an unchangeable dead river body. The Isar-Alliance started to beat the drum for their "utopian" vision by inviting artists, organizing expositions, staging events (including light shows, university campaigns, demonstrations, petitions and parliamentary queries, public discussions, and radio and TV interviews), by printing books, leaflets, and T-shirts, and providing other giveaways. All of these actions achieved a turnaround in the positions of the river management, which until then had been dominated by the hydropower lobby,[43] but now agreed to a new distribution of the water and to ecological measures.

The Isar Alliance started its political activity with the Mühltal project and the vision of a revitalized river. Mühltal is a ten-kilometer-long stretch to the south of Munich, where almost all the Isar water was injected into a hydropower canal and the river ran nearly dry. This had severe consequences for the alluvial areas and the species living there. At the same time, some governmental planners, too, had visions of an ecological river management. These were not in accordance with the economic interests of the hydropower lobby and remained unpublished. The Isar Alliance created political pressure for a change in the Mühltal stretch and later for implementing the new IsarPlan.[44] During these intensive discussions with the Isar Alliance the Bavarian government started publishing its first new Isar River management plan. Thus the Isar Alliance initiated a political process toward new perspectives for the Isar.

In Mühltal the first pioneer project ran from 1993 until 1999. The riverside woodland recovered slowly but remarkably, through guiding more water out of the canals into the side branches of the riverbed. Hydropower plants dispensed the necessary water to the drier area adjacent to the technically constructed concrete canals, and installed suitable measures for the amelioration of the heavily modified hydromorphology. A new appreciation for nature led to a new understanding of natural qualities, which were given more weight in the decision-making process regarding the maintenance of the river. The measures in Mühltal were highly appreciated in the "Isar community" and also publicly.[45] The Mühltal experience became a path-breaking and renowned example for all subsequent proceedings.[46]

So a first paradigm shift happened with the results of Mühltal. More than forty hydropower plants already existed along the Isar River and more than a hundred existing hydropower plants in Bavaria had licenses that were due to expire and had to be renewed for the next thirty years. The powerful hydropower lobbyists feared that new, ecologically sound but economically less profitable, guiding principles would become established as the new standard.[47] At the same time in Munich the administration tried to push through a nonpublic program enabling dams for flood protection, calling it the IsarPlan. The recently successful Isar Alliance interfered at this phase.[48] The Alliance proposed measures to widen the riverbed in combination with improvements for nature and leisure activities instead of higher dams. Since an EU flood management directive with new standards meant that some measures at the river were necessary, officials became willing to start discussions about this important topic. Roundtable circles were established, integrating the new experiences of a river management that included nature. After having found mutual ground among themselves, Isar Alliance members began a tough struggle with the authorities in favor of the Isar. Public relations campaigns—as mentioned above—continued, and intelligent communication strategies with all officials of the city and the state were initiated to promote these ideas with the aim of raising support at all levels, and of building political pressure. If crucial conflicts occurred between the planning group on one side, and districts, citizens, and the IsarPlan group on the other, they were reconciled by mediation. Weakening the old plans and establishing new ideas were thus a masterpiece in the "art of right communication." In 1998 Munich, together with the Bavarian government, published a leaflet with the title *New Life for the Isar*. The ideas that spread publicly provided much inspiration for other river restoration projects. The new ecologically influenced concept was politically prepared, and commonly accepted three years later by the city council. Based on this political decision, a long-term project was initiated, the New IsarPlan.

The New IsarPlan Munich

Construction works according to the New IsarPlan started in February 2000, beginning with remodeling eight kilometers of the riverbanks based upon ecologically oriented guiding principles. The New IsarPlan intended to bring together ideas in flood protection, sustainability, and recreation toward creating a new urbanity, while respecting the new environmental guiding principles. This was a pioneer project, since no one had any experience with comparable projects concerning public acceptance. Step-by-step changes were implemented, including complex preliminary works for urban technical infrastructures under flood meadows and the riverbed. The renaturation project was finished in 2011.

Instead of the former technically stabilized homogeneous floodplains combined with slopes restricted by inaccessible concrete or granite dump riprap, the urban restoration measures provided space for new accessible gravel shores, which cover underlying shore-fixing rocks. Continuous gravel banks and a morphologically heterogeneous riverbed were intended to increase biological activity. Flood protection could be improved through three techniques: replacing the granite concrete, widening and revitalizing the canalized riverbed into the higher floodplains, and—if still necessary—improving the dams.[49]

Munich's action for the Isar became an early holistic approach toward revitalizing natural river environments including multifunctional public and nature spaces in an urban area. Visitors are now able to experience the dynamic power of an Alpine river from the gravel shores within Munich. With

Table 15.1. Isar: Project Chronology

1992	Governmental and municipal scoping studies (not public) for flood protection measures at the Isar in Munich with the decision for new high dams
	Governmental studies for hydropower licenses and parallel for river restoration (not public)
1993	Foundation of the Isar Alliance with campaigns for sustainable river management
1994	Decision for a first river restoration at the Isar River in connection with a new license for an old hydropower plant
1995	New flood protection measures in Munich consisting exclusively of dam enlargement
	Intervention by the Isar Alliance
	The new interdisciplinary working group "New IsarPlan" is established (not public)
1997	Campaign for the Isar River water quality
1998	City council resolution for the integral IsarPlan with flood protection, measures for river ecology, and leisure activities
2000–2002	Publicly discussed construction stages 1–3
2002–2004	Construction stage 4
2003	Competition for construction stage 5 (without public discussions and public decisions)
2003–2004	Controversy over competition results
2004	Mediation and final consensus decision on construction stage 5
2005–2006	Realization of construction stage 5
2006–2011	Realization of final stages

the "improved health of the river" it complies with three demands: in addition to flood protection, it created a space within a densely populated area where recreational and leisure activities for people successfully coexist with flora and fauna. A total of 1.3 million metric tons of earth were removed and 385,000 metric tons of new stone were added for new, mostly hidden shoreline protection. In the process of remodeling the canalized river, nearly thirty groundsills were removed and modified as rock ramps for fish or canoes. The total cost was €35 million. The majority of the financial investment was for optimizing infrastructure measures. Project goals were met, increasing flood protection, providing accessible river shorelines for leisure activities, and improving conditions for river-specific flora and fauna.[50] Studies show that flora and fauna are reestablishing along the project area; however, for gravel-bound species and fish the observed improvement is less than was hoped. Recent flooding confirmed the technical quality of the IsarPlan.[51]

The Isar River was still contaminated with sewage water, other effluents, and coliform bacteria. These problems were caused by other municipalities along the river, so they were not subject to the influence of the IsarPlan of Munich, although the water quality was important for the public acceptance of the newly shaped and accessible river section in Munich. In a first project in Bavaria, which was also promoted by the Isar Alliance, a sewage plant was improved to have a cleaning rate of 99 percent and UV treatment of the water for elimination of coliform bacteria. After optimizing all sewage plants and after the addition of UV radiation units to twenty-seven plants along the Isar River, the water widely complies with the EU bathing water directive.[52]

Tamed and Lost, Revitalized and Regained: A Conclusion

The first restoration project for the Mühltal, decided upon in 1994, was a paradigm-changing first success. New water distribution rates between the river and adjacent turbines were agreed upon, and in addition a pathway for fish was installed beside the weirs. A first Isar management concept for gravel loads was initiated and implemented in the Mühltal. Concrete dams were removed and the lateral junctions to oxbows were opened. Where formerly only a rivulet (5 cubic meters per second) remained in the nearly dry riverbed, now a minimum of 12–17 cubic meters of water per second remain in the river. Before 1925 an average of 90 cubic meters per second was in the river.[53] These measures initiated a shifting of the riverbed and allowed wide gravel banks to settle. So, a process had begun, but further measures were necessary for achieving the predicted hydromorphological dynamic of a braided river. The alluvial hinterlands would allow a further furcation into the natural borders. In general, the Mühltal experience was a first step in revitalizing a disrupted area of the Isar, which was well accepted publicly and provided a good basis for further projects.

The implementation of the second project, the urban IsarPlan, is likewise perceived as a success. Whereas the rural Mühltal attracted comparably little interest, the urban Isar became a precious nature attraction for people. Sloping stone river shores in Munich were transformed to more than 50 percent accessible gravel shore, but regrettably, space limitations within the city would not permit broadening of the riverbed. Floodplains were lowered by taking off quantities of fine sandy accumulations; dams were stabilized and raised in some particular places for achieving the desired flood capacity of 1,100 cubic meters per second. Thus, flood protection measures were optimized with the revitalization of the river. The ecological revitalization of the Isar also found its response through the relationship of people to "their" Isar. Not only the river itself but this relationship was revitalized. Apart from urban dwellers' perception of the river's vital "inflow of fluvial energy" into the city, comparable to an aorta and its essential pulse, the restoration also resulted in the desirable spread of seeds of regional xeric gravel grasslands across the land, with good results for flora and fauna. Weirs and groundsills were reconstructed as rock ramps at least partially traversable by fish and canoes. Water quality was improved by optimizing sewage factories upstream, attaining the acceptable water quality for swimming that complied with the EU bathing water directive. The new Isar internationally gained the image of being a whitewater Alpine river with a visible swirling dynamic. Nature conservation and flood protection were combined in a harmonic way, so that the risk of flood catastrophes—as demonstrated in spring 2013—was reduced and the natural dynamic simultaneously regained.

Future actions to enhance the dynamic of the Isar are in discussion. The gravel river shores and waterbound habitats remain underdeveloped: in favor of the former floodplain, the riverbed obtained a gravel-covered reinforcement with narrow limits that inhibit natural branching and the reaggregation of gravel banks and islands. By inhibiting the self-forming dynamic of the river, technically constructed gravel shores reconvert to alluvial plains or willow bush land. Aquatic zones should receive more space for dynamic gravel movements with its characteristic sounds. Rock ramps have been formed with a boat passage for canoeing and swimming. These kinds of reshaping of the riverbed enhance the living conditions for fish and improve the river's biological cleaning capacity. An additional result is the sound of the water and of the swirling of gravel—an unplanned but pleasant side effect. The people love "their" new Isar, where they listen to sweeping water sounds filling the air, drowning out the noise of inner-city traffic. According to municipal estimations, up to forty thousand people relax there during sunny summer weekends.

Old willow trees, which formerly grew alongside the canalized river, were planned to be maintained on a new small willow island. This new willow island was fixed with rocks in the highly erosive river. This spot has become one of the most popular meeting places and sites for a short refreshing swim, just a few meters away from the city center. Not only has the Isar gained space, but coincidently the citizens of Munich have regained it as a new place to relax and to be. Breaking down the concrete dams also brought the people together.

FIGURE 15.8. The new Isar with the willow island beyond the bridge, 2014. Photo by Georg Jochum.

The Munich historian Christine Rädlinger says, "People were now able to enjoy their recreational time in 'natural surroundings' and wanted to seize that opportunity. Thanks to new technological approaches the citizens' wish for a natural river landscape could now be reconciled with flood protection."[54]

One central deficit is treated as a marginal point in the discussion. One-hundred-thousand-cubic-meter bed loads are retained in the Alpine Sylvenstein Reservoir and are lacking downstream. The recent reactivation of downstream lateral input of gravel loads (fifty thousand cubic meters) and an occasional truck transport of gravel loads surrounding the reservoir show good results in covering the riverbed with fine gravel, but are not enough to inhibit the erosion process. For experienced geology specialists it is clear that this is a crucial demand for the long-term viability of the Isar.

The New IsarPlan was subject to several restrictions. Other measures were dropped, even when complying with the common guiding principles. Technically redundant westward rip-rap dams were not removed over long distances. Lateral flood forests with old oxbows neighboring the dams of the actual flood zone were not integrated into the planning area. Therefore, the accepted dynamic of the river in the town remained too limited for a new furcation process with self-regulating gravel islands.

The historic review has shown that, after the foundation of the city of Munich in the late Middle Ages, a first sociotechnical system developed with a multifunctional use of the Isar River and its water. It was the basis of a daily and intensive relationship with local citizens. Despite this economic use, the river was still largely characterized by its natural circumstances. It is true that the economic use of the river throughout was controlled by the city of Munich and the Bavarian regents. Through industrialization a fundamental change took place. The river was domesticated and manipulated by building giant river dams and weirs with the aim of reducing dangers from flooding. By building over forty hydropower plants on the Isar, most of the river water was directed in order to electrify Bavaria through a huge power supply network. This led to a separation between the site of energy production and its consumption, and fundamentally changed the economic and social structures of Bavaria. As a result, a second sociotechnical system was established, disregarding the intrinsic logic of the river ecosystem and disturbing it—even destroying it—in many ways. It also had corresponding impacts on the lives of people around the Isar.

The second system focused on profit maximization for the electricity industry and can be regarded as one example of the early stages of the Anthropocene.[55] The goal of total control over nature with anthropocentric and primary economic values replaced all other conceptions and relationships between mankind and the environment. Ignoring natural laws had, and still has, severe effects on the ecosocioeconomic network of the Isar River.

In the last few years the contours of a new system have gained shape. It was found that the second man-made system did not ensure any sustainability and hence destabilized the system in its entirety. The New IsarPlan took into account the self-propelling dynamic of the river and integrated it into the renaturalized system, with the result that the Isar regained some of her former untamed and vital dynamic.

The shift from exploitation to a respectful interaction with nature also indicates a change in social mechanisms: hierarchical planning was replaced by a cooperative, collaborative process of diplomacy and agreement between all parties involved. As implied, the goals of re-creating the Isar and the aims of establishing civil participation processes in political decisions become, to some extent, a reality.

Now people of all ages with different leisure habits are dedicated to discovering the new Isar and enjoying the new dynamic in various ways. During hot summer weekends Munich's river shore is filled with new life and becomes a new inner-city highlight for visitors. The people have reacquired—in a different way from premodern times—a close emotional relationship with the river.

Taken as a whole, this process can be regarded as a transition to a new ecosociotechnical system with an exemplary importance. Following Ulrich Beck's theory of a transformation from the first, classical modernity, to a second, "reflexive modernity,"[56] increasing control over nature is no longer the only goal. The focus is now also on limiting the negative impact of human domination. Consideration of the intrinsic logic of nature and the effects upon the relationship between humans and their environment has become a main element in anthropogenic concepts. Within this context, the renaturalization of the Isar is a first step in a paradigm-shifting guiding vision.[57] Currently a municipal team is developing options (IsarPlan II) for the strictly canalized second municipal Isar section. It remains to be seen how further approaches (the citizens' wish for a natural river landscape) might be reconciled with already-existing urban structures and how the second step might be a continuity of the paradigm-shifting guiding vision.

Chapter 16

Union Is a Raging River, or Remembering Fez as the River Remembers

Shelley Hornstein

Union is a raging river running toward the sea, O Beloved, Be Like that to Me

—Rumi

This chapter considers an innovative river project in Fez, Morocco, one that mines history, memory, and place. The argument is that matter matters, as does the connection (or perhaps more aptly called the interface) between matter and the idea that matter is always connected to—never detached from—memory. The project that will unfold on these pages puts any theory about remembering into practice and considers an archeology of geographic place or matter, in a sense, where what transforms is a kind of opening up of the ground—the matter that matters—beneath our feet, both literally and metaphorically. Rather than a project that invites a peopled place to imagine a future after some traumatic past event that has fractured what they know of as their home, the example I will describe in Fez, Morocco, which has evolved organically for generations—a process of beneficial demolition without intentional destruction of any considerable consequence—has permitted a city that has negated its river, its history, and its source, to rediscover itself. Where once, quietly and strategically, this city opted to imagine itself cleaner and more efficient by covering up its river, what resulted in the process of attending to a problem as a knee-jerk reaction was that it destroyed the core of the place—it eviscerated its center of gravity, ironically, by stuffing its core: a continuous thick concrete slab sealed it over to hide its odorous and toxic detritus and thereby link—in a manner of speaking—its densely imbricated banks.

FIGURE 16.1. Historical bridge across the Fez River near the place Rcif, 2008. Aziza Chaouni, Aziza Chaouni Projects. Reproduced by permission of Aziza Chaouni Projects.

Condos, transportation hubs, monuments, parks, rivers, streets, and sidewalks are not simply the result of gluttonous neoliberal economies blindly ravaging nature or the "evil genius of capitalism and the traditional hegemony of engineering and architecture," as Richard Weller puts it when acknowledging and forgiving landscape architecture's inability to resolve entire configurations of cities and places.[1] Contemporary cities are tangled and denatured infrastructures "impossible to romance . . . as a collective work of art."[2] Yet I want to argue that while Weller's point is accurate, I think there is ample room for romance, at least at some moments, to consider how civically minded performers within the body politic of city space partake in a responsibility within what is a massive and complex and interconnected place called the city, and contribute to and affect (and here the term holds its twin meaning)—at least in part—the direction of its development. As such, romancing the city can result through performativity in cities, a factor often left out of the equation when discussing architecture, planning, and technologies of change. Civic governance notwithstanding, actors are present in city space, taking its pulse and recording its vital signs and evolving historical narratives. Citizens, through everyday ambulatory encounters, mark places in cities, places that recall other moments in those very places, places that serve as vectors to memory—perhaps increasingly so of late, often because the world of things is fast dissolving into the virtual, there is a prevailing worrisome sense by agents in city space about the loss of the past. No doubt, part of this anxiety must be, as many others have elaborated, that the fear of moving at an accelerated pace into an increasingly digitized world removes us from the tangible quality of objects and things—things we can hold on to in order to claim a place somewhere, or to determine an inventory of what we consider markers of our lives, actions, and memories. Jacques Derrida echoes this concern in *Archive Fever* when he points out the exponential speed of technology aggressively present in everyday life, which has pushed us to want to remember the past. The shift to ephemeral documents such as social media and virtual books, for example, instead of physical, material evidence, contributes to fears of what might disappear forever. Email for Derrida, for example, is digitization that dehumanizes; it is a trace that is subject to "iterability," and if something can be iterated indefinitely then it supplants the actual document, sending it further and further away from its origin.[3] Contemporary culture enables a fixation, arguably, with holding on to memory in the face of a potential and collective cultural amnesia. This dilemma is a signal that a memory boom of "unprecedented proportions" is in our midst, or what Andreas Huyssen refers to as "the twilight status of memory," where "twilight is that moment of the day that foreshadows the night of forgetting, but that seems to slow time itself, an in-between state in which the last light of the day may still play out its ultimate marvels."[4] He calls for a consideration of memory to "recover a mode of contemplation outside the universe of simulation and fast-speed information and cable networks, to claim some anchoring space in a world of puzzling and often threatening heterogeneity, non-synchronicity, and information overload."[5] Huyssen's plea is for us to consider how to remember places, to engage with the past while forging ahead without necessarily forgetting, in spite of the suggestion of ephemerality in a digitized world, or the endless, iterable path of virtual information. This makes for a compelling argument, as I will demonstrate.

The River of Trash

The Fez River Resuscitation project by architect Aziza Chaouni takes into consideration existing places and the lives within its compacted, overdensified, and severely deteriorating residential and commercial urban configuration as cooperative opportunities for an exercise in memory recovery in addition to a healthier lifestyle. She achieves this by recognizing and honoring—by excavating—the river that runs through it. Through that process of excavation, the river is then remediated by several interventions that remove the sewage and gives a newfound ecology and desperately needed open spaces back to this intramural city. Over time, Fez will recall, through the performativity of its citizens, the presence of its river.

Located in one of Morocco's most fertile valleys, the Sebou River basin, Fez will be revived and thrive with a genuine community life steeped in centuries of historical significance. Through this intense infrastructural examination and restoration, Chaouni places front and center the question: What can the city of Fez be today in light of the fact that it—or certain civic authorities—turned its back on its river and its past and without much resistance? The project incubated during a time when Chaouni's firm was called the Bureau of Ecological Architecture & Systems of Tomorrow (Bureau EAST), now known as Aziza Chaouni Projects. The office, then as now, is dedicated to pairing sustainability with design. Trained at Columbia University as a civil engineer and later at Harvard as an architect, Chaouni has consistently anchored her research in creating built environments that are sustainable, and with a strong emphasis in working in the developing world. In 2014 she was named a TED Fellow, an achievement that singles out trailblazers, and was listed in the Public Interest Design 100 Global list, a network of advocates, communicators, funders, makers, educators, policymakers, organizers, visualizers, and others who continue to contribute to effectively shaping the world. She is the recipient of the Progressive Architecture Award in 2007 for Hybrid Urban Sutures: Filling the Gaps in the Medina of Fez and the following year, in 2008, the Holcim Gold Award for Sustainable Construction for the Fez River remediation and urban development scheme, among others. For the Fez project specifically, the Bureau EAST initial project statement reiterates the challenge: "We choreographed a phased implementation strategy in which measures for enhancing water quality become both the locus and agent for addressing social and economic concerns."[6] As for her design approach, Chaouni's trademark is to highlight the often invisible or underrepresented spaces of cities or landscapes that are blithely ignored, always tethered to the investigation of viable and environmentally sound solutions. In fact, it is difficult—even useless—to divide city from landscapes in her work, as those traditional tropes suggest, because she concentrates often on the infrastructural components of a city and its natural habitat as a synchronized unit. Much like the recent literature on landscape urbanism (as opposed to landscape architecture), Chaouni uses the lens of ecology as well as critical issues of heritage conservation to understand spatiality, but not alone, as these issues are complicated by mul-

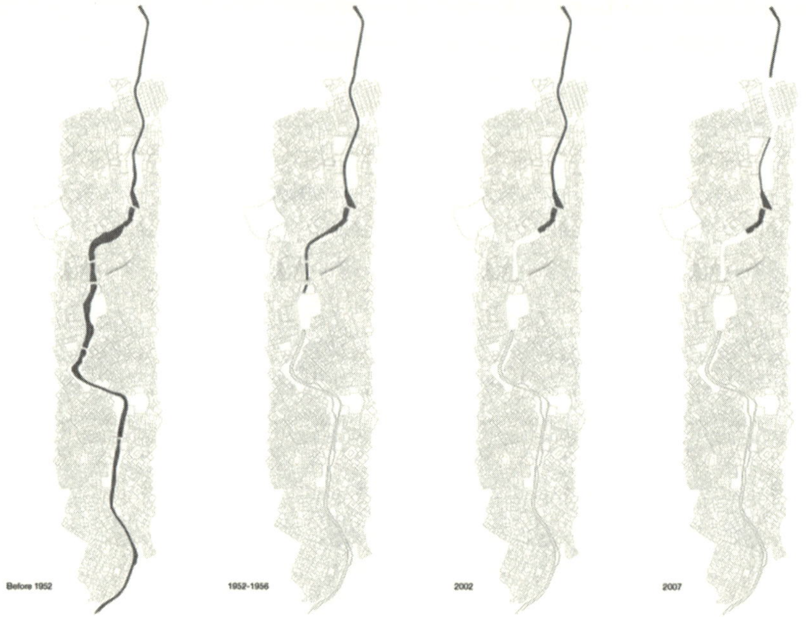

FIGURE 16.2. Diagram of the Fez River transformation, 2008. Aziza Chaouni and Takako Tajima, Bureau EAST Reproduced by permission of Aziza Chaouni Projects.

tiple overlapping issues. One of the earlier Bureau EAST project statements describes the challenge: "The complex interweaving of preservation initiatives is . . . under scrutiny both as a way to evaluate the past and to plan the future . . . new problems have emerged such as localized gentrification caused by tourism and environmental pollution linked to the industrialization of crafts."[7] With this in mind, what is the heritage to which Chaouni refers?

A UNESCO Heritage Site since 1981, Fez is cherished as one of the largest, densest, and most elaborate medieval morphologies and boasts, among other architectural treasures, the world's oldest university. The city teems with life and preserves its plethora of religious, civil, and military monuments. In point of fact, the city, as an aging and crowded medieval hub of activity, is congested and choking for open space, yet ironically thrives as an urban center in its comprehensive yet heterogeneous and overpopulated fabric. This is a city composed of different microcommunities. Founded on the eastern bank of the river in 789 CE by Prince Idris, a descendant of the Umayyad dynasty of Damascus, and about twenty years later a second settlement on the other bank by his son Idris II, the city was a unification of two separate and fortified quarters—the Umayyads of Cordoba on the right bank and on the left bank the Quarawiyyia who emigrated from Kairouan in the eleventh century—each with its center, divided naturally by the Fez wadi, or valley, and its river, the Oued, or Sebou Fes, or Fez River. In the twelfth century, the Almoravid dynasty succeeded in uniting the divided town. Only a century later, a new city grew out of these two and beyond the walls, and immediately to the west, following the conquest by the Marinid, under Abu Yusuf Yaqub, who ruled the Maghrib, and from the fourteenth century, the Mellah or Jewish quarter formed yet another annex to the increasingly crammed and stretched city.[8] In spite of the shift of the capital of Morocco from Fez to Rabat in 1912, it is widely accepted that Fez today remains the cultural and spiritual heart of the nation. Principal monuments in the medina are the madrasas (Islamic religious schools), *fondouks* (warehouses, stables, and inns for merchants), palaces, mosques, important private residences, and the elaborate and multiple fountains, most of which date from the Marinid Sultan period. Often referred to as "The City of a Thousand Fountains," it has over two hundred water sources for public use such as *hamams* and basins for ablutions, many of which are now largely dry.

The densification of the historic city more than doubled from about 100,000 inhabitants in 1900 to over 250,000 in the old city in 1975. This considerable cramming of residential, commercial, and religious space, along with increasing poverty, has exacerbated the network of streets and traffic, risks of flammable manufacturing materials, and the collapse of the irrigation and sewage system. Yet at the same time, Fez preserves the first centuries of Islamization of Morocco and is an exemplary pedestrian network with vernacular architecture that provides passive cooling in this extreme arid climate. These features are prized and valued, which has led to its designation as a UNESCO World Heritage site whose properties are considered to have outstanding and universal value in preserving cultural and natural heritage. This laudable designation is a blessing, as it will in turn allow the local and state resources to help preserve and cherish their tangible heritage legacy both for the citizens of Fez and of course for the burgeoning tourism industry.

This multistaged rehabilitation plan for the river runs through the center of the old city in Fez.[9] The history of the river's demise, and the city's amnesia of it, is linked to modernity, the acceleration of tourism, and subsequently the demand on the local leather and copper crafts industries. Once considered the fluid and raging spine of the city, the Oued Fes became increasingly unusable as a source of clean water when the population turned to it as a convenient sewage outlet not only for domestic but also for industrial waste.

FIGURE 16.3. Chouarra Tannerie, medina of Fez, 2010. Aziza Chaouni, Aziza Chaouni Projects. Reproduced by permission of Aziza Chaouni Projects.

Poisonous chemicals such as sulfur sulfate, formic acid, and liquid chrome from the leather tanning industry (the Chouarra tanneries, chiefly) and copper crafts were dumped into the once pristine source. Because these industries contributed to the history and tourism of the city for hundreds of years, the production only continued to accelerate and tragically, the City of a Thousand Fountains instead became the Oued Boukhrareb, or the River of Trash.

In order to obscure this eyesore in the center of the city, the river was paved over to become a road and parking lot. "The covered riverbed was seen as an alien intrusion into the organism of the old city, a foreign body as it were, which had to be absorbed and healed up in terms of the morphological rules of the historic urban fabric."[10] In 2004, UNESCO's World Heritage Committee took official note of a report "concerning the construction of a concrete slab, covering the Oued Boukhrareb, within the Medina of Fez," and its negative impact on this World Heritage property. It requested that the slab be demolished in order to "re-establish the integrity of the property."[11] The architectural firm's goal is to "enhance regional water quality while addressing the scarcity of open public space, overpopulation, and an aging infrastructure within the Medina of Fez. . . . This includes proposals for restructuring tannery operations and securing the local leather industry, enhancing health standards, and improving the general environmental quality of the Medina."[12] The Régie Autonome Distribution d'Eau et Electricité de Fes (RADEEF) commissioned Chaouni's firm to develop a rehabilitation plan for the river in tandem with their efforts to build two sewage treatment facilities to eliminate raw sewage from the river. For the first stage of the project they immediately eliminated the parking lot, or what became a parking lot as a result of the "concretization" of the river. To quote Chaouni, "For the river to reappear, this must disappear."[13] Above all, Chaouni sought aeration of the overcrowded space of the medina by the newly revealed river. This is not the first time a river has been erased from visibility. Partial or complete examples of buried rivers abound, particularly in modern industrial cities of the twentieth century. In Seoul, a four-mile freeway system dating from 1977 that paved over the Cheonggyecheon River and ran through the center of the city was removed.[14] In Toronto, many streams and creeks have been lost and diverted into underground sewage systems, and a soft movement is afoot to bring them to the attention of the public.[15] Successful cities that take their fluvial assets as central features, either as major or minor highways (the Danube as it services Belgrade, Budapest, and other cities), as tourist circuit options (Bateaux Mouches on the Seine in Paris or the Chicago Architecture Foundation River Cruise Tours on the Chicago River), or as fluid visual jewels (Venice, Strasbourg, Amsterdam), demonstrate what it means to be an aesthetically, socially, and economically river-centered place that supports local history and culture. Yet while thousands of urban bridges cross rivers and serve our cities daily connecting one side to the another, or even serve as viewing stations for pedestrians to enjoy the view, however rose-colored the lenses, still they remind us that the function of a river in many urban centers as the union between two sides or the metaphoric glue of a place consciously considered has been lost.

Creatively, the proposal by Chaouni's team in Fez identified the few vacant sites within the medina in order to introduce a tripartite intervention that would at once "enhance water quality, remediate contaminated sites and create open spaces."[16] In its place is a vast square that she likens to a sort of grand outdoor living room complete with benches made of recycled wood and riparian plants that will oxygenate runoff water. A future secondary phase of the elaborate zoning plan includes a playground with terraced wetland plantings. Finally, the third stage will include relocating the tanning facility to the new industrial zone outside the medina. In its place, the former dye vats will form part of the park concept where, among other activities, workshops will be held in a new Center for Leather Design. Each of these component parts constitutes one of the silent and unmonolithic aspects of design that churn up the existing patterns to repurpose the space.

FIGURE 16.4. Rendering view of proposed playground in front of the Chouarra Tanneries, 2008. Aziza Chaouni and Takako Tajima, Bureau EAST Reproduced by permission of Aziza Chaouni Projects.

Rather than deferring to unethical demolition solutions outright, this elaborate yet spare architectural intervention sets out to repair and remember—a key aspect of the project—its (pre)existing sites and newly discovered spaces of encounter. Instead of clearing a tract to make way for the new, Chaouni's project salvages what is present and dives into what might be considered the poetic DNA of Fez to return to the surface with something unexpected yet familiar. This is not only a river remediation; it is a renaissance project that takes up the riverfront and river as a unit. The process is marked by a sort of unfastening, as it were, of the hinge between one riverbank and the next. In a way, we can see this approach as one that invents the future by recycling the past and returning to what the city was—almost. At its root lies a powerful consideration of Chaouni's own civic responsibility dedicated to green and sustainable architecture, ecology, adaptive reuse combined with the historical practices of Fez and its ancient past. Rather than a simple proposal to study and rework the infrastructure or infill, this project disturbs and awakens with a surprisingly antimonumental project in what becomes visible, yet monumental in its scope and transformative qualities for the city to reactivate and revisit the past while setting the stage for its future.

In an article identifying water infrastructure and arid climates specifically, Chaouni and her University of Toronto colleague Liat Margolis point out that there has been co-optation by the design disciplines recently spearheaded by the likes of Norman Foster of Foster and Partners, for example, with his Masdar City in Abu Dhabi, arguably one of the most celebrated "carbon neutral" cities anywhere. Chaouni and Margolis note that this has promoted a "crossbreeding between different infrastructural systems where consideration of Middle East new city thinking has been circling around issues related to wastewater and drinking networks, among others."[17] Given the introduction of Western practices and infrastructural crossbreeding, Chaouni and Margolis see this as an attempt to develop innovative technologies for water infrastructures in particular. They ask: "What are the possible solutions that can improve the efficiency of existing water networks and at the same time generate spaces and structures that cities need?" Furthermore, they describe their research of current practices, which reveals that many projects "improve, subvert, opportunistically appropriate or rethink urban infrastructure as a legitimate ground for investigation and action in order to 'sustainably transform the city,' rather than take infrastructure as a given, immutable foundation." They continue by saying that "while infrastructure armatures are used as platforms for design innovations and speculations, often their related technologies are taken as de facto components rather than assets with auto-catalytic capacity, combinatory possibility and scalar transferability; characteristics that would in turn come to re-conceptualize their entire associated network."[18]

While these highly technical and engineering typologies of mapping water technologies form the backbone of Chaouni's work, what lies at the heart of this functional and pressing research is the ability for the cities, zones, and built environments in question to face their own rebirth. Put another way, while the braiding of architecture, water technologies, infrastructural investigations, and landscapes is essential in coming to terms with the complexities and the delicate nature of any urban fabric, culture, aesthetics, and context must always be part of the discursive system and never taken as isolated factors. Place, as some contend, is dynamic, never fixed, and is a process of working toward something rather than a formal, immutable, less eidetic idea. Tied to place is memory in that memory, too, works toward or within a process of working that is dynamic and that preserves the past in the present. Henri Bergson's concept of *durée*, or duration, in which he focuses on the problem of creating a distinction somehow between what is past and what is present, all the while recognizing the continuity of durational time, brings us to the important question he poses: how can we imagine a perception without the connection to memory, when "there is no perception which is not full of memories"?[19] I want to suggest that for the Fez River project, in addition to the stated goals that Chaouni argues are necessary to address, "not only the ecology of a river but also the social and economic concerns of a city," or the associated network to which Chaouni refers, although not explicitly stated, is rich within a cultured enterprise. In addition to the buildings and the urban infrastructure that makes use of the historic city of Fez are also the sociocultural values that these features represent to the community and how they assist in defining for the city dweller and visitor a sense of belonging and social cohesion. Sustainability, therefore, gels successfully when the tangible and intangible heritage are integrated.[20] Key here is the repetition in different publications by Chaouni to underscore the socioeconomic prerogatives while focusing on the technological innovations. If I raise this again it is to point out the critical importance for the contextual parameters to play a central role in the overall project where cultural locality and concern for the community interface is never to be bypassed. Indeed, in a separate publication, Chaouni and her former business partner, Takako Tajima, "state that culture is still a dimension of urbanism that is at best perceived to be the de facto byproduct of all that is tangible and calculable; the 'soft' outcome to the 'hard' inputs. We argue, however that culture serves [as] a useful and necessary starting point for interventions at the infrastructure scale. Through a thorough understanding of the culture of the place, we can make meaningful decisions that maximize the transformative power that infrastructure has to offer."[21]

Part of that thick consideration of locality is when, earlier on in her writing, Chaouni evokes the choreographic model, one that is clever and fitting. Choreography suggests a series of steps or sequences assembled as a whole, a thriving unit, indeed a unit dependent upon its parts to contribute to the project, and conversely the whole communicating in harmony. Chaouni's team notes the exemplary sustainable metropolis Fez has always been in its fluid pedestrian weaving of routes within the city as well as the vernacular architectural forms that always seek to exploit shade at every opportunity and subsequently become acquired strategies and contextually driven cooling technologies. With an eye to respecting these historic sustainable strategies, Chaouni thought it "only appropriate that a project that showcases sustainable planning and design concepts and innovations in energy-efficient construction methods be located in the Medina . . . [which] is constantly evolving to accommodate change. . . . This project repositions the Medina to not only overcome the negative impacts of its current crisis [of overcrowding and a toxic river] but to become a proactive component for improving the ecological quality of the region."[22] One of the exemplary (choreographed) aspects of the project is that it improves the quality of water, remediates contaminated sites, and engages the open city spaces in a sequencing pattern of open rooms that punctuate the tightly meshed ancient built forms within the medina. While the project was initiated to undo the damage to the river, the city as a whole benefits and is revived in the process as a domino effect program, since it is impossible to imagine one part of the urban fabric regenerated that does not ultimately regenerate another.

The rehabilitation plan the team proposed includes phytoremediation, an approach that is economically viable and turns to plants to cleanse the soil or water. Heralded as an environmentally sound technology, it can effectively "remove, detoxify or immobilize environmental contaminants in a growth matrix (soil, water or sediments) through the natural biological, chemical or physical activities and processes of the plants."[23] Phytoremediation is a means of placing plants in contaminated locations in order to restore these environments to their precontamination conditions by decreasing the total mass of contaminants.[24] What this implies is that the toxic waste found in the River of Trash can be treated through the uptake of plant roots. This technology underscores the design methodology Chaouni embraces, which, as she suggests, is "one of frugality in which we develop design solutions that maximize yield with minimum expenditure."[25] The approach, she argues, is two-pronged: not only is cost expenditure reduced but programmatic contributions can be implemented as well. The river project, therefore, addresses first and foremost the river, but in so doing it results in what Chaouni calls "multiperformability," because it "expands public space in the Medina, improves circulation, and restructures and revitalizes the historic leather industry, an integral but ailing component of the cultural heritage of the Medina."[26] The newly designed "great room," or open public square at R'cif Plaza, therefore serves as a gateway to the labyrinthine network of paths through the medina. It no longer welcomes vehicles but instead becomes a space with cafés, gardens that serve as a stormwater infiltration zone, an amphitheater, and a market. Due to its location it is also a dropoff and pickup area where donkeys—the main form of transportation within the medina—may be hired and loaded. The concept of performativity in these newly created public squares and community theaters of activity builds on the assumption that a pragmatic experience of public space contributes to the formation of social and cultural identities.[27]

A second site, at the location where the hides from the Chouarra tanneries are dried, will become a pedestrian zone as well, equipped with a playground and a constructed wetland. This riparian ecosystem will assist with the river cleaning as well as provide a new wildlife habitat, an important example of phytoremediation. Another is within the tanning pits currently polluted with chemical dyes such as chromium(III) oxide. With the proper species planted for seeding within these pits, the phytoremediation process would break down the chromium(III) oxide into nontoxic chromium(IV) oxide. This is typically known as a passive technique that is becoming increasingly popular as a "recognized pathway for contaminant removal from water and shallow soils and is an aesthetically pleasing, solar-driven . . . technique useful for remediation of shallow plumes with low to moderate levels of contamination."[28] It should be noted that embedded in the scientific literature on phytoremediation appears the reference to the above-mentioned and not to be ignored "aesthetically pleasing" process, underpinning the unassailable presence as part and parcel of place, the community, and the ever-present urban cultural form. Yet all too familiar is an approach that sees the trope of the psychologically interpreted evil, built urban environment pitted against rural—romanticized—bucolic landscapes. This inevitably sets up a posture of blame wherein the city and all that is associated with it is rebuked: the dystopic Metropolis of Fritz Lang is replete with civil unrest, derelict pockets of homelessness, crime, and broken civic life. Rather, I prefer to demonstrate how it is best to see how we, as performers and mobile actors in city space, are politic players, whether in passive or active mode, but engaged in the city writ large as a beneficial space of encounter and livability—now more than ever a hybrid digital and physical site.

Indeed, Chaouni's idea is predicated on civic engagement to make it work. As such, we can come to understand that the spaces and places we inherit through a language of action at all times has consequences, as much from inaction as from action. With the acknowledgment that municipal governance is multilayered and action is present on all levels, civic action is still possible. With civic action and awareness in mind, Hannah Arendt's "space of appearance" serves well as a guiding ideal. The *space of appearance*, Arendt's critical theory about citizenship, statelessness, and the base requirement for voice in community, is one way to consider Chaouni's work in Fez. If people are denied a voice, or denied the possibility to partake in and belong to the community, they are nothing. In *The Human Condition*, Arendt argues for the right to

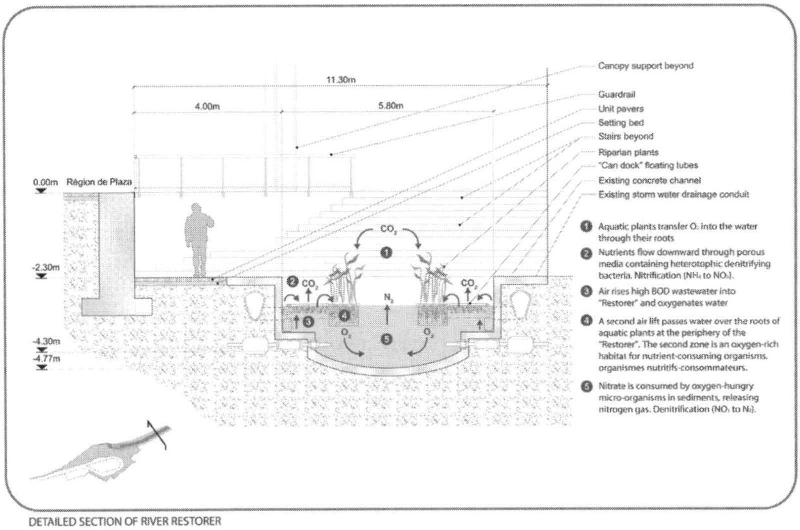

FIGURE 16.5. Detail section of the Fez River as proposed by the scheme, 2009. Aziza Chaouni and Takako Tajima, Bureau EAST Reproduced by permission of Aziza Chaouni Projects.

a space of appearance, where one belongs and is heard.[29] To be heard is to voice, to use language, where language stands for words as well as communication with other, nonverbal forms such as building materials or artistic constructions. Judith Butler, too, builds on this thought when she writes that language "is a name for our doing: both 'what' we do (the name for the action that we characteristically perform) and that which we effect, the act and its consequences."[30] As an architectural or artistic intervention in urban space, this project invites its public players to effectively undo and remake the city they knew. In the process, they are also invited to remember Fez, its history, its pathways, its people in everyday life, as a discursive system. By untangling some of the hidden networks of river and pedestrian activity as spatiality previously ignored, this resuscitation *event* challenges for citizens their spaces of appearance, their right to the city, their voice and sense of belonging in urban space. Put another way, by reevaluating and reeducating the river, its banks and connective tissue, this case study rehabilitates the sorely neglected *un-space* of city life.

To consider fully an *un-space,* let me add a sidebar to fully exploit the kind of work Chaouni introduces in Fez. Here, it is important to brush against the grain by focusing on the polar opposite of architectural symbols that are compelling and recognizable in physical space, particularly in cities, and instead turn to this un-space. Chaouni's work is what I would call *un-starchitecture.* What do I mean by this? Just as the river races deep within the wadi of historic Fez, Chaouni's project digs down, undoes and rebuilds through the existing and fragile fabric of the historic center. She is on a mission to have the local population relish in city space and face the water as the uniting force: the life-blood of the city. Her unorthodox approach to architecture renews the path of the river and its banks—as a circulation system for Fez. It is a radical urban transformation that celebrates nature and human agency and remembers the river's former life. Newly reidentified as the medina's iconic heartbeat and woefully misunderstood heritage, the river is quickly becoming its highly visible, audible, and functional central feature.

Memory and Place

As the late historian Tony Judt once suggested, the colorful graphic panels on the *autoroutes* of France succinctly deliver an immediate sense of the French past—a visual sound bite of sorts—to convey the richness of the nation and the memory of France. At excessive speeds only the most iconic symbols of monuments or sites on these panels can succeed in piquing one's interest. The picture might be of a mountain, a river, or a quaint village, followed perhaps a kilometer or so later by a brief account such as "you are now passing the vineyards of Burgundy, the cathedral at Reims, or the Mont Ste.-Victoire," and an arrow pointing to the location in the distance.[31] As Judt put it: "We are being led at seventy miles an hour through the Museum of France that is France itself." Once upon a time, sites were marked in relationship to and because of the "flowing fingers," or rivers and waterways, of connections etched into the landscape, not its highways. These fluid vectors determined in many ways the flows of migration and settlement. The rivers running through or around cities and towns once stood at the center of community life but have now fallen into the backdrop, silent, ignored, or even abused. Cast aside as irrelevant—increasingly so as goods rely less on river transport—and even dismissed outright as that which belongs to the unspoiled nature landscapes with no role to play in constructed city spaces, the river has been forsaken.

In *The Pain of Others*, as Aleida Assmann argues, Susan Sontag rejects the idea that there is such a thing as "collective memory" and boldly declares that all memory dies with the individual. On the other hand, Maurice Halbwachs, who first introduced the notion of "collective memory," understood the dilemma Sontag articulated and contextualized it by setting it within a framework of social space. He argued that memory is indistinguishable from a context made up of people, society, and a social setting. Furthermore, and as our invisible river in Fez proves irrevocably, Halbwachs emphasized that if those social frameworks in any way crumbled or disappeared, so too might memory.[32] Just as historical sites and monuments are or were initially intended as formal reminders, they too eventually disappear into a state of forgetting, otherwise known as memory's underbelly. "Once we assign monumental form to memory," as James Young famously wrote, "we have to some degree divested ourselves of the obligation to remember. . . . Under the illusion that our memorial edifices"—and here I want to insert the idea of natural geographic "objects" such as rivers—"will always be there to remind us, we take leave of them and return only at our convenience."[33]

Unlike the Seoul River project, Chaouni configures outdoor rooms taken from the river's lead. Once freshly exposed, the river weaves under, between, and through existing historic and contemporary structures in celebration of its new freedom and in homage to its past. Consequently, Chaouni's subtle but critical remediation invokes performativity by and with the community. Whereas starchitecture, or monolithic architectural buildings that make spectacular visual statements in city space, draw attention to themselves, this project speaks to a different kind of celebrity architecture that highlights a new sense of urban life and locates the river as the star architecture or the city's centerpiece. As such, the essence of Chaouni's subtle yet city-altering architectural work truly takes up Rumi's poetic suggestion that union is a raging river. Ultimately, the project considers the quintessential role of the river in the life of Fez, a river that was once considered a heterotopic—or negative—space. Her project challenges our concept of history and memory, monumentality and celebrity architecture, and circles back to the origins of the river's life force in the medina. Chaouni's remediation project in Fez is an almost invisible architectonic intervention as a key element for the city's renewal. The project relies on the invisible presence of the river becoming visible and thereby acting as a catalyst in the urban ecosystem in order to shake up a civic life and offer a place of transformation, a performative space of appearance. There is no question that unhinging the riverbanks by demolishing the concretization of the river itself in Fez is monumental in scale and therefore worthy of celebrity status, to be sure, but with a fundamental difference or what we might otherwise see as a reversal of the monumental: it involutes—that is, it hides behind and inside, swells, or reflects the city rather than builds on, up, and out, the way starchitecture typically does. There is a place for monolithic, self-contained, and absorbed starchitecture build-

FIGURE 16.6. Uncovering of the Fez river near the Rcif plaza during construction, 2011. Aziza Chaouni, Aziza Chaouni Projects. Reproduced by permission of Aziza Chaouni Projects.

ings, but this is not one of them. This project returns to what exists, the river and the river culture that has slipped away from view. In the process, the action of involution invites a community's agency, or self-reflexivity and self-awareness. This project unobtrusively inserts itself within these otherwise considered unsuitable spaces to alter patterns, behaviors, and contexts, and does so by using the geography to respatialize place. Rather than setting the stage for a spectacle to be played out, the project invites citizens to trace the contours of the place it thought it knew. Chaouni, who is from Fez and knows the history and culture of the place, now retells stories recounted to her by her grandmother about the role of the river in everyday life. She is not alone. The geographic return remaps the city and together with public participation the river is not just a reminder of the past and the locus of their memories but marks the future.

The Future of After

Whatever presence rivers have in our cities, however central they are or were to our everyday patterns in city spaces, all too many are, for the most part today, invisible. As a result of the intensification of urban construction the rivers, creeks, and brooks that were once discrete geographic markers of place

FIGURE 16.7. Uncovering of the Fez river near the Plaza all Yeddouna during the construction, 2013. Aziza Chaouni, Aziza Chaouni Projects. Reproduced by permission of Aziza Chaouni Projects.

have been buried under cities and therefore vanish from collective memory. The social frameworks that upheld the life of those waterways eventually crumbled from memory as well. Rarely are they the star attraction any longer, although a valiant effort is being made in the tourist industry to promote the river as a way to know a place.[34] Does it matter to revive those geographic sites as living memories of the city's origins? Can we return to recapture and restore a sense of the life the river gave of itself to the city? Rumi's poetic suggestion that union is a raging river aptly captures the essence of what it means for Aziza Chaouni to consider the quintessential role of the river in the life of Fez, a river that was once considered a heterotopic—or negative—space. Union comes from the river that rages through space and we, as actors in cities, are agents who bring about that union.

The project in Fez, Morocco, is an urban call to arms: a wakeup alert to memory and place, a mobilization of civic action through repair and revival. In Fez, the river rages once again, returning to its source in the center of its crumbling and alluring ancient medina. What was ignored and forgotten is now praised and remembered. Together with an environmental thrust and a reconsideration of landscape within and as part of the urban core, the local community is reengaged and active within a *space of appearance.*

The Fez resuscitation project takes a leap of faith and denies itself public authorship by creating, instead, a neutral zone where names of architects are unimportant for citizens as they partake of urban life, as they share words and deeds. The sharing of words and deeds was as Arendt imagined the "space of appearance." That space instead, as she framed it, was always in a state of being reinvented through action, or agency. Each recycles place into a *space of appearance* where the past is performed in the present tense. Each of these alternative manifestations acts as a signal to the civic imagination to recall the history of the sites. Through a language of architectonic action and reconfiguration, Fez, as now newly defined by its river, presents a space in which citizenship "takes" place and where citizens as agents respond by remembering place. These performances demonstrate how this architectural and landscape urbanism necessarily challenges our concept of monumentality and the spectacular: rather than reject the place itself, the city revels in place. An agent of change, this project is an emissary dispatching invitations for conversations hinged on the past, looking at the future squarely in the face, but lodged, all the same, in the present, to begin anew.

NOTES

INTRODUCTION

1. See Mark Cioc, *The Rhine: An Eco-Biography, 1815–2000* (Seattle: University of Washington Press, 2002); Joel Tarr, *The Search for the Ultimate Sink: Urban Pollution in Historical Perspective* (Akron, OH: University of Akron Press, 1996).

2. On rivers as resources see Dieter Schott, "Städte und ihre Ressourcen in der Geschichte: Blicke über und aus Europa," in *Raum und Zeit der Städte: Städtische Eigenlogik und jüdische Kultur seit der Antike*, ed. Andreas Hoppe (Frankfurt am Main: Campus 2011), 95–116.

3. See, for example, Jared Orsi, *Hazardous Metropolis: Flooding and Urban Ecology in Los Angeles* (Berkeley: University of California Press, 2004); Craig E. Colten, *Unnatural Metropolis: Wresting New Orleans from Nature* (Baton Rouge: Louisiana State University Press, 2005); Uwe Lübken, *Die Natur der Gefahr: Überschwemmungen am Ohio River im 19. und 20. Jahrhundert* (Göttingen: Vandenhoeck and Ruprecht, 2014); Lübken, "Rivers and Risk in the City: The Urban Floodplain as a Contested Space," in *Urban Rivers: Remaking Rivers, Cities, and Space in Europe and North America*, ed. Stéphane Castonguay and Matthew D. Evenden (Pittsburgh, PA: University of Pittsburgh Press, 2012), 130–44; Dieter Schott, "One City—Three Catastrophes: Hamburg from the Great Fire 1842 to the Great Flood 1962," in *Cities and Catastrophes: Coping with Emergency in European History*, ed. Geneviève Massard-Guilbaud, Harold Platt, and Dieter Schott (Frankfurt am Main: Peter Lang, 2002), 185–204.

4. See Martin Melosi, "The Place of the City in Environmental History," *Environmental History Review* 17, no. 1 (1993): 1–23; Dieter Schott, Bill Luckin, and Geneviève Massard-Guilbauld, eds., *Resources of the City: Contributions to an Environmental History of Modern Europe* (Burlington, VT: Ashgate, 2005); Joel Tarr and Clay McShane, *The Horse in the City: Living Machines in the Nineteenth Century* (Baltimore: Johns Hopkins University Press, 2007).

5. Chris Otter, "Locating Matter: The Place of Materiality in Urban History," in *Material Powers: Cultural Studies, History and the Material Turn*, ed. Tony Bennett and Patrick Joyce (London: Routledge, 2010), 38–59, 46.

6. The call for such rematerialization was a clear leitmotif at the European Conference on Urban History in Lisbon, September 2014.

7. Dieter Schott, "Stadt und Fluss: Flüsse als städtische Umwelten im 19. und 20. Jahrhundert," in *Beiträge zum Göttinger Umwelthistorischen Kolloquium 2004–2006*, ed.

Bernd Herrmann (Göttingen: Universitätsverlag Göttingen, 2007), 145–62; Ari Kelman, *A River and Its City: The Nature of Landscape in New Orleans* (Berkeley: University of California Press, 2003).

8. Tricia Cusack, ed., *Art and Identity at the Water's Edge* (Farnham, UK: Ashgate, 2012), 2.

9. Cusack, *Art and Identity*, 2; Eugen Trapp, "Beziehungs- und Grenzfragen: Regensburger Stadtbaugeschichte zwischen Aufklärung und Vormärz," in *1803—Wende in Europas Mitte*, ed. Peter Schmid and Clemens Unger (Regensburg: Schnell and Steiner, 2003), 291–94.

10. See Wolfgang Braunfels, *Abendländische Stadtbaukunst: Herrschaftsform und Baugestalt* (Köln: DuMont, 1977), 282–84.

11. Johann Heinrich von Thünen, *Von Thünen's Isolated State: An English Edition of Der isolierte Staat by Johann Heinrich v. Thünen.* Translated by Carla M. Wartenberg, ed. Peter Hall (Oxford: Pergamon, 1966).

12. See as a classic case the study on London's wood supply: James A. Galloway, Derek Keene, and Margaret Murphy, "Fuelling the City: Production and Distribution of Firewood and Fuel in London's Region, 1290–1400," *Economic History Review* 49, no. 3 (1996): 447–72. For the role of rivers at the beginning of the fossil fuel era see Dieter Schott, "Stadt und Fluss: Zentrale Dimensionen städtischer Umweltgeschichte," in *Umwelt und Klima im Geschichtsunter*richt, ed. Bärbel Kuhn and Astrid Windus (Saint Ingbert: Röhrig Universitätsverlag, 2013), 57–76.

13. See James A. Galloway, ed., *Trade, Urban Hinterlands and Market Integration c. 1300–1600*, Centre for Metropolitan History Working Papers, series no. 3 (London: Centre for Metropolitan History, Institute for Historical Research, 2000); Christopher Kennedy, John Cuddihy, and Joshua Engel-Yan, "The Changing Metabolism of Cities," *Journal of Industrial Ecology* 11, no. 2 (2007): 43–59; Gilles Billen, Josette Garnier, and Sabine Barles, "History of the Urban Environmental Imprint: Introduction to a Multidisciplinary Approach to the Long-Term Relationships between Western Cities and Their Hinterland," *Regional Environmental Change* 12, no. 2 (2012): 249–53.

14. See Jean-Pierre Poussou, "Towards a Definition of Town's Areas of Influence and Domination: The Large Hinterlands of French Ports and Their Development from the Middle of the 17th Century to the End of the 18th Century," in *Cities and Their Spaces: Concepts and Their Use in Europe*, ed. Michel Pauly and Martin Scheutz (Cologne: Böhlau, 2014), 235–49.

15. Richard W. Unger, "Feeding Low Countries Towns: The Grain Trade in the Fifteenth Century," *Revue belge de philologie et d'histoire* 77, no. 2 (1999): 329–58; Michael North, *Geschichte der Niederlande* (Munich: Beck, 2008).

16. Stephen Boyden, *The Ecology of a City and Its People: The Case of Hong Kong* (Canberra: Australian National University Press, 1981); Fridolin Krausmann, Heinz Schandl, and Rolf Peter Sieferle, "Socio-Ecological Regime Transitions in Austria and the United Kingdom," *Ecological Economics* 65, no. 1 (2008): 187–201; Friedrich Hauer,

Die Verzehrungssteuer 1829–1913 als Grundlage einer umwelthistorischen Untersuchung des Metabolismus der Stadt Wien, Social Ecology Working Papers 179 (Vienna: Institute of Social Ecology, IFF, 2010), http://www.uni-klu.ac.at/socec/downloads/WP129_web.pdf. See also the use of "urban metabolism" as a conceptual tool for a history of European urbanization in Dieter Schott, *Europäische Urbanisierung (1000–2000): Eine umwelthistorische Einführung* (Cologne: Böhlau, 2014), 17–20.

17. Marc Landry, "Water as White Coal," in *On Water: Perceptions, Politics, Perils*, ed. Agnes Kneitz and Marc Landry (Munich: Rachel Carson Center for Environment and Society, 2012), 7–11.

18. Eva Jakobsson, "Industrialization of Rivers: A Water System Approach to Hydropower Development," *Knowledge, Technology and Policy* 14, no. 4 (2002): 41–56.

19. Cioc, *Rhine*.

20. Verena Winiwarter, Martin Schmid, and Gert Dressel, "Looking at Half a Millennium of Co-Existence: The Danube in Vienna as a Socio-Natural Site," *Water History* 5, no. 2 (2013): 101–19.

21. Theodore Schatzki, "Nature and Technology in History," *History and Theory* 42, no. 4 (2003): 82–93; Marina Fischer-Kowalski and Helga Weisz, "Society as Hybrid between Material and Symbolic Realms: Toward a Theoretical Framework of Society-Nature Interaction," *Advances in Human Ecology* 8 (1999): 215–51. For the most recent summary of this approach see Winiwarter, Schmid, and Dressel, "Looking at Half a Millennium."

22. The results of the Viennese project have been published in a special issue of *Water History* 5, no. 2 (2013), and provide an impressive amount of material for a discussion of its scope and limits. See also for a discussion of the concept of socionatural sites Michael Toyka-Seid's contribution to this volume.

23. See Jeffrey H. Jackson, *Paris under Water: How the City of Light Survived the Great Flood of 1910* (New York: Palgrave MacMillan, 2010).

24. Stéphane Castonguay and Matthew Evenden, "Introduction," in Castonguay and Evenden, *Urban Rivers*, 3.

25. See Tarr, *Search for the Ultimate Sink*; Martin V. Melosi, *The Sanitary City: Environmental Services in Urban America from Colonial Times to the Present* (Pittsburgh, PA: Pittsburgh University Press, 2008).

26. "Endless Frontier," *Time*, July 30, 1951, quoted in Donald Worster, *The Wealth of Nature: Environmental History and the Ecological Imagination* (New York: Oxford University Press, 1993), 135. See also Leland R. Johnson, "Engineering the Ohio," in *Always a River: The Ohio River and the American Experience*, ed. Robert Reid (Bloomington: Indiana University Press, 1991), 180–209. For the Rhine see Christoph Bernhardt, *Im Spiegel des Wassers: Eine transnationale Umweltgeschichte des Oberrheins, 1800–2000* (Cologne: Böhlau, 2016); Martin Reuss, "Seeing like an Engineer: Water Projects and the Mediation of the Incommensurable," *Technology and Culture* 49, no. 3 (2008): 531–46.

27. John R. McNeill, *Something New under the Sun: An Environmental History of the Twentieth-Century World* (New York: W. W. Norton, 2000), 182.

28. Cioc, *Rhine*, 10, 16. The literature on the transformation of rivers by industrialization and urbanization is vast. See, most importantly, Castonguay and Evenden, *Urban Rivers*. For concrete examples, see Severin Hohensinner et al., "Changes in Water and Land: The Reconstructed Viennese Riverscape from 1500 to the Present," *Water History* 5, no. 2 (2013): 145–72; Sara B. Pritchard, *Confluence: The Nature of Technology and the Remaking of the Rhône* (Cambridge, MA: Harvard University Press, 2011); and Uwe Lübken, "Rivers and Risk in the City: The Urban Floodplain as a Contested Space," in Castonguay and Evenden, *Urban Rivers*, 130–44.

29. See Harold L. Platt, *Shock Cities: The Environmental Transformation and Reform of Manchester and Chicago* (Chicago: University of Chicago Press, 2005), 203, 393–94; Harriet Ritvo, *The Dawn of Green: Manchester, Thirlmere, and Modern Environmentalism* (Chicago: University of Chicago Press, 2009), 53–64.

30. Theodore Steinberg, *Down to Earth: Nature's Role in American History* (New York: Oxford University Press, 2002), 137, 166; David Soll, *Empire of Water: An Environmental and Political History of the New York City Water Supply* (Ithaca, NY: Cornell University Press, 2013).

31. Steinberg, *Down to Earth*, 166.

32. Stephen Halliday, *The Great Stink of London: Sir Joseph Bazalgette and the Cleansing of the Victorian Metropolis* (Stroud: Sutton, 1999), 71–75.

33. See Richard J. Evans, *Death in Hamburg: Society and Politics in the Cholera Years, 1830–1910* (Oxford: Clarendon, 1987); Tarr, *Search for the Ultimate Sink*; Melosi, *Sanitary City*; Susanne Frank and Matthew Gandy, eds., *Hydropolis: Wasser und die Stadt der Moderne* (Frankfurt am Main: Campus, 2006).

34. See Jennifer L. Bonnell, *Reclaiming the Don: An Environmental History of Toronto's Don River Valley* (Toronto: University of Toronto Press, 2014), 75–112; Lübken, *Natur der Gefahr*, 200; and the chapters by Vladimir Sánchez-Calderón and Nico Döring and Georg Jochum in this volume. For Hamburg, see Schott, "One City," 194.

35. For Pittsburgh, see Edward K. Muller, "River City," in *Devastation and Renewal: An Environmental History of Pittsburgh and Its Region*, ed. Joel A. Tarr (Pittsburgh, PA: University of Pittsburgh Press, 2003), 42: "With the shift to an industrial economy after the mid-nineteenth century, the role of the rivers changed and in the minds of many Pittsburghers diminished to a subordinate one in the city's economic world. Through technological means, the rivers were engineered and controlled, reduced in function to serve industry, and made less accessible, both physically and psychologically, to city residents."

36. Bonnell, *Reclaiming the Don*, 87.

37. Felix Fabri, *Fratris Felicis Fabri Tractatus de civitate Ulmensi, de eius origine, ordine, regimine, de civibus eius et statu*, vol. 186 (Stuttgart: Bibliothek des Litterarischen Vereins in Stuttgart, 1889), 50–51.

38. Achim Landwehr, "Kulturgeschichte, Version: 1.0," *Docupedia-Zeitgeschichte*, 14 May 2013, accessed 2 Nov. 2014, http://docupedia.de/zg/Kulturgeschichte.

39. See Schott, "Flüsse als städtische Umwelten," 147.

40. Reinhard Falter, "Natur als Landschaft und als Gott: Fluß- und Berggötter in der Spätantike," in *Natur-Bilder: Wahrnehmungen von Natur und Umwelt in der Geschichte*, ed. Rolf Peter Sieferle and Helga Breuninger (Frankfurt am Main: Campus, 1999), 148–55.

41. Cioc, *Rhine*, 14.

42. See, for example, John J. Audubon, "The Ohio," *The Museum of Foreign Literature, Science, and Art* 20, no. 117 (Mar. 1832): 334. This "melancholy sense of loss" was not limited to the experience of rivers, of course, but was witnessed in the face of many other modern transformations as well. For the United States, see Jackson Lears, *Rebirth of a Nation: The Making of Modern America, 1877–1920* (New York: HarperCollins, 2009), 137.

43. See, for example, Peter Danker-Carstensen, "Die Krückau—'Problemfluss' oder Lebensader einer Industriestadt," in *Flüsse in Norddeutschland: Zu ihrer Geschichte vom Mittelalter bis in die Gegenwart*, ed. Norbert Fischer and Ortwin Pelc (Neumünster: Wachholtz, 2013), 167–211. See also Steven Hawley, *Recovering a Lost River: Removing Dams, Rewilding Salmon, Revitalizing Communities* (Boston: Beacon, 2011). Jennifer Price, "Remaking American Environmentalism: On the Banks of the L.A. River," *Environmental History* 13, no. 3 (2008): 536–55, holds that "Angelenos have lost their fifty-one-mile river in plain sight—and can't see the essential daily connections from the river to their lives" (549). On the history and revitalization of the Los Angeles river see also Blake Gumprecht, *The Los Angeles River: Its Life, Death, and Possible Rebirth* (Baltimore: Johns Hopkins University Press, 1999). For the importance of the notion of loss within the field of environmental history, see the chapter "The Nature We Have Lost" in Worster, *Wealth of Nature*.

44. See the contributions by Geneviève Massard-Guilbaud on Nantes, by Nico Döring and Georg Jochum on Munich, and by Shelley Hornstein on Fez in this volume.

45. Keigo Nakamura, Klement Tockner, and Kunihiko Amano, "River and Wetland Restoration: Lessons from Japan," *BioScience* 56, no. 5 (2006): 419–29.

46. Anne Whiston Spirn, *The Granite Garden: Urban Nature and Human Design* (New York: Basic Books, 1984), 146.

47. Spirn, *Granite Garden*, 146. In Leicester, England, the large Abbey Park was opened in 1882 in the floodplain of the river Soar. Besides providing amenities for the local population and preventing further development at this flood-prone location, the park acted as planned spillway for floods and reduced the flood risk for adjacent districts close to the river. E. J. Emery, *The History of Abbey Park, Leicester* (Leicester: Leicester City Council, 1982), 19–21; Dieter Schott, "'The Taming of the Soar': Leicester Transforms Its River Environment," in *Leicester: A Modern History*, ed. Richard Rodger and Rebecca Madgin (Lancaster: Carnegie, 2016). See also chapter 6, "Restoration Is Ancient

History," of Ann Lawrence Riley's book *Restoring Streams in Cities: A Guide for Planners, Policy Makers, and Citizens* (Washington, DC: Island, 1998).

48. For the complicated story of the 1969 fire, its transformation by the media, the exaggeration of its importance, and a genealogy of its iconic status, see David Stradling and Richard Stradling, "Perceptions of the Burning River: Deindustrialization and Cleveland's Cuyahoga River," *Environmental History* 13, no. 3 (July 2008): 515–35.

49. See Cioc, *Rhine*, 109–10, 182; Thomas Lekan, "Saving the Rhine: Water, Ecology, and *Heimat* in Post–World War II Germany," in *Rivers in History: Perspectives on Waterways in Europe and North America*, ed. Christof Mauch and Thomas Zeller (Pittsburgh, PA: University of Pittsburgh Press, 2008), 110–36.

50. Nakamura, Tockner, and Amano, "River and Wetland Restoration," 419; Emily S. Bernhardt et al., "Synthesizing U.S. River Restoration Efforts," *Science* 308, no. 5722 (2005): 636–37.

51. Emily S. Bernhardt et al., "Restoring Rivers One Reach at a Time: Results from a Survey of U.S. River Restoration Practitioners," *Restoration Ecology* 15, no. 3 (2007): 482.

52. Richard Pinkham, *Daylighting: New Life for Buried Streams* (Snowmass, CO: Rocky Mountain Institute, 2000).

53. Riley, *Restoring Streams in Cities*, 2.

54. Riley, *Restoring Streams in Cities*, 2.

55. Myung-Rae Cho, "The Politics of Urban Nature Restoration: The Case of Cheonggyecheon Restoration in Seoul, Korea," *International Development Planning Review* 32, no. 2 (2010): 145–65; *Deconstruction/Construction: The Cheonggyecheon Restoration Project in Seoul*, ed. Joan Busquets (Cambridge, MA: Harvard University Graduate School of Design, 2011), and the review of this book by Tom Looser, *Journal of Korean Studies* 18, no. 2 (2013): 407–10.

56. Timothy J. LeCain, *Mass Destruction: The Men and Giant Mines that Wired America and Scarred the Planet* (New Brunswick, NJ: Rutgers University Press, 2009), 224.

Chapter 1. Rivers, Industrial Cities, and Hinterland Production in Quebec in the Nineteenth and Twentieth Centuries

A grant from the Social Sciences and Humanities Research Council of Canada made this research possible. Vincent Bernard, Jonathan Bernier, and Hubert Samson provided invaluable research assistance. I would like to thank the archivists of Hydro-Quebec, Martine Rodrigue and Caroline Rouleau.

1. Fridolin Krausmann, "Land Use and Industrial Modernization: An Empirical Analysis of Human Influence on the Functioning of Ecosystems in Austria, 1830–1995," *Land Use Policy* 18, no. 1 (2001): 17–26; "The Process of Industrialization from the Perspective of Energetic Metabolism: Socioeconomic Energy Flows in Austria 1830–1995," *Ecological Economics* 41, no. 2 (2002): 177–201; Jérome Buridant, "Flottage des bois et

gestion forestière: L'exemple du Bassin parisien, du XVIe au XIXe siècle," *Revue forestière française* 58, no. 3 (2006): 389–98; Sabine Barles, "Urban Metabolism and River Systems: An Historical Perspective—Paris and the Seine, 1790–1970," *Hydrology and Earth System Sciences*, special issue, "Man and River Systems: Long-Term Interactions between Societies and Nature in Regional Scale Watersheds" (2007): 1757–69.

2. Craig Colten, *Unnatural Metropolis: Wresting New Orleans from Nature* (Baton Rouge: Louisiana State University Press, 2005); Christopher G. Boone, "Language Politics and Flood Control in Nineteenth-Century Montreal," *Environmental History* 1, no. 3 (1996): 70–85; Jared Orsi, *Hazardous Metropolis: Flooding and Urban Ecology in Los Angeles* (Berkeley: University of California Press, 2004); Stéphane Castonguay, "The Construction of Flood as Natural Catastrophe: Extreme Events and the Construction of Vulnerability in the Drainage Basin of the St. Francis River (Quebec), Mid-Nineteenth to Mid-Twentieth Century," *Environmental History* 12, no. 4 (2007): 816–40; Harold L. Platt, "'The Hardest Worked River': The Manchester Floods and the Industrialization of Nature," in *Villes et Catastrophes: Réactions face à l'urgence dans l'histoire européenne*, ed. Geneviève Massard-Guilbaud, Harold L. Platt, and Dieter Schott (Frankfurt am Main: Peter Lang, 2002), 163–83.

3. Jacques Bethemont, *Les grands fleuves: Entre nature et société* (Paris: Armand Colin, 2002); François Molle, "River-Basin Planning and Management: The Social Life of a Concept," *Geoforum* 40, no. 3 (2009): 484–94; Shannon Stunden-Bower, "Natural and Unnatural Complexities: Flood Control along Manitoba's Assiniboine River," *Journal of Historical Geography* 36, no. 1 (2010): 57–67.

4. Among the numerous contributions to *Urban Rivers: Remaking Rivers, Cities, and Space in Europe and North America* (Pittsburgh, PA: University of Pittsburgh Press, 2012), edited by Stéphane Castonguay and Matthew Evenden, see the chapters by Craig Colten ("Fluid Geographies: Urbanizing Rivers Basins," 201–18) and Sabine Barles ("The Seine and Parisian Metabolism: Growth of Capital Dependencies in the Nineteenth and Twentieth Centuries," 95–112).

5. William Cronon, *Nature's Metropolis: Chicago and the Great West* (New York: W. W. Norton, 1991); Steven Stoll, *Fruits of Natural Advantage: Making the Industrial Countryside in California* (Berkeley: University of California Press, 1998); Joel Tarr and Clay McShane, "Urban Horses and Changing City-Hinterland Relationships in the United States," in *Resources of the City: Contributions to an Environmental History of Modern Europe*, ed. Dieter Schott, Bill Luckin, and Geneviève Massard-Guilbaud (Aldershot, UK: Ashgate, 2005), 48–62.

6. Eva Jakobsson, "Industrialization of Rivers: A Water System Approach to Hydropower Development," *Knowledge, Technology and Policy* 14, no. 4 (2002): 41–56.

7. René Hardy and Normand Séguin, *Histoire de la Mauricie* (Sainte-Foy: Presses de l'Université Laval, 2004).

8. *La Vallée du St-Maurice et les Avantages qu'elle Offre à l'Industrie, au Commerce et à la Colonisation* (Ottawa: Département de l'Agriculture, 1887).

9. Rapport du Commissaire des Travaux Publics pour 1865, Canada. *Documents de la Session*, 1866, vol. XXVI, no. 1 (no. I), app. 9, 29–30.

10. Rapport Général du Ministre des Travaux Publics du 30 juin 1867 au 1er juillet 1882, Canada, *Documents de la Session*, vol. 16, no. 10, annex. no. 21 (1883), 660.

11. René Hardy and Normand Séguin, *Forêt et société en Mauricie: La formation d'une région*, new ed. (Quebec: Septentrion, 2011), 74–76.

12. Jean-Pierre Kesteman, Peter Southam, and Diane Saint-Pierre, *Histoire des Cantons de l'Est* (Sainte-Foy: Presses de l'Université Laval, 1998), 382–85.

13. Jean-Pierre Kesteman, "Une bourgeoisie et son espace: Industrialisation et développement du capitalisme dans le district de Saint-François, Québec, 1823–1879" (PhD thesis, Université du Québec à Montréal, 1985), 432–39.

14. Library and National Archives of Canada, *Papiers BALC*, 8 Dec. 1834, 954.

15. John Derek Booth, *Railways of Southern Quebec*, vol. 1 (Toronto: Railfare, 1982), 12; Raoul Blanchard, *Le centre du Canada français: "Province de Québec"* (Montreal: Beauchemin, 1947), 315–23.

16. Jack Little, *Nationalism, Capitalism and Colonization in Nineteenth-Century Quebec: The Upper St. Francis District* (Kingston: McGill-Queen's University Press, 1989).

17. John Derek Booth, "Changing Forest Utilization Patterns in the Eastern Townships of Quebec, 1800–1930" (PhD thesis, McGill University, 1971).

18. National Archives of Canada, RG 4 C1 Civil Secretary's Correspondence (Incoming), no. 1947, vol. 442, "Pétition des Habitants de Weedon, Garthby, Stratford, Winslow, etc., Stratford, June 7 1858," quoted in Little, *Nationalism, Capitalism and Colonization*, 250n63; Archives of the City of Lennoxville, vol. GRP02, file 7/39, "Dam across St. Francis River," J. P. Cushing to Edward Chapman, August 28, 1856; Edward Chapman to J. P. Cushing, Inspector of Roads and Bridges, August 29, 1958; A. G. Woodward to the Municipal Council of Lennoxville, November 3, 1879.

19. Claude Bellavance, René Hardy, Pierre Lanthier, and Normand Séguin, *Shawinigan: Genèse d'une croissance industrielle au début du 20e siècle* (Trois-Rivières: Université du Québec à Trois-Rivières/Hydro-Québec, 1985).

20. Claude Bellavance and France Normand, "Bâtir une région industrielle: Le développement de l'arrière-pays de Trois-Rivières (1850–1930)," *L'Action nationale* 90, no. 1 (2000): 57–88.

21. Hardy and Séguin, *Histoire de la Mauricie*, 554; Kesteman, Southam, and Saint-Pierre, *Histoire des Cantons de l'Est*, 273.

22. Hydro-Quebec Archives, Paper of the Shawinigan Water and Power Company, F 18, vol. 4048, "Rate Study, 1949 and 1953."

23. Quebec Streams Commission, *Annual Report* (1912), 20–21.

24. Hydro-Quebec Archives, Paper of the Southern Canada Power Company, F 15, vol. 3477, file 30, A. S. Hurd to J. H. Turnbull, December 8, 1903. See also vol. 3473, file 27, *Preliminary Draft of Report on the Proposal to Form an Association for the Protection*

of the Lake Memphremagog Storage, n. d., City of Sherbrooke, Hydro-Sherbrooke Historical Archives, box 13, "Magog River Flow Control," n.d.

25. Quebec, Legislative Assembly, "Copie de Tous Documents, Correspondance, Options et Requêtes, Échangés entre Toutes Personnes, le Gouvernement de cette Province et les Officiers de la Commission des Eaux Courantes Relativement à l'Emmagasinement des Eaux de la Rivière Sainte-François et de ses Lacs et Rivières Tributaires: 1er mars 1915," *Sessional Papers* no. 32 (George V AD 1915), 3–4.

26. Castonguay, "Construction of Flood."

27. Serge Bouchard, *Mémoires d'un Simple Missionnaire: Le père Joseph-Étienne Guinard, 1864–1965* (Quebec City: Ministère des Affaires Culturelles, 1980).

28. Quebec, Legislative Assembly, "Copie de Tous Documents, Correspondance, Options et Requêtes, Échangés entre Toutes Personnes, le Gouvernement de cette Province et les Officiers de la Commission des Eaux Courantes Relativement à l'Emmagasinement des Eaux de la Rivière Saint-François et de ses Lacs et Rivières Tributaires: 15 Février 1916," *Sessional Papers* no. 55 (George V AD 1916), 3–4; Quebec, Legislative Assembly, *Debates*, 14th Parliament, 2nd Session (4 Dec. 1917), 3.

29. Olivier Lefebvre, "La Commission des Eaux Courantes et les Travaux de Régularisation du Régime des Rivières," *La Forêt et la Ferme* 2, no. 4 (Oct.–Nov. 1930): 291.

30. City of Sherbrooke Historical Archives, Plenary Commissions, box 75, "Report No. 10," 18 Jan. 1939. On the controversy surrounding the construction of a dam at Ulverton, see Jean-Pierre Kesteman, *La Ville Électrique: Sherbrooke 1880–1988* (Sherbrooke: Olivier, 1988), 165.

31. Kesteman, *La Ville Électrique*, 196.

32. René Verrette, *Les Idéologies de Développement Régional: Le Cas de la Mauricie, 1850–1950* (Quebec City: Presses de l'Université Laval, 1999).

33. Stéphane Castonguay and Dany Fougères, "Les Rapports Riverains de la Ville: Sherbrooke et ses Usages des Rivières Magog et Saint-François au 19e–20e Siècle," *Urban History Review/Revue d'histoire urbaine* 36, no. 1 (2007): 3–15.

34. Claude Allard, *Étude de la pollution des cours d'eaux de la province de Québec* (Montreal: Fédération des Associations de chasse et pêche du Québec, 1955); "Les Clubs de Chasse et Pêche de L'Estrie Veulent L'Épuration des Eaux du Saint-François," *Chasse et Pêche* 70, no. 1 (1955): 15 and 17.

35. Michel Philipponneau, *L'Avenir Économique et Social des Cantons de l'Est: Un Problème Type de Planification Régionale de la Province de Québec* (Quebec City: Ministère de l'Industrie et du Commerce, 1960).

36. Stéphane Castonguay, "Les Territoires de la Pollution: L'Environnement Comme Catégorie de l'Action Publique au Québec," in *Les Sciences du Territoire: Perspectives Québécoises*, ed. Guy Massicotte (Sainte-Foy: Presses de l'Université du Québec, 2008), 77–102.

37. J. H. Dubé, *Rapport sur la Qualité des Eaux de la Rivière Saint-François* (Quebec City: Régie des Eaux du Québec, 1970), 12.

38. Comité CHARM, Ville de Sherbrooke, *Aménagement de la rivière Magog* ([Sherbrooke]: Ville de Sherbrooke, 1980).

39. D. Laflamme, *Qualité des Eaux du Bassin de la Rivière Saint-Maurice, 1979 à 1992* (Quebec City: Ministère de l'Environnement et de la Faune, Direction des écosystèmes aquatiques), rapport QE-98, Envirodoq EN950251.

40. *La Saint-Maurice: Une Rivière en Danger* (Quebec City: Ministère de l'Environnement du Québec, 1985).

41. Louise Plante, "Bulletin de Santé de la Saint-Maurice," *Le Nouvelliste* [Trois-Rivières], 19 Aug. 1989, 22.

42. Stephen Mosley, *The Chimney of the World: A History of Smoke Pollution in Victorian and Edwardian Manchester* (Cambridge: White Horse, 2001).

43. Normand Brouillette, "Le Développement Industriel d'une Région du Proche Hinterland Québécois, la Mauricie, 1900–1975" (PhD thesis, McGill University, 1983).

44. Peter Brooke Clibbon, "The Nature, Evolution and Present Extension of the Ecumene in the Shield Section of the Saint-Maurice Valley," *Cahiers de Géographie du Québec* 13, no. 1 (1969): 30.

45. Corporation du Développement du Bassin de la Rivière Saint-Maurice, *Au Service de la Collectivité: La Rivière Saint-Maurice, une Inspiration, un Héritage, une Puissance: Mémoire* (1995).

Chapter 2. The Seine as a Parisian River

1. See, for example, André Guillerme, "Le testament de la Seine," *Revue de géographie de Lyon* 65, no. 4 (1990): 240–50; Isabelle Backouche, *La trace du fleuve, la Seine et Paris (1750–1850)* (Paris: éd. de l'EHESS, 2000).

2. See, for example, Joséphine Roulliard, Paul Benoit, and Raphaël Morera, *L'eau dans les campagnes du bassin de la Seine avant l'ère industrielle: Comprendre les paysages d'aujourd'hui* (Nanterre: AESN, 2011), accessed 29 June 2012, http://www.sisyphe.upmc.fr/piren/book/1325.

3. This study complements and extends a previous article: Sabine Barles, "The Seine and Parisian Metabolism: Growth of Capital Dependencies in the Nineteenth and Twentieth Centuries," in *Urban Rivers: Remaking Rivers, Cities, and Space in Europe and North America*, ed. Stéphane Castonguay and Matthew D. Evenden (Pittsburgh, PA: University of Pittsburgh Press, 2012), 95–112. In this first text, the emphasis was on the last two centuries and the focus was on urban metabolism in connection to river management; that is, on the supply and discharge side. It resulted in the introduction of the notion of "urban dependencies," as the material manifestation of the city outside its official limits. Pushing the analysis further, two needs came up: first, a better understanding of the prerevolutionary situation regarding river management and control; second, a better connection between urban dependencies, which had to be more precisely defined, and legal and power issues. This is what I try to do in the present text, limited to the supply side.

4. Marie-Hélène Bourquin-Simonin, *L'approvisionnement de Paris en bois de la Régence à la Révolution (1715–1789)* (Paris, 1969; Clamecy: Confrérie Saint-Nicolas de Clamecy, 2006), 198.

5. *Extrait et rapport d'un mémoire du procureur du Roi et de la Ville de Paris sur les moyens 1° d'augmenter les ressources de l'approvisionnement des combustibles destinés à la capitale; 2° de perfectionner et d'entretenir la navigation des rivières par lesquelles se font les transports de ces approvisionnements*, 1788, Archives Nationales F12 653. All quotes from French sources in this text have been translated by the author.

6. *Approvisionnement de Paris en bois. Rapport*, n.d. [1789–90], Archives Nationales F12 653.

7. Jean Boissière, "Une source pour l'histoire du commerce des forêts et du bois: Les enquêtes du bureau de la ville," in "La forêt en Île-de-France," special issue, *Mémoires de la fédération des sociétés historiques et archéologiques de Paris et de l'Île-de-France* 28 (1977): 119.

8. Military engineer and architect Sébastien Le Prestre de Vauban (1633–1707), usually referred to as Vauban, played a major role in France: he designed numerous fortifications, conducted many sieges, advised Louis XIV on the consolidation of French borders, and wrote several essays on various topics, from the art of siege to political reform.

9. Vauban, *Mémoire sur la navigation des rivières* [1698], in *Mémoires des intendants sur l'état des généralités dressés pour l'instruction du Duc de Bourgogne*, vol. 1, *Mémoire sur la généralité de Paris*, ed. A. M. de Boislisle (Paris, 1881), 404. The other "nursing mother" at the time would have been the Marne.

10. *Approvisionnement de Paris en bois. Rapport*, n.d. [1789–90], Archives Nationales F12 653.

11. *Approvisionnement de Paris en bois. Rapport*, n.d. [1789–90], Archives Nationales F12 653.

12. Paul Benoit, "Les aménagements hydrauliques liés au flottage du bois, leur impact sur le milieu fluvial XVIe–XVIIIe siècles," in *PIREN-Seine, Rapport d'activité 2002* (Paris: Programme Interdisciplinaire de Recherche sur l'Environnement de la Seine, 2003), 5, http://www.metis.upmc.fr/piren/?q=book/713.

13. Henri Louis Duhamel du Monceau, *Du transport, de la conservation & de la force des bois . . .* (Paris, 1767), 19.

14. Benoit, "Les aménagements liés au flottage du bois," 19.

15. Ordonnance des eaux et forêts, August 1669, titre XXVII, art. XLI, in *Commentaire sur l'ordonnance des eaux et forêts, du mois d'août 1669*, ed. Daniel Jousse (Paris, 1772), 357. Jousse specified that a declaration by the king in April 1683 added islands, islets, and fluvial deposits.

16. Ordonnance des eaux et forêts, August 1669, titre XXVII, art. XLII, in Jousse, *Commentaire sur l'ordonnance*, 358–59.

17. Ordonnance des eaux et forêts, August 1669, titre XXVII, art. XLIV, in Jousse, *Commentaire sur l'ordonnance*, 361.

18. Ordonnance des eaux et forêts, August 1669, titre XXVIII, art. VII, in Jousse, *Commentaire sur l'ordonnance*, 366.

19. Ordonnance des eaux et forêts, August 1669, titre I, art. III, in Jousse, *Commentaire sur l'ordonnance*, 10.

20. This order did for Paris what the 1669 Order did for waters and forests: it was both a summary of earlier texts and a foundational document that would serve as a reference throughout the eighteenth century and even after the Revolution.

21. Jousse, *Commentaire sur l'ordonnance*, 13.

22. See in particular Bonaventure Chailland, *Dictionnaire raisonné des eaux et forêts . . .*, vol. 1 (Paris, 1769), 516.

23. Ordonnance concernant la juridiction de l'Hôtel de Ville, December 1672, ch. 1, in Nicolas De La Mare, *Traité de la Police*, vol. 3 (Paris, 1719), 873.

24. Ordonnance concernant la juridiction des prévôt des marchands et échevins de la ville de Paris, 1672, ch. 1, art. XI, in *Code du commerce de bois et de charbon pour l'approvisionnement de Paris . . .*, vol. 1, ed. André Marie Dupin (Paris, 1817), 96.

25. Préfecture du département de la Seine, *Extrait des registres des procès-verbaux des séances du conseil général du département de la Seine, faisant office de conseil municipal de la ville de Paris*, séance du 13 juill. 1827, Archives de la Préfecture de Police DA 21.

26. Ordonnance concernant la juridiction de l'Hôtel de Ville, December 1672, ch. 17, 874.

27. Ordonnance concernant la juridiction des prévôt des marchands et échevins de la ville de Paris, vol. 1, ch. 3, art. II, December 1672, 96.

28. Ethys De Corny, *Extrait du rapport des moyens généraux de rétablir et de maintenir la navigation des rivières affluentes de Paris, pour l'approvisionnement de la capitale, et notamment d'exécuter les travaux à faire à la vanne d'Anglure sur la rivière Aube . . .*, 17 July 1788, Archives Nationales F12 653.

29. Antoine Furetière, *Dictionnaire universel . . .* (La Haye, 1690).

30. Furetière, *Dictionnaire universel.*

31. Antoine-François Prost de Royer, *Dictionnaire de jurisprudence et des arrêts, ou nouvelle édition du dictionnaire de Brillon*, vol. 6 (Lyon, 1781–1788), 774–75.

32. *Arrêt de la Cour de Parlement qui ordonne l'exécution d'une ordonnance rendue par le Prévôt des Marchands & Échevins de la Ville de Paris, pour la coupe & l'exportation des bois destinés à l'approvisionnement de Paris*, 30 déc. 1785, Archives de la Préfecture de Police DB 112, my emphasis.

33. Préfecture du département de la Seine, *Extrait des registres des procès-verbaux des séances du conseil général du département de la Seine, faisant office de conseil municipal de la ville de Paris*, séance du 13 juill. 1827, Archives de la Préfecture de Police DA 21.

34. Frédéric Graber, "Diverting Rivers for Paris, 1760–1820: Needs, Quality, Resistance," in Castonguay and Evenden, *Urban Rivers*, 183–200.

35. Eugène Belgrand, *Recherches statistiques sur les sources du bassin de la Seine qu'il est possible de conduire à Paris . . .*, vol. 1 (Paris, 1854), 26.

36. Graber, "Diverting Rivers," 190.

37. Henri Nadault de Buffon, *Traité des eaux de sources et des eaux thermales . . .* (Paris, 1870), 388.

38. Belgrand, *Recherches statistiques.*

39. In his memoirs, Haussmann claims to have been the originator of this choice. Baron Haussmann, *Mémoires*, ed. Françoise Choay (Paris: Seuil, 2000), 946.

40. Armand-Jacques Le Roy de Saint-Arnaud, in "Sénat. Extrait des comptes rendus de la séance du 28 juin 1862," *Le Temps*, 2 July 1862.

41. Belgrand, *Recherches statistiques*, 29, 28.

42. Belgrand, *Recherches statistiques*, 1, 75.

43. Haussmann, *Mémoires*, 1016.

44. Auguste Mathieu, *L'expropriation pour cause d'utilité publique et les eaux de la Somme-Soude et de la Dhuis, du Sourdon et du Surmelin* (Épernay, 1862), 9.

45. Mathieu, *L'expropriation pour cause d'utilité publique*, 12, 15, 7, 25.

46. Haussmann, *Mémoires*, 1016. Haussmann was actually more favorable to using springs in lower Burgundy, "whose diversion does not excite such strong opposition."

47. Le Roy de Saint-Arnaud, in "Sénat. Extrait des comptes rendus de la séance du 28 juin."

48. An infrastructural project is declared as being of public utility when it is deemed useful to the inhabitants of a territory or area. The declaration provides the legal means to expropriate land and buildings if necessary for the implementation of the project.

49. Le Roy de Saint-Arnaud, in "Sénat. Extrait des comptes rendus de la séance du 28 juin."

50. Victor Tourangin, in "Sénat. Extrait des comptes rendus de la séance du 19 mai," *Le Temps*, 21 May 1862.

51. Tourangin, in "Sénat. Extrait des comptes rendus de la séance du 28 juin."

52. Tourangin, in "Sénat. Extrait des comptes rendus de la séance du 19 mai."

53. Tourangin, in "Sénat. Extrait des comptes rendus de la séance du 19 mai." Article 643 of the Code Napoléon "forbids the owner of a source the right to divert it, if it supplies a community of inhabitants." Nadault de Buffon, *Traité des eaux de sources*, 30–31.

54. Tourangin, in "Sénat. Extrait des comptes rendus de la séance du 19 mai."

55. Le Roy de Saint-Arnaud, in "Sénat. Extrait des comptes rendus de la séance du 28 juin."

56. Noël Lefebvre-Duruflé, in "Sénat. Extrait des comptes rendus de la séance du 28 juin."

57. Tourangin, in "Sénat. Extrait des comptes rendus de la séance du 27 juin," *Le Temps*, 30 June 1862; Jean-Baptiste Dumas, in "Sénat. Extrait des comptes rendus de la séance du 27 juin," *Le Temps*, 1 July 1862.

58. Pierre Magne, in "Sénat. Extrait des comptes rendus de la séance du 28 juin."

59. I. Feyrnet, "Chronique," *Le Temps*, 31 July 1868.

60. Feyrnet, "Chronique."

61. Belgrand, *Mémoire sur l'avant-projet de dérivation des sources de la Vanne* (Paris, 1866), 21.

62. "Conseil municipal de Paris. Séance du 19 février," *Le Temps*, 21 Feb. 1880.

63. A. Vernier, "Causerie scientifique," *Le Temps*, 31 Jan. 1893.

64. Adrien Carpentier and Georges Frérejouan Du Saint, *Répertoire général alphabétique du droit français . . .*, vol. 29 (Paris: Librairie de la Société du recueil Sirey, 1902), 686.

65. It covered for the fact that "the City of Paris guarantees the town of Nemours 800 m^3 of springwater per day and proposed to all communes with inhabitants that held the sources from which water was to be diverted, treaties in which Paris committed itself to provide them with an equivalent amount of water, and also to carry out and pay for the construction works . . . should the communes so wish." Carpentier and Frérejouan Du Saint, *Répertoire général alphabétique*, 687.

66. Purchases of the sources of the Seine by the city of Paris in 1864 should also be mentioned.

67. H. Gilbert, "La direction technique du port de Paris," *Travaux*, hors série (1958): 239.

68. These sources alone were not enough to supply Paris, which also pumped water from the Seine and Marne.

69. Gilbert, "La direction technique," 239.

70. Gilles-Antoine Langlois, "Henri Chabal et l'aménagement du bassin de la Seine," *Centraliens*, no. 562 (May 2005): 41.

71. "Problème de la prévention des inondations dans le bassin de la Seine," *Journal officiel de la République française: Avis et rapports du Conseil économique*, no. 1 (1956): 9.

72. *Proposition tendant à inviter l'administration à pousser ses études en cours sur les questions d'approvisionnement en eau et de protection contre les inondations . . .* (Paris, 1925), 7.

73. "Problème de la prévention," 9.

74. Gilles-Antoine Langlois, *Pannecière* (Paris: Somogy/IIBRBS, 2003), 24.

75. Conseil général de la Marne, *2e session ordinaire de 1831: Rapport de M. Ch. Magny, préfet du département, et* délibérations du conseil (Châlons-sur-Marne: Imprimerie-librairie de l'Union Républicaine, 1931), 1075.

76. Langlois, *Pannecière.*

77. Langlois, *Pannecière*, 37–40.

78. "Problème de la prévention," 9.

79. "Problème de la prévention," 15.

80. Gilbert, "La direction technique," 240.

81. Translation of "levée de bouclier," "Problème de la prévention," 13.

82. Trans., "levée de bouclier," 12.

83. Service des barrages-réservoirs (Préfecture de la Seine), *Notice sur le réservoir "Seine,"* 2nd ed. (Paris: Imprimerie municipale, 1963), 3.

84. As of 1930, however, the Academic Society of Agriculture, Sciences, Arts, and Literature of the Aube (Département Société académique d'agriculture, des sciences, arts et belles-lettres du département de l'Aube) wanted to "ensure the failure of the project for creating an artificial lake in the forests of the Grand-Orient."

85. See Gilles-Antoine Langlois, *Le lac d'Orient* (Paris: Somogy/IIBRBS, 2004).

86. "Problème de la prévention," 14.

87. And from 2011 onward Établissement Public Territorial de Bassin Seine Grands Lacs.

88. See, for instance, Guillerme Barles, "Paris: A History of Water, Sewers, and Urban Development," in *A History of Water*, series 3, vol. 1, *From Jericho to Cities in the Seas: A History of Urbanization and Water Systems*, ed. Terje Tvedt and Terje Oestigaard (London: IB Tauris, in press).

89. See Barles, "Seine and Parisian Metabolism."

90. For other examples see, for instance, Bertrand Lemoine, ed., *Paris en Île-de-France, histoires communes* (Paris: Picard/Du Pavillon de l'Arsenal, 2006).

Chapter 3. Watershed Democracy or Ecological Hinterland?

Thanks to the Economic and Social Research Council for Research Grant ES/I031502/1, University of Greenwich, Sarah Palmer, Yvonne Rydin, Sue Tapsell, the editors of this volume, two anonymous referees, the Rachel Carson Center, Munich, Dominic Martyn (Environment Agency), and Action for River Kennet.

1. For more on the Water Act, see the section headed "Democracy, Local Government, and River Basins," later in this chapter.

2. W. J. Luckin, *Pollution and Control: A Social History of the Thames in the Nineteenth Century* (Bristol: Hilger, 1986); Christopher Hamlin, *A Science of Impurity: Water Analysis in Nineteenth Century Britain* (Bristol: Hilger, 1990); Richard H. Trainor, "The 'Decline' of British Urban Governance since 1850: A Reassessment," in *Urban Governance: Britain and Beyond since 1750*, ed. Robert J. Morris and Richard H. Trainor (Aldershot, UK: Ashgate, 2000), 28–46.

3. On ecological hinterland, see, e.g., Graham Haughton, Ian Rowe, and Colin Hunter, "The Thames Gateway and the Re-Emergence of Regional Strategic Planning: The Implications for Water Resource Management," *Town Planning Review* 68, no. 4 (1997): 407–22.

4. Donald Worster, "Watershed Democracy: Recovering the Lost Vision of John Wesley Powell," *Journal of Land, Resources, and Environmental Law* 23, no. 1 (2003): 65; Olivier Graefe, "River Basins as New Environmental Regions? The Depolitization of Water Management," *Procedia—Social and Behavioral Sciences* 14, no. 3 (2011): 26; David Marquand, *The Decline of the Public: The Hollowing-Out of Citizenship* (Cambridge: Polity, 2004).

5. See, e.g., Neil Brenner, "Metropolitan Institutional Reform and the Rescaling of State Space in Contemporary Western Europe," *European Urban and Regional Studies* 10, no. 4 (2003): 299.

6. David Kinnersley, *Troubled Water: Rivers, Politics and Pollution* (London: Hilary Shipman, 1988), 2; Barry M. Doyle, introduction to *Urban Politics and Space in the Nineteenth and Twentieth Centuries: Regional Perspectives* (Newcastle upon Tyne: Cambridge Scholars, 2007), 16. See important earlier environmental histories such as William Cronon, *Nature's Metropolis: Chicago and the Great West* (New York: W. W. Norton, 1991); and Stéphane Castonguay and Matthew Evenden, eds., *Urban Rivers: Remaking Rivers, Cities, and Space in Europe and North America* (Pittsburgh, PA: University of Pittsburgh Press, 2012).

7. José E. Castro, Maria Kaïka, and Erik Swyngedouw, "London: Structural Continuities and Institutional Change in Water Management," *European Planning Studies* 11, no. 3 (2003): 287.

8. Donald Snow, "The Persistence of Powell: The Idea of Watersheds and Participatory Democracy," *Journal of Land, Resources, and Environmental Law* 23, no. 1 (2003): 31–40; Erwin C. Hargrove, *Prisoners of Myth: The Leadership of the Tennessee Valley Authority, 1933–1990* (Princeton, NJ: Princeton University Press, 1994). See, however, François Molle, "River-Basin Planning and Management: The Social Life of a Concept," *Geoforum* 40, no. 3 (2009): 484–94; W. A. Maloney and J. Richardson, *Managing Policy Change in Britain: The Politics of Water* (Edinburgh: Edinburgh University Press, 1995); Judith Petts, "Learning about Learning: Lessons from Public Engagement and Deliberation on Urban River Restoration," *Geographical Journal* 173, no. 4 (2007): 300–311.

9. Vanessa Taylor and Frank Trentmann, "Liquid Politics: Water and the Politics of Everyday Life in the Modern City," *Past and Present* 211 (2011): 199–241; John Sheail, "Government and the Perception of Reservoir Development in Britain: An Historical Perspective," *Planning Perspectives* 1, no. 1 (1986): 45–60.

10. Matthew Flinders, "MPs and Icebergs: Parliament and Delegated Governance," *Parliamentary Affairs* 57, no. 4 (2004): 768; Ben Pimlott and Nirmala Rao, "Metropolitan Miasma: Blurred Accountabilities in the Governance of London," *London Journal* 29, no. 2 (2004): 33–45; Ben Page, "Results of the Case Study on the Water Supply in the Region of London, UK," in *Achieving Sustainable and Innovative Policies through Participatory Governance in a Multi-Level Context: Final Report (Water)*, ed. Eric Swyngedouw, Ben Page, and Maria Kaïka, annexed to Hubert Heinelt et al., *Achieving Sustainable and Innovative Policies through Participatory Governance in a Multi-level Context: Final Report* (EU Project No. HPSE-CT-1949-00028, 2002), 232.

11. John Davis, *Reforming London: The London Government Problem, 1855–1900* (Oxford: Clarendon, 1988); John Davis, "London Government 1850–1920: The Metropolitan Board of Works and the London County Council," *London Journal* 26, no. 1 (2001): 47–56; Bill Luckin, "The Metropolitan and the Municipal: The Politics of Health and Environment in London, 1860–1920," in *Cities of Ideas: Civil Society and Urban*

Governance in Britain, 1800–2000, ed. Robert Colls and Richard Rodger (Aldershot, UK: Ashgate, 2004), 46–66.

12. Flinders, "MPs and Icebergs," 771; William Robson, *The Government and Misgovernment of London* (London: Allen and Unwin, 1939), 114–20, 135–38.

13. Luckin, "Metropolitan and the Municipal," 66; James Gillespie, "Municipalism, Monopoly and Management: The Demise of 'Socialism in One County,' 1918–1933," in *Politics and the People of London: The London County Council 1889–1965*, ed. Andrew Saint (London: Hambledon, 1989), 105, 109; Stephen Inwood, *A History of London* (London: Macmillan, 1998), 709. For public participation, see also A. Skeffington, Committee on Public Participation in Planning, *People and Planning: Report of the Committee on Public Participation in Planning* (London: HMSO, 1969); John Davis, "'Simple Solutions to Complex Problems': The Greater London Council and the Greater London Development Plan, 1965–1973," in *Civil Society in British History: Ideas, Identities, Institutions*, ed. José Harris (Oxford: Oxford University Press, 2003), 249–74.

14. J. Abbernethy, evidence, *Report from the Select Committee of the House of Lords [HL] on Conservancy Boards, &c*, 1877 (371), Q. 423; Pugh, "River Conservancy and Floods Prevention Bill," House of Commons [HC] Debate, Hansard, vol. 260, cc. 958–62, 7 Apr. 1881.

15. "The Water Supply of London," *Times*, 19 July 1892, 8. See also: J. Prestwich, *A Geological Inquiry Respecting the Water-bearing Strata of the Country around London . . .* (London, 1851); "Mr. Bailey Denton's Plan," 1868–69 [4169] Royal Commission on Water Supply, Report, Part II, no. 68; R. Etheridge, "On the Rivers of the Cotteswold Hills . . . as Supply to the Main River and the Metropolis," *Proceedings of the Cotteswold Naturalists' Field Club* 11, part 1 (1892–93): 49–101; John Broich, "London's and Surrey's 'Fight for the River': Politics, Perceptions and the Thames, 1897," *Surrey History* 6, no. 2 (2000): 75–90; John Sheail, "Underground Water Abstraction: Indirect Effects of Urbanization on the Countryside," *Journal of Historical Geography* 8, no. 4 (1982): 395–408. For an 1894 Thames basin map see *Researching the River Thames*, accessed 14 Nov. 2015, http://www2.gre.ac.uk/about/faculty/ach/research/centres/maritime/research/case-studies/runningriverthames/timeline/maps/river-thames-and-thames-basin.

16. Elizabeth Porter, *Water Management in England and Wales* (Cambridge: Cambridge University Press, 1978), 19–23.

17. John Sheail, "Planning, Water Supplies and Ministerial Power in Inter-War Britain," *Public Administration* 61, no. 4 (1983): 386–95; John Hassan, "The Water Industry 1900–1951: A Failure of Public Policy?," in *The Political Economy of Nationalisation in Britain 1920–1950*, ed. Robert Millward and John Singleton (Cambridge: Cambridge University Press, 1995), 189–211. For RWAs' failure to integrate with local authority planning, see Yvonne Rydin, *Urban and Environmental Planning in the UK*, 2nd ed. (Basingstoke: Palgrave Macmillan, 2003), 162–63.

18. Metropolitan Water Board [MWB], *London's Water Supply, 1903–1953* (London: Staples, 1953), 311–12. The National Archives [TNA]: HLG 50/2250: Departmental Com-

mittee on Greater London Water Supplies [DCGLWS]: Verbatim Notes 1947: Maxwell Fyfe, 10 Mar. 1947, 6; Sir Cyril Radcliffe: 21 Apr. 1947, 126. HLG 50/2264: DCGLWS: Review of Technical Evidence 1949: MWB, "A Review of the Resources and Consumption of Water in the Great London Area by P.G.H. Boswell," 4.

19. TNA: HLG 50/2263: Ministry of Health, *Report of the Departmental Committee on Greater London Water Supplies* (London: HMSO, 1948), 8, 11. HLG 50/2261: "DCGLWS. Statement of Evidence by the Metropolitan Boroughs' Standing Joint Committee" (20 June 1947), 2; LCC, "DCGLWS. Statement of Evidence" [undated], 5. HLG 50/2248: DCGLWS: Report. Memo: Mr Armer to Sir John Wrigley, 7 July 1948; MWB, *London's Water Supply*, 316; Asok Kumar Mukhopadhyay, "The Politics of London Water," *London Journal* 1, no. 2 (1975): 212–17.

20. Glyn, "Water Resources," HL Debate, Hansard, vol. 204, c. 99, 29 May 1957; B. J. Hardcastle, "From Concept to Commissioning," in *Thames Ground Water Scheme*, ed. ICE (London: ICE, 1978), 5; "Boring May Solve London's Water Problem," *Times*, 24 June 1963, 6; "Topping Up the Thames," *Times*, 17 Feb. 1967, 2.

21. *Newbury Weekly News*, 28 Oct. 1965, 22; Letter from Lord Hurd, *Times*, 7 Oct. 1965, 13; Letter from Richard Nugent, *Times*, 8 Oct. 1965, 13; "Water Plan for London Makes Farmers Angry," *Times*, 23 Oct. 1965, 5; "Correction," *Times*, 27 Oct. 1965, 12; Lord Nugent, "Opening Address," in *Thames Ground Water Scheme*, ed. ICE, 2–3; "Cirencester Public Inquiry 1980," *Wilts and Gloucestershire Standard*, 27 June 1980, 1; 5 Dec. 1980, 1; "No. 2007 Water, England and Wales, The Thames Water Authority (Gatehampton Farm Boreholes) Order 1987"; "Gatehampton Borehole Upgrade," accessed 14 Nov. 2015, http://www.gpsuk.com/casestudy/8/6/gatehampton-borehole-upgrade.html.

22. Porter, *Water Management*, 25.

23. Berkshire Record Office [BRO]: D/EX 423/6/25: Letter from (Secretary) Ministry of Housing and Local Government [MHLG] to Town Clerk, Swindon Corporation, 30 Mar. 1960 [Result of Public Inquiry].

24. See, e.g., "Water Supplies of England: Lord Desborough's Warning," *Times*, 15 Nov. 1927, 13; Thames Conservancy, *The Thames Conservancy, 1857–1957* (London: Thames Conservancy, 1957), 46.

25. BRO: D/EX 423/6/25: Prof. H. L. Hawkins, "Report to the Kennet Valley Protection Association on the Proposal by Swindon Corporation . . . " (July 1958), 3.

26. BRO: D/EX 423/6/25: Letter from (Secretary) MHLG to Town Clerk, Swindon Corporation, 30 Mar. 1960; Note by Godwin, Bembridge and Clifton, 8 June 1959; Water Act, 1948, 11 & 12 Geo 6. Ch. 22, S. 8; Ancram, "River Kennet, Wiltshire," HC Debate, Hansard, vol 312, c. 579, 14 May 1998. For earlier abstractions in the Kennet, see: W. S. Atkins Consultants Ltd, *Kennet and Coln: River Levels Study, Final Report, Volume 1: River Kennet. Report to National Rivers Authority, Thames Region* (1992), accessed 14 Nov. 2015, http://ea-lit.freshwaterlife.org/archive/ealit:3650/OBJ/20001620.pdf. For key towns: MHLG/Minister for Welsh Affairs, *The South-East Study* (London: HMSO, 1964), "Appendix 1: Water Supplies in South-East England," 54–59.

27. TNA: HLG 127/1212: "Report of the MHLG on the Port of London Bill: Preamble" (7 Apr. 1964)," no. 7.

28. Castro, Kaïka and Swyngedouw, "London," 287. The RWAs took on the duties of 1,580 former bodies. Porter, *Water Management*, 28; Daniel A. Okun, *Regionalization of Water Management: A Revolution in England and Wales* (London: Applied Science Publishers, 1977), 147.

29. Water Resources Board, *Water Supplies in South East England* (London: HMSO, 1966), 11–12; A. G. Jordan, J. J. Richardson, and R. H. Kimber, "The Origins of the Water Act 1973," *Public Administration* 55, no. 3 (1977): 319–20; Hassan, *History of Water*, 123–24.

30. TNA: HLG 127/1073: "Water Reorganisation: Note of a Meeting Chaired by Lord Kennet on 10th May 1968," no. 3(f, g). Local government was also being reconsidered at this time. *Report of the Royal Commission on Local Government* [Cd. 4040] (London HMSO, 1969); Local Government Act 1972, Ch. 70.

31. TNA: HLG 127/1506: "Implementation of CAWC Report, Note of a Meeting Held on 3 March 1971," nos. 1, 2(i); Department of the Environment (DoE), "Water Reorganisation" (undated, c. March 1971), No. 4(iv); "Implementation of CAWC Report. Notes of a Meeting Held on 26 Feb 1971," No. 4(v). For hydrosocial cycles: Karen Bakker, *An Uncooperative Commodity: Privatizing Water in England and Wales* (Oxford: Oxford University Press, 2003), 28–29; Erik Swyngedouw, "The Political Economy and Political Ecology of the Hydro-Social Cycle," *Journal of Contemporary Water Research & Education* 142, no. 1 (2009): 56–60. For the construction of regions, see Allan Cochrane, "Making Up a Region: The Rise and Fall of the 'South East of England' as a Political Territory," *Environment and Planning C: Government and Policy* 30, no. 1 (2012): 95–108.

32. HLG 127/1073: Lord Kennet, "Water Reorganisation," [Kennet Report] 1 June 1969, no. 8; "Paper 13. Water Organisation: Relationship to Local Government" (undated), part III, no. 29. See also MHLG, "Local Government Reform Water Functions and Organisation," 16 Apr. 1969 (Memo by M. Stevenson).

33. Jordan, Richardson, and Kimber, "Origins of the Water Act," 324. On technology and silences see Bruno Latour, *Science in Action* (Cambridge, MA: Harvard University Press, 1987), 2–3, 21.

34. TNA: HLG 127/1506: J. E. Beddoe memo, "CAWC Report on the Organisation of the Water Industry in England and Wales" (Feb. 1971), no. 17.

35. See, e.g., Gloucestershire Archives [GA]: D6804/1/7: Cotswold Water Board Minutes, 1972–74: "British Waterworks Association: Vice-Chairman's Report on Proceedings . . . 20th–23rd June 1972," 2; Surrey History Centre [SHC]: 6128/1/84: East Surrey Water Company; *Annual Report and Accounts for the Year ended 25th March 1971*, Chairman's Statement, 7; Okun, *Regionalization of Water Management*, 95.

36. TNA: MT 88/327: D. Plummer (GLC) letter to P. Walker (DoE), 9 Feb. 1972; HLG 29/1244. E. Griffiths (DoE) letter to D. Plummer, 15 Aug. 1972, 1–2. For Greater London and neighboring boundaries, see PLA, "Port of London Act 1964" map at *Researching*

the River Thames," accessed 14 Nov. 2015, http://www2.gre.ac.uk/about/faculty/ach/research/centres/maritime/research/casestudies/runningriverthames/timeline/maps/map-of-port,-river-and-greater-London.

37. TNA: MT 88/327: "Draft Paragraphs on Water Reorganisation . . . " [Lord Greenwood's Speech to River Thames Society, 21 September 1973]; Okun, *Regionalization of Water Management*, 69, 86–88, 148–49.

38. The GLC was responsible for land drainage in the "London Excluded Area" (approx. 402 square meters/1,040 square kilometers). Edmund C. Penning-Rowsell, Dennis J. Parker, and D. M. Harding, *Floods and Drainage: British Policies for Hazard Reduction, Agricultural Improvement and Wetland Conservation* (London: Allen and Unwin, 1986), 45.

39. Michael Keating and Malcolm Rhodes, "Politics or Technocracy? The Regional Water Authorities," *Political Quarterly* 52, no. 4 (1981): 488–89; Howell, "Water Authorities (Constitution)," HC Debate, Hansard, vol. 965, c. 1322, 4 April 1979; Wilts and Swindon History Centre [WSHC]: K7/130/1: Thames Water, *Annual Report and Accounts for the Year Ended 31st March 1978*, 5, 41. In 1978 TWA had fifty-seven members: thirty-six local government representatives (twenty from London) and twenty-one government appointees.

40. My emphasis. [WSHC]: K7/130/1: *Thames Water, Annual Report and Accounts . . . for the Year Ended 31st March 1984*, 3. For citizens, consumers, and customers, see Bakker, *Uncooperative Commodity*, 26; Taylor and Trentmann, "Liquid Politics."

41. Local Government Act 1985, Ch. 51; Local Government Act 1985 (Land Drainage Functions) Order 1986; Thames Barrier and Flood Prevention Act 1972 (Amendment) Order 1986.

42. Mukhopadhyay, "Politics of London Water," 224; Jordan, Richardson, and Kimber, "Origins of the Water Act," 317–34; Graham Taylor, *State Regulation and the Politics of Public Service: The Case of the Water Industry* (London: Mansell, 1999), 78–80; Bakker, *Uncooperative Commodity*, 60; Vanessa Taylor, "Whose River? London and the Thames Estuary, 1960–2014," *London Journal* 40, no. 3 (2015): 252.

43. Museum of London Docklands [MOLD], Port and River Collection: London Rivers Association papers [Aspinall Collection]; Marjorie Mayo, Ines Newman, and Nick Sharman, "Rebuilding London's Docklands: The Strategic Plan—Two Years On," *Community Development Journal* 15, no. 1 (1980): 68; P. Pinch, "Waterspace Planning and the River Thames in London," *London Journal* 40, no. 3 (2015): 272–92.

44. SHC: 6304/box16: Molesey Residents' Association; 3696/19/12: Flood Victims Campaign; 8416: Ember Action Group. See also Thames Preservation Society (1890s); Underground Water Protection Society (1902); Thames Tributaries Power Users Association (1920s); Pure Rivers Society (1930s).

45. Gresham-Cooke, "River Thames," HC Debate, Hansard, vol. 749, cc. 1224-25, 3 July 1967; BRO: D/EX 1465, TPAF Records; MOLD: British Ports Association 95: PLA Inquiry, 7, 8, 9 November 1966 (Proposed Revision Order) Verbatim Report: Third Day,

RTS evidence, 10–14; MOLD: Proposed Reconstitution of the PLA: Public Inquiries and Working Papers (1966–1974) (box 1.1): Port of London Revision Order Public Inquiry, 22–25 July 1974: Third Day, RTS evidence, 28–29.

46. *Wilts and Gloucestershire Standard*, 2 May 1980, 2; 27 June 1980, 1; 4 July 1980, 14; GA: DA22/132/132: TPAF Minutes, 26 June 1968, 1. For river basin protest, see Yorkshire WaterWatch: Graham Haughton, "Private Profits—Public Drought: The Creation of a Crisis in Water Management for West Yorkshire," *Transactions of the Institute of British Geographers* 23, no. 4 (1998): 419–33. Comparing watersheds and "polycentric . . . subwatershed communities of interest": William Blomquist and Edella Schlagerb, "Political Pitfalls of Integrated Watershed Management," *Society and Natural Resources* 18, no. 2 (2005): 101–17. On problems of a "generally applicable typology" for state/nonstate actor relationships, see Rydin, *Urban and Environmental Planning*, 116.

47. Jan Kooiman, ed., *Modern Governance: New Government-Society Interactions* (London: Sage, 1993); M. Goldsmith and J. Garrard, "Urban Governance: Some Reflections," in Morris and Trainor, *Urban Governance* 15–27; Nigel Watson, Hugh Deeming, and Raphael Treffny, "Beyond Bureaucracy? Assessing Institutional Change in the Governance of Water in England," *Water Alternatives* 2, no. 3 (2009): 448–60; Maureen G. Reed and Shannon Bruyneel, "Rescaling Environmental Governance, Rethinking the State: A Three-Dimensional Review," *Progress in Human Geography* 34, no. 5 (2010): 646–53.

48. Bakker, *Uncooperative Commodity*; Jeremy Purseglove, *Taming the Flood: History and Natural History of Rivers and Wetlands* (Oxford: Oxford University Press, 1988).

49. Frances Craig and Phil Craig, *Britain's Poisoned Water* (Harmondsworth: Penguin Books, 1989), 118. EU Water Framework Directive, accessed 12 Aug. 2015, http://ec.europa.eu/environment/water/water-framework/index_en.html. For recent consultation on river basin management plans, see Environment Agency, *Water for Life and Livelihoods: Consultation on the Draft Update to the River Basin Management Plans—Summary Response Document* (2014), https://web.archive.org/web/20151009042106/https://www.gov.uk/government/uploads/system/uploads/attachment_data/file/452286/Water_for_life_and_livelihoods_-_Summary_response_document.pdf.

50. On postprivatization consumer/environment conflicts: Bakker, *Uncooperative Commodity*.

51. BRO: D/EX 423/6/25: Borough of Swindon: Proposed Water Abstraction—Kennet Valley, Notes of a Meeting at Civic Offices, Swindon, 16 June 1958; KVPA Minutes, 23 Oct. 1959; Letter from Jepson Turner to KVPA, 9 Dec. 1959.

52. WWF, *WWF-UK Case Study: The Impact of Over-Abstraction on the River Kennet* (WWF/Natural England, undated; c. 2011), 3, citing "Axford Licence Variation Application, Environmental Report" (Thames Water Utilities Ltd: August 2007), accessed 9 Feb. 2014, http://assets.wwf.org.uk/downloads/case_study_kennet_final.pdf; Water Briefing, "Thames Water Wins Two Year Extraction Licence Extension," 9 May 2011, accessed 11 Feb. 2014, http://www.waterbriefing.org/index.php/home/company-news/item/4176

-thames-water-wins-two-year-extraction-licence-extension; *Action for the River Kennet*, http://www.riverkennet.org/. See also W. S. Atkins Consultants, *Kennet and Coln*.

53. Thanks to Dave Webb, then Biodiversity Team Leader, SE Area, Thames Region, EA, Oral History Interview: 15 Feb. 2011.

54. On NGOs' influence and limits, see Matthew Hilton, James McKay, Nicholas Crowson, and Jean-François Mouhot, *The Politics of Expertise: How NGOs Shaped Modern Britain* (Oxford: Oxford University Press, 2013).

55. For London and the region, see also Castro, Kaïka, and Swyngedouw, "London," 287; Defra, *Catchment Based Approach: Improving the Quality of Our Water Environment*, May 2013, accessed 14 Nov. 2015, https://www.gov.uk/government/uploads/system/uploads/attachment_data/file/204231/pb13934-water-environment-catchment-based-approach.pdf.

56. See, e.g. the creation of LDDC (1981), abolition of metropolitan counties (1986), and recent announcement on fracking: Adam Vaughan, "Government Will Step in if Councils Don't Fast-Track Fracking Applications," *Guardian* (UK), 13 Aug. 2015, accessed 13 Aug. 2015, http://www.theguardian.com/environment/2015/aug/13/government-will-step-in-if-councils-dont-fast-track-fracking-applications.

57. Heinelt et al., *Achieving Sustainable and Innovative Policies through Participatory Governance in a Multi-Level Context*, 49, accessed 14 Nov. 2015, http://cordis.europa.eu/documents/documentlibrary/82608091EN6.pdf.

Chapter 4. The City Whose Rivers Disappeared

1. This outline of the estuary history is based on the records of the Chambre de commerce of Nantes (series 1ET), preserved at the Archives départementales de Loire-Atlantique (hereafter ADLA), the records of the local Ponts-et-chaussées (also at ADLA, listing in progress) and the many printed sources available at the Nantes central library.

2. The Loire Improvement Act (24 Dec. 1903) authorized the deepening of the river up to eight meters.

3. 17 July 1913 Act of Enlargement and Improvement of Nantes Harbor.

4. The Ponts-et-chaussées (literally "bridges and roads") engineering body was one of the most highly respected bodies of State engineers.

5. Conseil régional des Pays de la Loire, "Politiques régionales: Environnement," Région pays de la *loire*, www.paysdelaloire.fr.

6. Archives nationales (hereafter AN) F14 12604.

7. Archives municipales de Nantes (hereafter AMN) 1298 W2.

8. Literally: "Damned Bridge"! The origin of this name is unknown. It has no connection with this collapse.

9. ADLA 509 S 39 to 42.

10. ADLA 509 S 39 to 42, 21 Mar. 1928.

11. The Pont de Pirmil, for instance, was the only bridge allowing tramways, cars, and walkers to get to the south bank, where sat a smaller part of Nantes and the town of Rezé. The second bridge visible on fig. 4.2 is a railway bridge.

12. AMN sous-série 3 O, carton 4, dossiers 15 to 19.

13. AMN O3 4/16.

14. Although they did so in their own meetings (AN F14 12604).

15. AMN 1298 W2.

16. This was a first victory of the Ponts-et-chaussées, as the city architect responsible for the design of the plan had always been (and would remain) opposed to the filling.

17. *Architectes et urbanistes à Nantes, 1892–1947,* Nantes, CERMA rapport no. 45, rapport de fin d'étude remis au ministère de l'Urbanisme et du Logement, 1985, 63.

18. The boats should have passed through a lock, but there was a lock already at the entrance of the river Erdre to protect it from the tide and keep it afloat.

19. AMN 1298 W2.

20. AMN 1298 W2.

21. AMN 1298 W2.

22. ADLA 509 S 39 to 42.

23. AN F14 12604.

24. AN F14 12604.

25. Technical blueprint in *Et si la Loire océane . . .* (Nantes: Loire Estuaire, 2005), 15. See also Guireg Migaud, "Inventaire des ressources bentho-démersales dans la masse d'eau de transition de la Loire" (master's thesis, Université Paul Verlaine, 2006).

26. *Nature*, I admit, is rather inaccurate a word, as the Saint Félix branch had been fitted out by Félix, bishop of Nantes, in the sixth century and was not really "natural."

27. AN F14 12604.

28. AMN 1298 W23.

29. The Pirmil branch and the south of the city are not visible on these pictures, which show only three of the four branches the Loire still had at that time.

30. On this topic see Yannick Le Marec, *Le Temps des capacités, les diplômés nantais à la conquête du pouvoir dans la ville* (Paris: Belin, 2000).

31. Source: "Des villages de Cassini aux communes d'aujourd'hui," Database *EHESS*, http://cassini.ehess.fr/cassini/fr/html/1_navigation.php.

32. On the history of the relationship between Nantes and Saint-Nazaire see Gabriel Le Bras, "Les ports de la basse Loire," *Annales d'histoire économique et sociale* 4, no. 15 (1932): 308–11. See also "Avis à M. Le ministre du Commerce," in *Exposé des travaux de la Chambre de commerce de Nantes* (1860), 63; and "Lettre à monsieur le préfet de Loire inférieure," *Exposé des travaux de la Chambre de commerce de Nantes* (1871), 91. These documents are available at the Heritage section of the Médiathèque Jacques Demy in Nantes.

33. See, for instance, William Arnous Rivière, *Notice sur l'amélioration de la Basse-Loire et ses conséquences pour Nantes, d'après les résultats obtenus sur d'autres fleuves* (Nantes, 1850); AN F14 12601 to 12607: "Amélioration et extension des grands ports maritimes: Rapports imprimés de quelques comités d'études, 1883–1943."

34. On the history of the port and the estuary, along with the local board of trade and the administrative records, the main printed sources are: Méderic-Clément Lechalas, *Transformation de la Basse Loire et du port de Nantes* (Nantes, 1868); *Critique du projet*

d'amélioration de la Loire présenté par M. l'ingénieur en chef Léchalas (Nantes, 1870); Maurice Sibille, *Rapport fait au nom de la Commission des Travaux publics . . . chargée d'examiner le projet de loi ayant pour objet l'amélioration et l'extension du port de Nantes par la Chambre des députés* (Paris: Martinet, 1912); Maurice Kauffman, *La Loire et le port de Nantes* (Paris: A. Dumas, 1915).

35. *Ouest-Éclair,* Jan. 1929 and Jan. 1930.

36. Louis Lefèvre-Utile, *La Loire: Un programme de destruction et d'infection de la ville de Nantes,* booklet available in AMN, and the many letters from the same available in the public records and the press.

37. Brunelière, Amieux, Billard, Harden, and others.

38. No good history of the Ponts-et-chaussées is available yet. For a first and rather uncritical approach, see Bertrand Lemoine, *Construire, aménager, équiper: La France de Ponts en chaussées* (Paris: Gallimard, coll. Découverte, 2004).

39. AN F14 12604.

40. Direction régionale de l'Environnement, de l'Aménagement et du Logement, "Pourquoi intervenir dans le lit ?," 22 June 2010, http://www.centre.developpement-durable.gouv.fr/pourquoi-intervenir-dans-le-lit-a99.html.

Chapter 5. The New Cuyahoga

1. Cleveland Chamber of Commerce, *The New Cuyahoga: River Straightening Recommendations by the Committee on River and Harbor Improvement of the Cleveland Chamber of Commerce* (Cleveland, OH: Cleveland Chamber of Commerce, 9 Feb. 1917).

2. In theory the old river bed would be filled in after creation of the new channel, so as to prevent the creation of stagnant water and to create new development opportunities in the flats. See *Cuyahoga River, Ohio, Letter from the Secretary of War,* 63rd Congress, 2nd Session, House of Representatives, Document No. 707, 69, 70.

3. Chamber of Commerce minutes from the early decades of the twentieth century include committee membership lists. They are found in the Greater Cleveland Growth Association Records (MS 3471) at the Western Reserve Historical Society (WRHS).

4. On rivers in history see Mark Cioc, *The Rhine: An Eco-Biography, 1815–2000* (Seattle: University of Washington Press, 2002); Stéphane Castonguay and Matthew Evenden, eds., *Urban Rivers: Remaking Rivers, Cities, and Space in Europe and North America* (Pittsburgh, PA: University of Pittsburgh Press, 2012); Christof Mauch and Thomas Zeller, eds., *Rivers in History: Perspectives on Waterways in Europe and North America* (Pittsburgh, PA: University of Pittsburgh Press, 2008).

5. Cleveland Chamber of Commerce, *Annual Report* (Cleveland, OH: Cleveland Chamber of Commerce, 1912), 134.

6. Cleveland Chamber of Commerce, *Annual Report* (Cleveland, OH: Cleveland Chamber of Commerce, 1910), 148.

7. On this topic I've been influenced by the work of David Harvey, especially through the essays in *The Urban Experience* (Baltimore: Johns Hopkins University Press, 1989).

8. William Cronon, *Nature's Metropolis: Chicago and the Great West* (New York: W. W. Norton, 1991).

9. *The City of Cleveland v. The Standard Bag and Paper Company,* 72 Ohio St. 324; 74 N. E. 206; 1905 Ohio Lexis 132; "Sanitary Condition of the Cleveland Water Supply," *Cleveland Medical Journal* (Nov. 1912): 792–809; Winthrop Pratt, "Sewage Disposal Investigations at Cleveland," *Engineering News* 69, 13 Feb. 1913, 287–89; Howell Wright, *Pollution of the Public Water Supply* (Cleveland, OH: 5 Mar. 1925), pamphlet held by Cleveland Public Library.

10. Cleveland City Plan Commission, *Lorain-Carnegie Bridge and River Improvement* (Cleveland, OH: City Plan Commission, 1926), 44. Perhaps the most important "improvement" to the river for navigation purposes had occurred naturally before the arrival of white settlers, when the Cuyahoga broke through its former bank and emptied into Lake Erie nearly a mile east of its former outlet. The residual channel, which eventually lost its opening to the lake through siltation, became known as the Old River, and it served as an important part of the city's inner harbor. *Cuyahoga River, Ohio, Letter from the Secretary of War,* 63rd Congress, 2nd Session, House of Representatives, Document No. 707, 12.

11. For shipping data see *Cuyahoga River, Ohio, Letter from the Secretary of War,* 34–35.

12. Newton Baker's comments at a public hearing on proposed Cuyahoga improvements, 29 Oct. 1913, as quoted in *Cuyahoga River, Ohio, Letter from the Secretary of War,* 74.

13. Cleveland Chamber of Commerce, *New Cuyahoga*, 10; *Report of Charles Keller to E. J. Kulas, Chairman, River & Harbor Committee of the Cleveland Chamber of Commerce on the Changes Feasible to Be Made in the Channel of the Cuyahoga River . . .* (Cleveland, 1929), 10.

14. Memorandum, 13 May 1909; River and Harbor Improvement Committee Minutes, 18 Oct. 1909, Greater Cleveland Growth Association Records (MS 3471), con 13, vol. 2, WRHS.

15. *Cuyahoga River, Ohio, Letter from the Secretary of War,* 15.

16. Cleveland Chamber of Commerce, *Annual Report* (Cleveland, OH: Cleveland Chamber of Commerce, 1911), 77.

17. Although it is quite old, the best single volume on the American conservation movement remains Samuel P. Hays, *Conservation and the Gospel of Efficiency* (1959; repr., Pittsburgh, PA: University of Pittsburgh Press, 1999). The best recent works on conservation concern only rural topics, leaving urban conservation largely unexamined. See Louis S. Warren's *The Hunter's Game: Poachers and Conservationists in Twentieth-Century America* (New Haven, CT: Yale University Press, 1997) and Karl Jacoby, *Crimes against Nature: Squatters, Poachers, Thieves, and the Hidden History of American Conservation* (Berkeley: University of California Press, 2001). Kendrick Clements's *Hoover, Conservation, and Consumerism: Engineering the Good Life* (Lawrence:

University Press of Kansas, 2000) tracks conservationism through the career of Hoover, and it thereby crosses the traditional topical boundary between industrial efficiency and natural resource conservation. Tellingly, however, Clements addresses oil pollution control, a major environmental issue in the mid-1920s, largely through the lens of the conservation of fisheries rather than as a significant urban problem; see 61–69.

18. Cleveland Chamber of Commerce, *Annual Report* (1910), 150.

19. Cleveland Chamber of Commerce, *Annual Report* (1912), 131.

20. *Reports of the Cleveland River and Harbor Commission of the Improvement of the Cuyahoga River* (Cleveland, 1913), 18. In a similar vein, the Chamber expressed the desire to see harbor facilities "subject to some degree of public supervision or control, so as to prevent any possible hindrance to their best utilization as harbor facilities." Cleveland Chamber of Commerce, *Annual Report* (1912), 137.

21. Memorandum, 6 Oct. 1911, Greater Cleveland Growth Association Records, con 15, vol. 2, WRHS.

22. Chamber of Industry, *Annual Report* (Cleveland, OH: Cleveland Chamber of Industry, 1913), 50.

23. *Cuyahoga River, Ohio, Letter from the Secretary of War*, 8–9.

24. *Cleveland Press*, 25 and 26 Mar. 1913; *Laws of Ohio* 104 (Springfield, OH: Springfield Publishing, 1914).

25. *Cleveland Plain Dealer*, 27 Mar. 1913; Cleveland Chamber of Commerce, *Conservancy and the Cuyahoga: River Improvement Recommendations by the Committee on River and Harbor Improvement of the Cleveland Chamber of Commerce* (Cleveland, OH: Cleveland Chamber of Commerce, 1919), 5–6; *Plain Dealer*, 4 and 27 Mar. 1919; Minutes of the River and Harbor Improvement Committee Meeting, 29 Mar. 1919, Greater Cleveland Growth Association Records, con 23, vol. 2, WRHS.

26. "City Growth about Flats Is like Tree with Hollow Heart," *Cleveland Press*, 23 Feb. 1925; "Where the City's Future Is Threatened," *Cleveland Press*, 16 Feb. 1925.

27. City Plan Commission, *Lorain-Carnegie Bridge*, 13, 40–45.

28. *Report of Charles Keller*, 11.

29. Created in 1930, Republic Steel brought together several extant companies, including Corrigan-McKinney and Upson Nut in Cleveland. See "The Republic Steel Corp.," *Encyclopedia of Cleveland History*, http://ech.case.edu/cgi/article.pl?id=RSC.

30. Clifford F. Hood, *Easing the Cuyahoga: The Cleveland Chamber of Commerce's 100-Year Project* (Cleveland, OH: Cleveland Chamber of Commerce, 1947), 6, 16; *Plain Dealer*, 5 Oct. 1937. See also C. Langdon White, "After 100 Years of Vacillation Cleveland Solves Its River Problem," *Annals of the Association of American Geographers* 30, no. 3 (Sept. 1940): 195–209.

31. *Plain Dealer*, 6 Apr. 1928.

32. *Plain Dealer*, 7 Feb. 1937.

33. On the water pollution, the Cuyahoga fires, and deindustrialization, see David Stradling and Richard Stradling, *Where the River Burned: Carl Stokes and the Struggle*

to Save Cleveland (Ithaca, NY: Cornell University Press, 2015), and "Perceptions of the Burning River: Deindustrialization and Cleveland's Cuyahoga," *Environmental History* 13, no. 3 (2008): 515–35.

34. Mabel Walker, "The American City Is Obsolescent," *Vital Speeches of the Day* 13 (1946-1947): 697–99.

Chapter 6. A "Slum River"

This chapter is part of the dissertation "Tunjuelo: Un río del Sur. Desigualdad urbana en Bogotá (Colombia) a mediados del siglo XX," sponsored by COLCIENCIAS. I want to thank the UMBRA Collective and its coordinator, Prof. Marta Herrera, advisor of the dissertation, and Prof. Stefania Gallini, member of the Committee Staff, for all their insightful and valuable suggestions throughout this process.

1. Raúl Fernando Proaño López, "Diagnóstico de emergencia 1479—9 de Junio de 2002" (Technical report, Dirección para la Prevención y Atención de Emergencias—DPAE and Secretaría de Gobierno de Bogotá, typescript, Bogotá, 2002), http://www.sire.gov.co/sirepro/informes/verArchivo.jsp?id=13188.

2. The first press report found about disasters caused by the Tunjuelo River is "Obras de defensa contra inundaciones solicitarán los vecinos de Tunjuelito," *El Espectador* (Bogotá), 21 July 1955.

3. "Inundaciones en los barrios del sur—más de 500 familias damnificadas," *El Tiempo* (Bogotá), 12 July 1968; "Se desborda el río Tunjuelito," *El Espectador*, 9 June 1975; "1,800 damnificados en Bogotá," *El Tiempo*, 16 May 1996.

4. Francisco Sabatini et al., "Residencial Segregation in Santiago: Scale-Related Effects and Trends, 1992-2002," in *Urban Segregation and Governance in the Americas*, ed. Bryan Roberts and Robert Wilson (New York: Palgrave Macmillan, 2009), 121–44; Bryan Roberts and Robert Wilson, "Residential Segregation and Governance in the Americas: An Overview," in Roberts and Wilson, *Urban Segregation and Governance*, 1-20; Martin Smolka and Laura Mullahy, eds., "Uso del suelo y desarrollo urbano," in *Perspectivas urbanas: Temas críticos en políticas de suelo en América Latina* (Cambridge, MA: Lincoln Institute of Land Policy, 2007), 259–334; Francisco Sabatini, *La segregación social del espacio en las ciudades de América Latina* (Santiago de Chile: Inter-American Development Bank, 2006); Samuel Jaramillo, "Segregación socio-espacial y mercado de suelo en las ciudades de América Latina," in *Producción, uso y consumo de ciudad*, ed. Fundación Hábitat Colombia (Bogotá: Fundación Hábitat Colombia, 1996), 61–75; Samuel Jaramillo, "El papel del mercado en la configuración de algunos rasgos socioespaciales de las ciudades latinoamericanas," *Territorios* no. 2 (July 1999): 107–29; John Harold Córdoba Aldana, "Formes et Facteurs de Ségrégation Socio-Spatiale à Bogotá et dans l'Amérique Latine" (master's thesis, Université de Rennes 2, 2008); Teresa Pires do Rio Caldeira, *Ciudad de muros* (Barcelona: Gedisa, 2007).

5. Felipe Hernández and Peter Kellett, "Introduction: Reimagining the Informal in Latin America," in *Rethinking the Informal City: Critical Perspectives from Latin Amer-*

ica (New York: Berghahn, 2010), 1–19; Carlos Alberto Torres Tovar, *Ciudad informal colombiana: Barrios construidos por la gente* (Bogotá: Universidad Nacional de Colombia, Facultad de Artes, 2009); Samuel Jaramillo, *Urbanización informal: Diagnósticos y políticas: Una revisión al debate latinoamericano para pensar líneas de acción actuales* (Bogotá: CEDE, Universidad de los Andes, 2012). In this article *informal settlements* and *slums* are used as synonyms, although it can be debated.

6. Nik Heynen, Maria Kaika, and Erik Swyngedouw, "Urban Political Ecology: Politicizing the Production of Urban Natures," in *In the Nature of Cities: Urban Political Ecology and the Politics of Urban Metabolism*, ed. Nik Heynen, Maria Kaika, and Erik Swyngedouw (New York: Routledge, 2006), 10; Erik Swyngedouw, *Social Power and the Urbanization of Water: Flows of Power* (Oxford: Oxford University Press, 2004); Maria Kaika, *City of Flows: Modernity, Nature, and the City* (New York: Routledge, 2005); Marco Armiero, "Enclosing the Sea: Remaking Work and Leisure Spaces on the Naples Waterfront, 1870–1900," *Radical History Review* 109 (2011): 13–35.

7. Some examples about river degradation due to urban growth are Ari Kelman, *A River and Its City: The Nature of Landscape in New Orleans* (Berkeley: University of California Press, 2006); William Steele, "The History of the Tama River: Social Reconstructions," in *A History of Water*, vol. 1, *Water Control and River Biographies*, ed. Terje Tvedt and Eva Jakobsson (London: I. B. Tauris, 2006), 217–38; Charles E. Closmann, "Holding the Line: Pollution, Power, and Rivers in Yorkshire and the Ruhr, 1850–1990," in *Rivers in History: Perspectives on Waterways in Europe and North America*, ed. Christof Mauch and Thomas Zeller (Pittsburgh, PA: University of Pittsburgh Press, 2008), 89–109; Andrew C. Isenberg, "Banking on Sacramento: Urban Development, Flood Control, and Political Legitimization, 1848–1862," in *The Nature of Cities: Culture, Landscape, and Urban Space*, ed. A. Isenberg (Rochester, NY: University of Rochester Press, 2006), 103–21; Stuart Oliver, "The Thames Embankment and the Disciplining of Nature in Modernity," *Geographical Journal* 166, no. 3 (2000): 227–38.

8. Antonio Brailovsky, *Buenos Aires, Ciudad Inundable: Por qué está condenada a un desastre permanente* (Buenos Aires: Capital Intelectual, 2012); Bibiana Andrea Preciado Zapata, *Canalizar para Industrializar: La domesticación del Río Medellín en la primera mitad del siglo XX* (Bogotá: Universidad de los Andes, 2015); Janes Jorge, *Tiete, O Rio Que a Cidade Perdeu: São Paulo, 1890–1940* (São Paulo: Alameda, 2006); Simón Castillo, "El Mapocho Urbano del Siglo XIX," *Arquitectura* (Chile) 72 (2009): 46–49; Mike Davis, *Planet of Slums* (New York: Verso, 2003), ch. 3 and 5.

9. Some works that analyze Bogotá's urbanization process from an environmental history persepctive are Germán Palacio, ed., *Historia ambiental de Bogotá y la Sabana, 1850–2005* (Leticia: Universidad Nacional de Colombia, Instituto Amazónico de Investigaciones, 2008); Jair Preciado Beltrán, Robert Orlando Leal Pulido, and Cecilia Almanza Castañeda, *Historia ambiental de Bogotá, Siglo XX: Elementos históricos para la formulación del medio ambiente urbano* (Bogotá: Fondo de Publicaciones Universidad Distrital Francisco José de Caldas, 2005); Julián Osorio, *El río Tunjuelo en la historia de Bogotá: 1900–1990* (Bogotá: Alcaldía Mayor de Bogotá, Secretaría de Cultura, Recreación y

Deporte, 2007); Stefania Gallini, Laura Felacio, Angélica Agredo, and Stephanie Garcés, "The City's Currents: A History of Water in 20th-Century Bogotá," 2014, accessed 3 Dec. 2015, http://www.environmentandsociety.org/exhibitions/water-bogota, are examples of recent literature that focus on the environmental conditions of Bogotá's urbanization process.

10. Departamento Administrativo Nacional de Estadística, *Proyecciones nacionales y departamentales 2005–2020* (Bogotá: Departamento Administrativo Nacional de Estadística, 2010), 47.

11. Julián Vargas Lesmes, *Historia de Bogotá.*, vol. 1, *Conquista y Colonia*, 2nd ed. (Bogotá: Villegas editores, Alcaldía Mayor de Bogotá, 2007), ch. 2 and 3.

12. Germán Mejía Pavony, *Los años del cambio: Historia urbana de Bogotá, 1820–1910* (Bogotá: CEJA, 1998), 230 and 340.

13. Julián Osorio, "Los cerros y la ciudad: Crisis ambiental y colapso de los ríos de Bogotá al final del siglo XIX," in *Historia ambiental de Bogotá y la Sabana 1850–2005*, ed. Germán Palacio (Leticia: Universidad Nacional de Colombia, 2008), 169–93.

14. Benjamín Villegas, ed., *El agua en la historia de Bogotá*, vol. 1, *1538–1937* (Bogotá: Villegas, 2003), 530–40; Benjamín Villegas, ed., *El agua en la historia de Bogotá*, vol. 2, *1938–1985* (Bogotá: Villegas, 2003); Gallini et al., "City's Currents."

15. Villegas, ed., *El agua*, 2:57–61; Osorio, *El río Tunjuelo*, 67.

16. Fabio Zambrano, *Historia de Bogotá, Siglo XX* (Bogotá: Villegas, 2007); Marco Cortés Díaz, *La anexión de los 6 municipios vecinos a Bogotá en 1954* (Bogotá: Universidad Nacional de Colombia, 2006); Osorio, *El río Tunjuelo*. The Tunjuelo River has a very long history of human interaction that goes back to pre-Hispanic time. Ana María Falchetti, *El territorio de los muiscas a la llegada de los españoles* (Bogotá: Universidad de los Andes, Comité de Publicaciones, 1973); Santiago Martínez Medina, *Poderes de la mimesis: Identidad y curación en la comunidad indígena muisca de Bosa* (Bogotá: Universidad de Los Andes, Facultad de Ciencias Sociales, CESO, Departamento de Antropología, 2009); Zambrano, *Historia de la localidad*, ch. 1. During the colonial period and the nineteenth century the Tunjuelo basin was occupied mainly by large landowners from Bogotá, including some religious orders like Dominicans, and it was mainly devoted to grain cultivation and recreational and seasonal housing uses. Zambrano, *Historia de la localidad*, ch. 2; Santiago Luque Torres, *Historia del patrimonio rural y urbano del Colegio Mayor de Nuestra Señora del Rosario: Contexto cartográfico y valoración sociopolítica y económica de los habitantes de la ciudad, el altiplano y la vertiente, 1650–1870* (Bogotá: Universidad del Rosario, 2009), ch. 5; Juan Carrasquilla Botero, *Quintas y estancias de Santafé y Bogotá* (Bogotá: Banco Popular, Fondo de Promoción de la Cultura, L. Fas Producciones Editores, 1989); Darío Fajardo and Centro de Invetigaciones para el Desarrollo, *Estudio socio-económico del valle alto del río Tunjuelo* (Bogotá: Centro de Invetigaciones para el Desarrollo, Universidad Nacional de Colombia, 1975).

17. Zambrano, *Historia de Bogotá*, 104–6; Villegas, *El Agua*, 2:28; Osorio, *El río Tunjuelo*, 34.

18. Villegas, *El Agua*, 2:29–34. In Colombia's Andean region, El Niño tends to diminish the precipitation levels affecting the amount that dams can store. IDEAM and Alcaldía Mayor de Bogotá, *Estudio de la caracterización climática de Bogotá y cuenca alta del río Tunjuelo* (Bogotá: Alcaldía Mayor de Bogotá, 2007).

19. Villegas, *El Agua*, 2:59.

20. Villegas, *El Agua*, 2:101–14.

21. Universidad Nacional de Colombia, Departamento de Geografía and Secretaría Distrital de Ambiente (UN-SDA), "Informe fase de diagnóstico—Convenio 040 de 2007 'Aunar esfuerzos técnicos, humanos, administrativos y económicos para la formulación del Plan de Ordenación y Manejo de la cuenca del río Tunjuelo en el perímetro urbano de Bogotá'" (Technical report, Secretaría Distrital de Ambiente, Bogotá, unpublished typescript, 2008).

22. José Luis Romero, *Latinoamérica: Las ciudades y las ideas* (Medellín: Universidad de Antioquia, 1999), ch. 6 and 7; Luis Mauricio Cuervo and Luz Josefina González, *Industria y ciudades en la era de la mundialización: Un enfoque socioespacial* (Bogotá: COLCIENCIAS—CIDER—Tercer Mundo, 1997), ch. 8.

23. Alberto Saldarriaga, *Bogotá siglo XX: Urbanismo, arquitectura y vida moderna* (Bogotá: Alcaldía Mayor de Bogotá, 2000), 87. One hectare = 2.47 acres.

24. Cyrille Simonnet, *Hormigón, historia de un material* (Donostia-San Sebastian: Nerea, 2009); Amy E. Slaton, *Reinforced Concrete and the Modernization of American Building, 1900–1930* (Baltimore: Johns Hopkins University Press, 2001); Robert Courland, *Concrete Planet: The Strange and Fascinating Story of the World's Most Common Man-Made Material* (Amherst, NY: Prometheus, 2011), intro. and ch. 1.

25. Carlos Sanz de Santamaría, *Historia de una gran empresa* (Barcelona: Gaudí, 1983); Fabio Zambrano and Carolina Castelblanco Castro, *El kiosko de la luz y el discurso de la modernidad* (Bogotá: Instituto Distrital de Cultura y Turismo, 2002); Instituto Distrital de Cultura y Turismo, *La ciudad de la luz: Bogotá y la Exposición Agrícola e Industrial de 1910* (Bogotá: Alcaldía Mayor de Bogotá, Corporación La Candelaria, 2005).

26. Technically, aggregates are divided in three main size groups: sand (0.06–2 mm), gravel (2–60 mm), and crushed stone (60–150 mm). M. R. Smith and L. Collis, eds., *Aggregates: Sand, Gravel and Crushed Rock for Construction Purposes*, 1st ed. (London: Geological Society, 1993), ch. 8; Hernando Vargas C., "Materiales, procesos y productos," in *Cincuenta años en la construcción de Colombia, 1957–2007*, ed. Cámara Colombiana de la Construcción (Bogotá: Panamericana, 2007), 169–216.

27. Smith and Collis, *Aggregates*, 199.

28. Hernando Vargas Caicedo, "De la tapia pisada a la piedra líquida," in *La construcción del concreto en Colombia: Apropiación, expresión, proyección*, ed. Asociación Colombiana de Productores de Concreto (Bogotá: Panamericana, 2006), 62.

29. Hernando Parra and José Royo y Gómez, "Informe sobre las explotaciones rocosas en el municipio de Bogotá," in *Compilacion de los estudios geologicos oficiales en Colombia, Tomo VII* (Bogotá: Servicio Geológico Nacional, 1946), 285–315.

30. Pardo Restrepo y Santamaría Ltda., *Edición conmemorativa de los 50 años de su fundación* (Bogotá: OP Gráficas, 1988). Civil engineer, son of a cofounder of Central de Mezclas, interview with the author, Dec. 2012.

31. Alberto Pardo y Pardo, ed., *Instrucciones para el uso del concreto* (Bogotá: Compañía Distribuidora de Cementos Colombianos Diamante y Titán, Ed. Cromos, 1943), 253–67; Sanz de Santamaría, *Historia*; Pardo Restrepo y Santamaría, *Edición conmemorativa*; Alberto Escovar Wilson-White and Miguel Dario Cárdenas, "Hitos y protagonistas," in *La construcción del concreto en Colombia*, 44–123. Civil engineer, son of a cofounder of Central de Mezclas, interview with the author, Dec. 2012.

32. Preciado et al., *Historia ambiental de Bogotá*, 167–72.

33. UN-SDA, "Informe," ch. 1; Osorio, *El río Tunjuelo*, 58.

34. Alix Maritza Bolaños Busto and Silvia Patricia Forero Valencia, "Génesis del cono del río Tunjuelo—Sureste de Bogotá, Cundinamarca" (undergraduate thesis, Universidad Nacional de Colombia, 2003); Víctor Luna Torres and Otto Rodríguez Sánchez, "Análisis sedimentológico del cono del Río Tunjuelito Al S.E. de la ciudad de Bogotá" (undergraduate thesis, Universidad Nacional de Colombia, 1990).

35. Bolaños and Forero, "Génesis del cono." For size ranges of the particles the rivers transport see Richard Huggett, *Fundamentals of Geomorphology*, 2nd ed. (New York: Routledge, 2007), 105.

36. According to Smith and Collis, aggregates can also be irregular, flaky, or elongated, but those shapes have technical problems in the making of concrete. Smith and Collis, *Aggregates*, 158.

37. Archivo de Bogotá, Fondo Secretaría de Obras Públicas, box 25, folder 11.

38. Civil engineer, son of a cofounder of Central de Mezclas, interview with the author, Dec. 2012. In the report made by the Servicio Geológico Nacional mentioned above, most of the *areneras* (sand quarries) reported the use of dynamite.

39. Juan Carrasquilla Botero, "La tenencia de la tierra en Bogotá desde 1539 hasta 1939" (working paper, Fundación para la promoción y el avance de la tecnología, Bogotá, unpublished typescript, 1987); Luque Torres, *Historia del patrimonio rural.*

40. Fundación San Antonio, *Fundación San Antonio, 1894–1994: Un siglo al servicio de los más necesitados* (Bogotá: Uribe Sáenz Asociados, 1994).

41. Besides interpretation made during this research, aerial photographs have been used also by Secretaría Distrital de Ambiente, "Geología y geomorfología de la zona del Parque Minero-Industrial Tunjuelo: Contrato 680 de 2008 y Contrato 525 de 2007" (Technical report, Secretaría Distrital de Ambiente, Bogotá, unpublished typescript, 2008); Nohra León Rodríguez and Vladimir Sánchez, "Caracterización socioambiental de la localidad de Tunjuelito: Bogotá" (Technical report, Alcaldía Local de Tunjuelito, Bogotá, unpublished typescript, Bogotá, 2004); Esther A. Galeano Tibaquirá, "Inundaciones en la cuenca del río Tunjuelo (Bogotá, Colombia): Evidencias ambientales para la geología forense" (undergraduate thesis, Universidad Nacional de Colombia, 2004).

42. Saldarriaga, *Bogotá Siglo XX*, ch. 2 and 3; Zambrano, *Historia de Bogotá*, ch. 4 and 7; Silvia Arango, *Historia de la arquitectura en Colombia*, 1st ed. (Bogotá: Universidad Nacional de Colombia, Centro Editorial y Facultad de Artes, 1989), ch. 7; Lauchlin Currie, *Una política urbana para los países en desarrollo* (Bogotá: Tercer Mundo, 1965); Ricardo Arias Trujillo, *Historia de Colombia contemporánea (1920–2010)*, 1st repr. (Bogotá: Universidad de los Andes, 2013).

43. Camila Aguilar, Nicolás León, and Marcela Meléndez, *El sector de materiales de construcción en Bogotá-Cundinamarca* (Bogotá: Fedesarrollo, 2006). Civil engineer, son of a cofounder of Central de Mezclas, interview with the author, Dec. 2012; Former chief of Central de Mezclas, interviews with the author, Apr. and Nov. 2012.

44. Vargas, "De la tapia pisada"; Zambrano, *Historia de Bogotá*; Sanz de Santamaría, *Historia*. Civil engineer, son of a cofounder of Central de Mezclas, interview with the author, Dec. 2012; Former chief of Central de Mezclas, interview with the author, Nov. 2012.

45. Secretaría Distrital de Ambiente, *Geología y geomorfología*, 8. The sinuosity index (SI) estimates the *ratio* of the channel length and the downvalley length. If SI is near 1.0 then the river course is straight.

46. Ro Charlton, *Fundamentals of Fluvial Geomorphology* (New York: Routledge, 2008), 171–75.

47. Galeano, "Inundaciones."

48. Cortés Díaz, *La anexión*; Zambrano, *Historia de Bogotá*; Saldarriaga, *Bogotá siglo XX*, 82 and 87.

49. Jaramillo, "El papel del mercado"; Alan Gilbert, *La ciudad latinoamericana* (México: Siglo XXI, 1997).

50. Jaramillo, "Segregación socio-espacial"; Sabatini, *La segregación social*.

51. The analysis of the origin and consolidation of this spatial process has been studied by several authors, among them Saldarriaga, *Bogotá siglo XX*; Zambrano, *Historia de Bogotá*; Peter Amato, "An Analysis of the Changing Patterns of Elite Residential Areas in Bogotá" (PhD diss., Cornell University, 1968); Adriana María Suárez Mayorga, *La ciudad de los elegidos: Crecimiento urbano, jerarquización social y poder político, Bogotá (1910–1950)* (Bogotá: Guadalupe, 2006); Juan Carlos del Castillo Daza, *Bogotá, el tránsito a la ciudad moderna, 1920–1950* (Bogotá: Universidad Nacional de Colombia, Facultad de Artes, 2003).

52. Carlos Alberto Torres Tovar, "La comunidad, el Estado y la iniciativa privada: Agentes sociales en la consolidación de la ciudad periférica informal—El caso de Ciudad Bolívar" (master's thesis, Universidad Nacional de Colombia, 1993); Lucy Riveros, "Las políticas urbanas del estado y las urbanizaciones clandestinas en Bogotá D.E., El caso de la alcaldía de Usme" (undergraduate thesis, Universidad Nacional de Colombia, 1976); UN-SDA, "Informe"; Torres, *Ciudad informal*; Jaramillo, *Urbanización informal*.

53. Hernández and Kellett, "Introduction."

54. Torres, *Ciudad informal*; Jaramillo, *Urbanización informal*.

55. Torres, *Ciudad informal*, 108–12; UN-SDA, "Informe," ch. 4. Bogotá is divided into twenty *localidades*, which are the basic political-administrative units. Tunjuelo's urbanized watershed includes also, partially, *localidades* of Kennedy and Bosa.

56. Barrios El Carmen (1946), Fátima (1946), Tunjuelito (1946), Abraham Lincoln (1946), Meissen (1946), San Vicente (1948), and San Carlos (1949). Zambrano, *Historia de la localidad*; Juan David Mob [pseud.] and Jak [pseud.], "En la hacienda del 'Maisen': 50 años de lucha silenciosa, historia del barrio Meissen" (Unpublished typescript, Departamento Administrativo de Acción Comunal Distrital, Bogotá, 1998).

57. Zambrano, *Historia de la localidad*, 69.

58. UN-SDA, "Informe," ch. 3.

59. Gerardo M. Santafe, "Usme y su historia: Generalidades regionales y locales a través del tiempo." (Unpublished typescript, Departamento Administrativo de Acción Comunal Distrital, Bogotá, 1998); Zambrano, *Historia de Bogotá*; Torres, *Ciudad informal*; UN-SDA, "Informe," ch. 3.

60. Gilbert, *La ciudad latinoamericana*; Jaramillo, *Urbanización informal*; Torres, "La comunidad"; Alfonso Torres Carrillo, *La ciudad en la sombra: Barrios y luchas populares en Bogotá, 1950–1977* (Bogotá: CINEP, 1993).

61. Gilbert, *La ciudad latinoamericana*; Torres Carrillo, "Luchas barriales"; Torres, "La comunidad"; Juan David Mob and Jak, "En la hacienda"; Mahe [pseud.], "Historia del barrio Tunjuelito" (Unpublished typescript, Departamento Administrativo de Acción Comunal Distrital, Bogotá, 1997); Gonzalo Jiménez de Quesada [pseud], "El pensador de mi barrio o historia del barrio San Carlos" (Unpublished typescript, Departamento Administrativo de Acción Comunal Distrital, Bogotá, 1998); Luis Ortiz G., "Los relatos de la Isla del Sol," in *Bogotá, Historia común: Ganadores del concurso de historias barriales y veredales* (Bogotá: Departamento Administrativo de Acción Comunal, 1997), 73–121; and interviews conducted in the area of study.

62. Interviews were conducted in November and December 2012 and September through December 2013. Twenty-seven inhabitants were interviewed, mainly from the four neighborhoods that constituted the central area of study: Meissen, México, Tunjuelito, and San Benito (shown in fig. 6.2). This is a very common topic in "informal" settlements all over Latin America. See, for example, Erik Swyngedouw, *Social Power and the Urbanization of Water: Flows of Power* (Oxford: Oxford University Press, 2004).

63. Mahe, "Historia del barrio Tunjuelito"; Zambrano, *Historia de Bogotá*; Jiménez de Quesada, "El pensador."

64. Woman, eighty-three years old, barrio México, settler since 1968, interview with the author, Dec. 2012.

65. Woman, eighty-three years old, barrio México, settler since 1968, interview with the author, Dec. 2012.

66. Osorio, *El río Tunjuelo*; Zambrano, *Historia de la localidad*; Jiménez de Quesada, "El Pensador"; Juan David Mob and Jak , "En la hacienda."

67. Woman, eighty-seven years old, barrio Abraham Lincoln, inhabitant since 1946, interview with the author, Sept. 2013.

68. Villegas, *El Agua*, 2:195–210; Zambrano, *Historia de la localidad*, 113, 150–60.

69. Man, fifty-five years old, barrio Meissen, interview with the author, Sept. 2013.

70. Interview (8), Woman, eighty-two years old, barrio Meissen, inhabitant since 1948, interview with the author, Sept. 2013; Woman, sixty years old, barrio Tunjuelito, son of a founding settler, interview with the author, Oct. 2013.

71. "La desesperación no se va de San Benito," *El Tiempo* (Bogotá), 16 May 1996; "Inundaciones," *El Tiempo*, 16 May 1996.

72. León and Sánchez, *Caracterización socio-ambiental*, 43–46.

73. Zambrano, *Historia de Bogotá*, 161–65; Osorio, *El río Tunjuelo*, 60–62; "La peor inundación en 27 Años," *El Tiempo*, 15 May 1996; "La Regadera inundó a Tunjuelito," *El Tiempo*, 7 May 1994; "Obras de defensa contra inundaciones solicitarán los vecinos de Tunjuelito," *El Espectador* (Bogotá), 21 July 1955; "Inundaciones anoche en el sur," *El Tiempo*, 12 July 1968; "Inundado el Meissen al desbordarse el río Tunjuelito," *El Tiempo*, 4 Oct. 1969.

74. IDEAM and Alcaldía Mayor de Bogotá, *Estudio de la caracterización*, 34.

75. Proaño López, "Diagnóstico"; UN-SDA, "Informe," 989–91.

76. Agencia Nacional de Minería, *Expediente 8151*, folder 1–3; Archivo Central e Histórico de la CAR, *Expediente 048*; Archivo de Bogotá, *Fondo Secretaría de Obras Públicas*, box 25, folder 11; Autoridad Nacional de Licencias Ambientales, Expediente 530, folder 1.

77. Torres Carrillo, *La ciudad en la sombra*, 25.

78. Jaramillo, *Urbanización informal*; Jaramillo, "El papel del mercado"; Torres, *Ciudad informal*; Torres Carrillo, *La ciudad en la sombra.*

79. "Inundaciones en los barrios del sur."

80. "Inundaciones anoche en el sur."

Chapter 7. Urbanizing a River in a Bicultural Border Region

1. Febvre, *Der Rhein und seine Geschichte*, trans. and with an afterword by Peter Schöttler (Frankfurt am Main: Campus, 1994); see also André Traband, *Villes du Rhin: Strasbourg et Mannheim-Ludwigshafen* (Paris: Belles lettres, 1966). The issues developed in this chapter are discussed in a wider context in my book *Im Spiegel des Wassers: Eine transnationale Umweltgeschichte des Oberrheins (1800–2000)* (Cologne: Böhlau Köln, 2016).

2. The broad literature on border cities and border rivers cannot be presented here in extenso. While David Blackbourn, *The Conquest of Nature* (London: Pimlico, 2006), set a milestone in the analysis of border rivers, Helga Schultz, *Stadt-Grenze-Fluss: Europäische Doppelstädte* (Berlin: Berliner Wissenschafts, 2005), collected articles on European cities at border rivers with a socioeconomic focus. See also Christoph Mauch and Thomas Zeller, *Rivers in History: Perspectives on Waterways in Europe and North Ameri-*

ca (Pittsburgh, PA: University of Pittsburgh Press, 2008); Christoph Bernhardt, "Grenzregionen im Zeitalter des Kalten Krieges: Regionalentwicklung und Umweltpolitik an Rhein und Oder nach 1945," in *Europas Grenzen*, ed. Sabine Penth et al. (St. Ingbert: Röhrig, 2006), 205–28; and Terje Tvedt and Terje Oestigaard, eds., *A History of Water*, vol. 3, *The World of Water* (London: Tauris, 2006). For Strasbourg see Traband, *Villes du Rhin*, n1; and Christoph Cornelissen, Stefan Fisch, and Annette Maas, *Grenzstadt Straßburg: Stadtplanung, kommunale Wohnungspolitik und Öffentlichkeit 1870–1940* (St. Ingbert: Röhrig, 1997), n36.

3. Traband, *Villes du Rhin*, 20.

4. Traband, *Villes du Rhin*, 22.

5. Strasbourg's decision dates from 1814. See Mathieu Wehrle, "La maîtrise d'un patrimoine forestier, Strasbourg et sa forêt rhenane (1800–1870)," *Chantiers historiques en Alsace* (2002): 91–102, esp. 98.

6. See F. J. Bär, *Die Wasser- und Straßenbau-Verwaltung in dem Großherzogthum Baden* (Karlsruhe, 1870).

7. See Arthur Valdenaire, "Das Leben und Wirken des Johann Gottfried Tulla," *Zeitschrift für die Geschichte des Oberrheins* n.s., 42 (1929): 594.

8. See Protocol du Conseil municipal from 22 Dec. 1838 in Archives Nationales Paris, F14-6694.

9. See Wehrle, "La maîtrise d'un patrimoine," 95.

10. M. Defontaine, "Des travaux du Rhin," *Annales des ponts et chaussées* (1833), vol. 2, 177.

11. Defontaine, "Des travaux du Rhin."

12. "Extrait des registres de la commission mixte, séance du 12.6.1843," Archives Nationales Paris, F14-6694.

13. Jean-Jacques Schwien et al., "De sources en nappe: Une histoire des eaux souterrains de Strasbourg," *Saison d'Alsace* 101 (1988): 31–32.

14. See "La correction du Rhin et ses effets désastreux pour Strasbourg," in Pierre Buntz, "Rhin nouveau, ville nouvelle," *Saison d'Alsace* 101 (1988): 59.

15. Protokolle der Rheinischen Strombefahrungs-Commission von 1874, Mannheim 1875, Schlussprotokoll 2–4.

16. Buntz, "Rhin nouveau," 59.

17. Buntz, "Rhin nouveau," 59.

18. Traband, *Villes du Rhin*, 28.

19. *Bericht an den Gemeinderath betreffend die Wasserstrasse Mannheim-Strassburg und die Schiffahrt auf dem Oberrhein* (Strasbourg, 1894), 3.

20. Max Honsell, *Die Wasserstraße zwischen Mannheim-Ludwigshafen und Kehl-Strassburg—Canal oder Freier Rhein?* (Berlin, 1890), 5.

21. Gaston Haelling, Jean Majorelle, *Le Rhin: Politique, Economique, Commercial* (Paris: Eyrolles, 1930), 81.

22. See his study *Kanal oder freier Rhein* (1890).

23. Buntz, "Rhin nouveau," 60.

24. Sarrazin and Hofsfeld, "Der neue Hafen von Straßbourg," *Zentralblatt der Bauverwaltung* (1891): S. 251.

25. Sarrazin and Hofsfeld, "Der neue Hafen," 251.

26. Buntz, "Rhin nouveau," 62.

27. Sarrazin and Hofsfeld, "Der neue Hafen," 250.

28. See Karlheinz Hillenbrand, *100 Jahre Rheinhafen Kehl: 1900–2000* (Kehl: Hafenverwaltung, 2000).

29. Traband, *Villes du Rhin*, 37.

30. Sarrazin and Hofsfeld, "Der neue Hafen," 251.

31. Sarrazin and Hofsfeld, "Der neue Hafen," 252.

32. Joseph Krieger, *Topographie der Stadt Strassburg nach ärztlich-hygienischen Gesichtspunkten bearbeitet* (Strasbourg, 1889), 95–98.

33. Buntz, "Rhin nouveau," 62. See also Otto Back, *Bericht an den Gemeinderath betreffend die Wasserstrasse Mannheim-Strassburg und Schiffahrt auf dem Oberrhein* (1894), 3.

34. Buntz, "Rhin nouveau," 62.

35. Buntz, "Rhin nouveau," 61.

36. See Cornelissen, Fisch, and Maas, *Grenzstadt Straßburg.*

37. Viviane Claude, "La Germanisation de Strasbourg après 1871," *Les Annales de la recherche urbaine* 37 (1988): 38–45.

38. See *Verwaltungsbericht der Stadt Straßburg 1919–35*, 238, 245; *Verwaltungsbericht der Stadt Straßburg 1900–1910*, 137–38.

39. *Verwaltungsbericht der Stadt Straßburg 1919–35*, 238, 245; Claude, "La Germanisation," S. 44; Vivian Claude, "Strasbourg 1850–1914, assainissement et politiques urbaines" (Thèse 3ème cycle, École des Hautes Études en Sciences Sociales, 1985), 299.

40. *Verwaltungsbericht Straßburg 1900–1910*, 266.

41. Paulssen, "Entwässerung der Stadt Strasbourg," in *Drittes Heft des XV. Bandes des Archivs für öffentliche Gesundheitspflege in Elsaß-Lothringen für die Monate Oktober bis Dezember 1893*, 98; Claude, "Strasbourg 1850–1914," 321.

42. See Paulsen, "Entwässerung," 142.

43. Paulsen, "Entwässerung," 98.

44. Claude, "Strasbourg 1850–1914," 322.

45. Gemeinderath der Stadt Strassburg, *Bericht der Commission zur Berathung des Entwurfs für die Entwässerung der Stadt Straßburg* (Strasbourg, 1891), 5.

46. Claude, "Strasbourg 1850–1914," 326.

47. Paulsen, "Entwässerung," 140.

48. D. Goldschmidt, *Le tout à l'égout et les champs d'épandage* (Strasbourg, 1890), 10.

49. *Besprechung über die Schwemmkanalisation in der Generalversammlung des Straßburger Eigenthümer-Vereins vom 26.2.1892 durch dessen Vice-Präsidenten Herrn Fastinger*, Straßburg 1892, 5, 4–5, 12–13.

50. Jörg Lange, "Zur Geschichte des Gewässerschutzes am Ober- und Hochrhein" (PhD diss., University of Freiburg, 2002), 37.

51. Claude, "Strasbourg 1850–1914," 303, 308–10.

52. Gemeinderath der Stadt Straßburg, *Bericht der Commission zur Berathung des Entwurfs für die Entwässerung der Stadt Straßburg* (Strasbourg, 1893).

53. Gemeinderath der Stadt Straßburg, *Bericht der Commission* (1893), 5–6. Lange, "Zur Geschichte des Gewässerschutzes," 23–24.

54. Gemeinderath der Stadt Strassburg, *Bericht der Commission* (1893), 11; Claude, "Strasbourg 1850–1914," 304.

55. *Jahrbuch der Medizinalverwaltung Elsaß-Lothringen* (Strasbourg, 1909), 144; *Verwaltungsbericht für die Stadt Strassburg 1900–1910*, 277–78.

56. Febvre, *Der Rhein*, 49–50.

57. François Walter, *Les figures paysagères de la nation: Territoire et paysage en Europe (16e–20e siècle)* (Paris: École des Hautes Études en Sciences Sociales, 2004), 416, 456.

58. Emmanuel de Martonne, *Conditions physique et économiques de la navigation rhénane* (Paris: Imprimerie nationale, 1918), 7, 22.

59. Ermin Berninger, *La situation économique de l'Alsace et de la Lorraine* (Strasbourg: Comité Alsacien d'Etudes et d'Informations, 1922–23), 189–93.

60. Gaston Haelling, *Le port de Strasbourg* (Paris: Association française pour le développement des travaux publics, 1924), 24–27.

61. *Rapport presenté au conseil municipal sur le développement des ports de Strasbourg* (1922–23), 208–9.

62. The editorial in the first issue of the journal *La Navigation du Rhin* reflects very clearly this strategy of the French state and its instruments. See Le comité de redaction, *Au Lecteur* (1922), 1–2.

63. Haelling, *Le port de Strasbourg*, 24.

64. *Rapport presenté au conseil municipal*, 212.

65. *Rapport presenté au conseil municipal*, 211.

66. *Rapport presenté au conseil municipal*, 213.

67. Horst-Johannes Tümmers, *Der Rhein: Ein europäischer Fluss und seine Geschichte* (Munich: C. H. Beck, 1994), 168.

68. See, e.g., Mark Cioc, *The Rhine: An Eco-Biography, 1815–200* (Seattle: University of Washington Press, 2002).

69. Bernhardt, *Im Spiegel*, 153–81.

Chapter 8. Path Dependencies Managing the River Elbe and the Requirements of Hamburg's Open Tidal Seaport

1. James Mahoney, "Path Dependence in Historical Sociology," *Theory and Society* 29, no. 4 (2000): 507–48.

2. Scott E. Page, "Path Dependence," *Quarterly Journal of Political Science* 1, no. 1 (2006): 88.

3. Further important literature on path dependency: Lars Magnusson and Jan Ottosson, eds., *The Evolution of Path Dependence* (Cheltenham: Edward Elgar, 2009); Martin V. Melosi, "Path Dependence and Urban History: Is a Marriage Possible?," in *Resources of the City: Contributions to an Environmental History of Modern Europe*, ed. Dieter Schott, Bill Luckin, and Geneviève Massard-Guilbaud (Aldershot, UK: Ashgate, 2005).

4. Peter von Allwörden, Nikolaus Ruhl, and Christian C. Schmidt, *Erlebnis Elbe: Eine Entdeckungsreise durch die maritime Landschaft von Hamburg nach Helgoland* (Drochtersen: MedienContorElbe, 2004), 10.

5. K. Heinrich Altstaedt, *Der Hafen von Harburg: Schifffahrt, Handel und Hafenleute an der Süderelbe* (Harburg: Helms Museum, 2011), 31.

6. Ortwin Pelc and Susanne Grötz, eds., *Konstrukteur der modernen Stadt. William Lindley in Europa und Hamburg 1808–1900* (Hamburg: Dölling und Galitz, 2008).

7. Dieter Maass, *Der Ausbau des Hamburger Hafens 1840–1910* (Hamburg: Hansa, 1990), 21.

8. Hans J. Teuteberg, "Die Entstehung des modernen Hamburger Hafens 1866–1896," *Tradition*, no. 17 (1972): 285.

9. This is currently called "24/7" and means service that is available any time (24 hours) and every day (7 days).

10. The Warehouse District (together with the adjacent Office House District) became a World Heritage in 2015. Frank Hesse, ed., *Urban Development towards Modernism: The Birth of Metropolitan Harbour and Commercial Districts* (Berlin: Bässler, 2012).

11. Richard J. Evans, *Tod in Hamburg: Stadt, Gesellschaft und Politik in den Cholera—Jahren 1830–1919* (Reinbek bei Hamburg: Rowohlt, 1990), 367. Evans's study is also an example of environmental history and the relationship between city and river.

12. Dirk Schubert, "Die Herausbildung der Regional- und Landesplanung im Raum Hamburg," *Forum Stadt* 40, no. 3 (2013): 215.

13. Dirk Schubert and Hans Harms, *Wohnen am Hafen: Leben und Arbeiten an der Wasserkante* (Hamburg: VSA, 1993), 128.

14. Christoph Strupp, "Im Bann der 'gefährlichen Kiste': Wirtschaft und Politik im Hamburger Hafen," in *19 Jahre Hamburg Ereignisse und Entwicklungen der Stadtgeschichte seit den fünfziger Jahren*, ed. Forschungsstelle für Zeitgeschichte Hamburg (Hamburg: Dölling und Galitz, 2012), S. 132.

15. Dirk Schubert, "The Great Fire of Hamburg 1842: From Catastrophe to Reform," *Flammable Cities: Urban Conflagration and the Making of the Modern World*, ed. Greg Bankoff, Uwe Lübken, and Jordan Sand (Madison: Wisconsin University Press, 2012), 212–23; Evans, *Tod in Hamburg*.

16. Herbert Hötte, ed., *Die grosse Flut, Katastrophe, Herausforderung, Perspektiven [Hamburg und die Sturmflut 1962]* (Hamburg: Landeszentrale für politische Bildung, 2012), 12.

17. Dieter Schott, "One City—Three Catastrophes: Hamburg from the Great Fire 1842 to the Great Flood 1962," *Cities and Catastrophes / Villes et catastrophes: Coping with Emergency in European History / Réactions face à l'urgence dans l'histoire européenne*, ed. Geneviève Massard-Guildbaud (Frankfurt am Main: Peter Lang, 2002), 185–204; Martina Heßler and Christian Kehrt, eds., *Die Hamburger Sturmflut von 1962: Risikobewußtsein und Katastrophenschutz aus der zeit-, technik- und umweltgeschichtlichen Perspektive* (Göttingen: Vandenhoeck and Ruprecht, 2014).

18. Kern was also responsible for commissioning an aluminum mill for Hamburg with government subsidies. It was sold soon after the start of production, went bankrupt, and later was nicknamed "Onkel Helmuth's Hütte," which in German has a double meaning of "Uncle Helmuth's" "mill" and "hut."

19. Gemeinsamer Landesplanungsrat Hamburg/Schleswig-Holstein, Raumordnung an der Niederelbe, Methoden und Ziele, Kiel/Hamburg/Hannover 1968, 7-28.

20. Helmut Kern, *Ein Modell für die wirtschaftliche Entwicklung der Region Unterelbe* (Hamburg: Behörde für Wirtschaft und Verkehr, 1972).

21. Ernst Schmacke, ed., *Hamburg auf dem Weg in das Jahr 2000* (Düsseldorf: Droste, 1970), 10.

22. Wilhelm Wortmann, "Das Gebiet Hamburg—Unterelbe im Raum der norddeutschen Bundesländer," in *Zwischen Stadtmitte und Stadtregion*, ed. Deutsche Akademie für Städtebau und Landesplanung (Stuttgart und Bern: Krämer, 1970), 29.

23. Hans Laucht, "Aspekte der Planung des Tiefwasserhafens Neuwerk/Scharhörn," in *Landespflegerische Probleme an der Region Unterelbe*, Deutscher Rat für Landespflege (Bonn: Schriftenreihe des Deutschen Rates für Landespflege, 1976), 276–84; Hans Laucht, *Hafenprojekt Scharhörn: Eine Planung im Spiegel der Zeit (1948 bis 1980)* (Aumühle: Selbstverlag, n.d.).

24. The Wadden Sea mudflat area with tidal flatlands and wetlands was inscribed in the UNESCO World Heritage List in 2009 and 2011 because of its biologically diverse ecosystem.

25. Michael Grüttner, *Wem die Stadt gehört: Stadtplanung und Stadtentwicklung in Hamburg 1965–1975* (Hamburg: Association, 1976), 58.

26. Jürgen Oßenbrügge, *Umweltrisiko und Raumentwicklung: Wahrnehmung von Umweltgefahren und ihre Wirkung auf den regionalen Strukturwandel in Norddeutschland* (Berlin: Springer, 1993), 111.

27. Michael Bose, "Wirkungsanalyse eines stadtregionalen Siedlungsstrukturkonzeptes und Ansätze für eine Neuorientierung, Das Entwicklungsmodell Hamburg und sein Umland" (diss., Technischen Universität Hamburg-Harburg, 1994), 248.

28. Lenkungsgruppe Regionales Entwicklungskonzept für die Metropolregion Hamburg, *REK Regionales Entwicklungskonzept für die Metropolregion Hamburg: Leitbild und Orientierungsrahmen* (Hamburg, 1994), 3.

29. Steffen Bukold and Petra Thinnes, eds., *Boomtown oder Gloomtown? Strukturwandel einer deutschen Metropole: Hamburg* (Berlin: Edition Sigma, 1991), 36.

30. Officially the deepening of the fairway is called adjustment of the fairway. When this paper was finished in November 2015 there was no final legal judgment.

31. Dirk Schubert, "Seaport Cities: Phases of Spatial Restructuring and Types and Dimensions of Redevelopment," in *Port Cities, Dynamic Landscapes and Global Networks*, ed. Carola Hein (New York: Routledge, 2011), 58.

Chapter 9. Rivers as Prisms of Urban Imagining

1. The first rail line connecting Chongqing was Chengdu–Chongqing Railway (Cheng Yu tielu 成渝鐵路), built between December 7, 1949, and July 1, 1952. Zheng Jingdong 鄭敬東, Zhu Peilin 朱培麟, and Fu Zhou 傅舟, *Changjiang Sanxia jiaotong wenhua* 長江三峽交通文史 [Culture of transportation in the Yangzi Three Gorges region] (Beijing: Zhongguo wenshi chubanshe, 2005), 293.

2. Paul J. Smith, "Commerce, Agriculture, and Core Formation in the Upper Yangzi, 2 A.D. to 1948," *Late Imperial China* 9, no. 1 (1988): 1–78.

3. Wang Di 王笛, "Qingdai Chongqing renkou yu shehui zuzhi" 清代重慶人口與社會組織 [Population and social organization in Qing dynasty Chongqing], in *Chongqing chengshi yanjiu* 重慶城市研究 [Urban history of Chongqing], ed. Wei Yingtao 隗瀛濤 (Chengdu: Sichuan daxue chubanshe, 1989), 310–78.

4. Zhou Lisan 周立三, Hou Xuetao 侯學燾, and Chen Siqiao 陳泗橋, *Sichuan jingji dituji shuoming* 四川經濟地圖集說明 [Explanations to the collected economic maps of Sichuan] (Beibei: Zhongguo dili yanjiusuo, 1946), 118–24.

5. D. J. Dwyer, "The Development of China's Inland Waterways," *Geography* 46, no. 2 (1961): 165–67.

6. Louis Audemard, *Les Jonques Chinoises*, vol. 5, *Haut Yang Tse Kiang* (Rotterdam: Museum voor Land- en Volkenkunde en het Maritiem Museum "Prins Hendrik," 1963), 9–17, 27–31; G. R. G. Worcester, *The Junks and Sampans of the Yangtze* (Annapolis, MD: Naval Institute Press, 1971), 111–17, 473–81. For a detailed account on the social and economic history of Sichuan boatmen see Igor Iwo Chabrowski, *Singing on the River: Sichuan Boatmen and Their Work Songs, 1880s–1930s*, China Studies 32 (Leiden: Brill, 2015), 53–105.

7. Ted Gioia, *Work Songs* (Durham, NC: Duke University Press, 2006), 8–12. Definitions of *haozi*: Chen Xinming 陳鑫明, "Chuanjiang haozi" 川江號子, *Xungen* 尋根 (2009): 100–101; Deng Xiao 鄧曉, "'Chuanjiang haozi' de wenhua neihan" '川江號子'的文化内涵, *Zhongguo wenhua luntan* 中華文化論壇 [Chinese culture forum] 1 (2005): 17–18; Han Kuo-Huang, "Folk Songs of the Han Chinese: Characteristics and Classifications," in "Chinese Music Theory," special issue, *Asian Music* 20, no. 2 (1989): 113–16; Kuang Tianqi 匡天齊, "Sichuan hanzu mingede fenlei yu leibie tedian" 四川漢族民歌的分類與類別特點 [Particular classification and categorization of the folk songs of the Sichuan Han ethnicity], *Yinyue tansuo* 音樂探索 [Musical explorations] 3 (1984): 48–49; Sichuansheng difangzhi bianzuan weiyuanhui 四川省地方志編纂委員會, *Sichuansheng zhi: Minsuzhi* 四川省志: 民俗志 [Sichuan province gazetteer: folk customs] (Chengdu: Sichuan renmin chubanshe, 2000), 621–24.

8. These functions are well exemplified by the differences between particular songs. In order to understand them, ethnomusicologists adopted a variety of classifications. For a well-established one see Sichuansheng difangzhi bianzuan weiyuanhui, *Sichuansheng zhi: Minsuzhi*. A much more sophisticated descriptive classification was offered by Duan Ming 段明 and Hu Tiancheng 胡天成, *Chuanjiang haozi* 川江號子 [Work songs of the Sichuan Rivers] (Guizhou: Guizhou renmin chubanshe, 2007), 113–26.

9. Duan and Hu, *Chuanjiang haozi*, 371–82.

10. Chabrowski, *Singing on the River*, 21–41, 106–31.

11. Gioia, *Work Songs*.

12. See Frank Kidson, "Sailors' Songs: A Paper Read before the Folk-Song Society, November 23rd, 1899," *Journal of the Folk-Song Society* 1, no. 2 (1900).

13. For example, Polish *flisak* vocabulary was already noticed by the late sixteenth-century writer Sebastian Fabian Klonowic (1545–1602). He devoted to these boat workers a long poem titled *Flis, to jest Spuszczanie statków Wisłą i inszymi rzekami do niej przypadającymi* [Flis, that is sailing boats down the Vistula and the other rivers that join her], 1595. On *burlaks* see Henri Troyat, *Daily Life in Russia under the Last Tsar* (Stanford, CA: Stanford University Press, 1999). A very interesting comparison of the culture and representation of the *burlaks* and the Mississippi boatmen is given by Dorothy Zeisler-Vralsted, *Rivers, Memory, and Nation-Building: A History of the Volga and Mississippi Rivers* (New York: Berghahn, 2014), 6–9, 55, 65–69.

14. Mark Edward Lewis, *China between Empires: The Northern and Southern Dynasties* (Cambridge, MA: Belknap Press of Harvard University Press, 2009), 94–102.

15. See, for example, Li Daoyuan 酈道元 (Northern Wei, 386–534), *Shuijingzhu* 水經注, Shan E 單鍔 (1031–1110), *Wuzhong shuili shu* 吳中水利書, Šansi/Šakši 贍思/沙克什 (1278–1351), *Hefang tongyi* 河防通議, Wang Xi 王喜 (Yuan dynasty, 1279–1368), *Zhihe tulüe* 治河圖略, Fu Zehong 傅澤洪 (probably flourished during the Kangxi reign 康熙, 1662–1722), *Xingshui jinjian* 行水金鑒, etc.

16. The river view was especially practiced in paintings; see "Qingming shanghetu" 清明上河圖 or "Changjiang wanlitu" 長江萬里圖.

17. Daria Berg, "Marvelling at the Wonders of the Metropolis: Perceptions of Seventeenth-Century Chinese Cities in the novel *Xingshi yinyuan zhuan* . . . ," in *Town and Country in China: Identity and Perception*, ed. David Faure and Tao Tao Liu (Houndmills: Palgrave in association with St Anthony's College, Oxford, 2002), 17–40; Sun Zhongquan 孫忠銓, "Chongqing juan" 重慶卷, in *Zhonghua zhuzhici quanbian* 中華竹枝詞全編 [Full edition of Chinese bamboo branch poetry], edited by Qiu Liangren 丘良任, Pan Chao 潘超, Sun Zhongquan 孫忠銓, and Qiu Jin 丘進 (Beijing: Beijing chubanshe, 2007), 6:523–61.

18. Henry George Woodhead, *Yangtsze and Its Problems* (Shanghai: Mercury Press, 1931), 50. For a description of Chongqing docks see George C. Basil and Elizabeth Foreman Lewis, *Test Tubes and Dragon Scales* (Philadelphia: John C. Winston, 1940), 65–67.

19. Nie Yunlan 聶雲嵐, Peng Weijin 彭維金, and He Dashun 賀大舜, eds., *Zhongguo geyao jicheng Chongqingshi juan* 中國歌謠集成重慶市卷 [Grand compendium of Chi-

nese folksongs, volume on Chongqing Municipality] (Chongqing: Kexue jishu wenxian chubanshe Chongqing fenshe, 1989), 1–2. Hereafter referred to as ZGYJC.

20. According to one source the name Nanjimen was related to a phrase of a poem from *Shijing* 詩經 [Book of songs], Xiaoya 小雅, Siyue (Fourth month) 四月, line 19–20: "滔滔江漢，南國之紀， 盡瘁以仕，寧莫我有" (Grandly flow the Jiang and the Han, Regulators of the southern States. Worn out as I am with service, He yet takes no notice of me). See "Nanjimen 南紀門," *Baidu Online Encyclopedia*, accessed 27 Sept. 2014, http://baike.baidu.com/view/1232001.htm. "Siyue 四月," *Book of Poetry*, analysis and explanations in modern Chinese, accessed 27 Sept. 2014, http://www.yshin.com/Article/fenleijingdian/200609/554.html. "Siyue 四月," Chinese Text Project, *Book of Poetry*, accessed 27 Sept. 2014, http://ctext.org/book-of-poetry/si-yue.

21. Worcester, *Junks and Sampans*, 527. However scanty, archival evidence from the wartime (Feb.–Mar. 1945) points out that the areas of the main city gates of Chongqing were still a typical place of residence for the unqualified workers of the shipping companies. Chongqing Municipal Archive, Chongqingshi shehuiju 重慶市社會局 [Chongqing Municipal Bureau of Social Affairs]: 0060/9/81–1, 432–46.

22. ZGYJC, 5–8.

23. Peng Botong 彭伯通, *Chongqing diming qutan* 重慶地名趣談 [Amusing stories of Chongqing toponyms] (Chongqing: Chongqing chubanshe, 2001), 92–93. Rebellion started in early 1599 and ended with suppression by Li Hualong's troops and Yang Yinglong's suicide on July 15, 1600. Peter Allan Lorge, *War, Politics and Society in Early Modern China, 900–1795* (New York: Routledge, 2005), 136.

24. Peng, *Chongqing diming qutan*.

25. Sichuan junkmen's *haozi* described whole lengths of river valleys with much shorter songs.

26. A paste used for makeup.

27. Jin Xinagming 金祥明, Zhong Youling 鐘游玲, Ran Zhuang 冉庄, and Yang Zhongjin 楊忠全, eds., *Bayu minjian wenxue huicui* 巴渝民間文學薈萃 [Distinguished collection of popular literature from BaYu] (Chengdu: Sichuan wenyi chubanshe, 1992), 63–65. Ba and Yu are historical names of Chongqing.

28. On local gods see Stephan Feuchtwang, *Popular Religion in China: The Imperial Metaphor* (Richmond, Surrey: Curzon, 2001).

29. C. A. S. Williams, *Chinese Symbolism and Art Motifs: An Alphabetical Compendium of Antique Legends and Beliefs, as Reflected in the Manners and Customs of the Chinese*, 3rd ed. (Shanghai: Kelly and Walsh, 1932; repr., Rutland, VT: Charles E. Tuttle, 1989), 403–406.

30. Meaning not clear.

31. Cao Xuequan 曹學佺, in the geographical treaty published in 1643, stated, "Today the circuit capital has seventeen gates, nine opened and eight closed; they resemble Nine Palaces and Eight Trigrams, or the turtle's head turned toward Mountain Tu" (今府城為門十有七， 九開八閉， 以象九宮八卦， 或曰象龜首， 乃繇塗山). Cao Xuequan 曹學佺, *Shuzhong guangji* 蜀中廣記 [Extensive records of Sichuan], 1643, in *Qinding*

Siku quanshu 欽定四庫全書 [Complete imperial library in four sections], 1773–1782, Shibu 史部, Mingshengji di shiqi (17), Shang Chuandong dao, Chongqingfu, Baxian fuguo. Peng Botong 彭伯通, *Gucheng Chongqing* 古城重慶 [Old city of Chongqing] (Chongqing: Chongqing chubanshe, 1981), 17; Zhou Yong 周勇, *Chongqing tongshi* 重慶通史 [General history of Chongqing] (Chongqing: Chongqing chubanshe, 2002), 147–50; Peng, *Chongqing diming qutan*, 57–58.

32. Duan and Hu, *Chuanjiang haozi*, 850.

33. Most probably today's Taihe township, north of Hechuan City in Hechuan District, Chongqing Municipality (Chongqingshi, Hechuanqu Taihezhen 重慶市合川區太和鎮). Suining is north of Hechuan in today's Sichuan Province (Sichuansheng Suiningshi 四川省遂寧市).

34. Sichuan opera title.

35. Duan and Hu, *Chuanjiang haozi*, 853.

36. Wei Yingtao 隗瀛濤, *Jindai Chongqing chengshi shi* 近代重慶城市史 [History of modern Chongqing] (Chengdu: Sichuan daxue chubanshe, 1991), 190.

37. Bamanzi: legendary hero of the Ba Kingdom, a polity covering an area of today's Chongqing Municipality obliterated by the state of Qin 秦 in 316 BCE. Möngke (1209–1259): fourth Great Khan of the Mongol Empire and a grandson of Genghis Khan, ruled from 1251 until 1259.

38. See *Sengni niehai* 僧尼孽海 [Monks and nuns in the sea of sins], in Howard S. Levy et al., *Two Chinese Sex Classics* (Taipei: Oriental Cultural Service, 1975). This motif is also present in the classical novels such as *Shuihuzhuan* 水滸傳 [Water Margin] and *Jinpingmei* 金瓶梅 [The Plum in the Golden Vase].

39. Duan and Hu, *Chuanjiang haozi*, 850–51.

40. Joshua Howard, *Workers at War: Labor in China's Arsenals, 1937–1953* (Stanford, CA: Stanford University Press, 2004).

41. Chabrowski, *Singing on the River*, 81–86.

42. For a record on the disappearance of the Three Gorges' social, historical, and natural world see, for example, Deirdre Chetham, *Before the Deluge: The Vanishing World of the Yangtze's Three Gorges* (New York: Palgrave Macmillan, 2002).

43. Such opinion still persists in the monumental studies of Wei Yingtao and Zhou Yong.

44. Such as ZGYJC and Duan and Hu, *Chuanjiang haozi*. The latter volume is a comprehensive study of Sichuan boatmen work songs. In contrast to the previous collections, it is also provided with a social, cultural, and historical analysis of the songs as well as biographies of the singers.

45. Deng Ping 鄧平, "Chuangong, chuangong haozi, chuangong jingshen—guanyu Chongqing xingxiang yu jingshende sikao he jianyi" 船工, 船工號子, 船工精神—關於形象與精神的思考和建議 [Boat workers, boat workers' haozi, boat workers' spirit—reflections and recommendations on the Chongqing's imagery and spirit], *Chongqing shehui zhuyi xueyuan xuebao* 重慶社會主義 [Journal of Chongqing Institute of Socialism] 2 (2002): 59–61; Li Liangpin 李良品, "Chuanjiang haozide xingcheng, neirong yu wenhua

jingshen" 川江號子的形成, 內容與文化精神 [The form, content and cultural spirit of the Sichuan *haozi*], *Fuling shifan daxue xuebao* 涪陵師範學院學報 [Journal of the Fuling Normal University] 19, no. 2 (2003): 58–61.

46. For a similar case in Wuhan see Robin McNeal, "Constructing Myth in Modern China," *Journal of Asian Studies* 71, no. 3 (2012): 679–704.

47. Such as the previously mentioned works of Peng Botong or recent collections of articles and sources on local history. See Zhonggong Chongqingshi weiyuan xuanbu 中共重慶市委員宣傳部, ed., *Chongqing duben* 重慶讀本 [Chongqing reader] (Chongqing: Chongqing chubanshe, 2009), hereafter *Chongqing duben*. On Great Yu see Anne Birrell, *Chinese Mythology: An Introduction* (Baltimore: John Hopkins University Press, 1993), 81–83, 146–59. For a full overview with accurate pictures of the sculpture park, see "Pictures of the Sculpture Park on the Three Gorges Square, Shapingba, Chongqing," accessed 13 Apr. 2014, http://blog1.poco.cn/myBlogDetail-htx-id-5785337-userid-56387375-pri-n-0.xhtml.

48. "Pictures of the Three Gorges Square, Shapingba, Chongqing," accessed 13 Apr. 2014, http://dcbbs.zol.com.cn/61/84_604737.html.

49. See interview with the construction engineer in *Die chinesischen Schuhe: Aus der Vergangenheit flussaufwärts in eine Gegenwart rasloser Veräderung*, dir. Tamara Wyss (Berlin: Good!Movies, 2004).

50. Wang Chuanping 王川平, "Jiedu 'Chongqing Zhongguo Sanxia Bowuguan'" 解讀"重慶中國三峽博物館 [Decoding "Chongqing Chinese Three Gorges Museum"] in *Chongqing duben*, 544–47.

51. Wang, "Jiedu 'Chongqing Zhongguo Sanxia Bowuguan,'" 544–47.

52. "Chongqing Sanxia bowuguan 重慶三峽博物館," *Baidu Online Encylopedia*, accessed 11 Aug. 2015, http://baike.baidu.com/view/478839.htm.

53. "Pictures of the Three Gorges Museum exhibition," accessed 13 Apr. 2014, http://www.mafengwo.cn/i/975495.html.

54. This was substantiated rather tenuously by an excavated wooden boat from the Ba Kingdom era (?–316 BCE).

Chapter 10. The Ganges as an Urban Sink

Epigraph: National Archives of India (hereafter NAI), Home, Municipalities, A, 15–16, June 1901, noting by L. D. Spencer, Officiating Sanitary Commissioner with the Government of India, 19 Apr. 1901.

1. In an interesting development, the government of India has recently renamed its Ministry of Water Resources as the Ministry of Water Resources, River Development and Ganga Rejuvenation. Ganga and Ganges are different names for the same river.

2. The Ganges River basin is the largest river basin in India, covering more than a quarter of the country's land area, and hosting about 43 percent of its population.

3. On exceptional occasions, such as the days of the Kumbh festival (below), the size of the congregation at holy sites along the rivers numbers in millions.

4. David Arnold, "Pollution, Toxicity and Public Health in Metropolitan India, 1850–1939," *Journal of Historical Geography* 42 (2013): 128.

5. In its most general sense, the idea of colonial difference contends that there was an inherent moral deficiency of the Oriental such that "a normative proposition of supposedly universal validity did not apply to the East." See Partha Chatterjee, *The Black Hole of the Empire: History of a Global Practice of Power* (Delhi: Permanent Black, 2013), 34. In this essay I draw upon this idea to look at how the understandings of material life and properties of nature were also inflected by the idea of colonial difference.

6. M. K. Gandhi, "Ganga and Jumna," *Young India*, 23 Dec. 1926, reproduced as "The Use We Make of Our Rivers," in Gandhi, *Social Service, Work and Reform*, ed. V. B. Kher (Ahmedabad: Navajivan, 1976), 115–16.

7. Kelly D. Alley, *On the Banks of the Gaṅgā: When Wastewater Meets a Sacred River* (Ann Arbor: University of Michigan Press, 2002).

8. GoI, *Annual Report on the Sanitary Administration of the Punjab, 1868* (Lahore, 1869), 36; see also Bernard Cohn, "Representing Authority in Victorian India," in Cohn, *An Anthropologist among the Historians and Other Essays* (Delhi: Oxford University Press, 1987); Veena Talwar Oldenburg, *The Making of Colonial Lucknow, 1856–1877* (Delhi: Oxford University Press, 1989).

9. The literature on this subject is too vast, but on rivers in India specifically, see Rohan D'Souza, *Drowned and Damned: Colonial Capitalism and Flood Control in Eastern India* (New Delhi: Oxford University Press, 2006); Christopher Hill, *River of Sorrow: Environment and Social Control in Riparian North India, 1770–1994* (Ann Arbor, MI: Association for Asian Studies, 1997); David Gilmartin, "Modes of Hydraulic Environment: Colonial Irrigation, State Power and Community in the Indus Basin," in *Nature, Culture, Imperialism: Essays on the Environmental History of South Asia*, ed. David Arnold and Ramchandra Guha (Delhi: Oxford University Press, 1996), 210–36; and John Broich, "Engineering the Empire: British Water Supply Systems and Colonial Societies, 1850–1900," *Journal of British Studies* 46, no. 2 (2007): 346–65.

10. NAI, Home, Public, A, 47–49, 27 Feb. 1869, *Destruction of Fish*.

11. See Alley, *On the Banks of the Gaṅgā*.

12. NAI, Home, Sanitary, 25–33, July 1876.

13. NAI, Home, Sanitary, no. 52 and 53, Feb. 1888.

14. "Fair at Hurdwar," *Asiatic Journal*, n.s., 13 (Apr. 1834): 245. On the threat to the sanctity of the river, see also Kama Maclean, *Pilgrimage and Power: The Kumbh Mela in Allahabad 1765–1954* (New York: Oxford University Press, 2008), 131.

15. "Fair at Hurdwar."

16. This did not rule out the fact that religious fairs, especially those connected with bathing festivals at the rivers such as the Ganges, Jumna, and Godavari, were important sites for Christian missionary activities, much resented by the native publics.

17. The literature on the revolt of 1857 is quite large. See for instance, Mahmood Farooqui, *Besieged: Voices from Delhi, 1857* (New Delhi: Penguin Viking, 2010); Eric Stokes, *The Peasant Armed: The Indian Revolt of 1857* (New York: Oxford University

Press, 1986) and Rudranghsu Mukherjee, *Awadh in Revolt, 1857–1858: A Study of Popular Resistance* (Delhi: Oxford University Press, 1984).

18. "Proclamation by the Queen in Council to the Princes, Chiefs and the People of India," 1858.

19. For this distinction between the sovereign inner domain and a colonized outer world see Partha Chatterjee, *The Nation and Its Fragments: Colonial and Postcolonial Histories* (Delhi: Oxford University Press, 1994).

20. *Kumbh* referred to the exceptionally large bathing festival that took place in cycles of twelve years.

21. Maclean, *Pilgrimage and Power.*

22. There was little information regarding Kumbh fairs previous to 1867, wrote the sanitary commissioner in 1879, but that which remained was nothing if not a record of disease and death. NAI, Home, Public, 67–72, Jan. 1892; G. Hutchenson, *Report on the Sanitary Arrangements of the Hardwar Great Fair, or Kumbh Mela of 1891*; A. B. Patterson, *Memorandum of Hardwar Kumbh, 1891.*

23. NAI, Home, Public, 67–72, Jan. 1892; Hutchenson, *Report*; Patterson, *Memorandum.*

24. NAI, Home, Public, 67–72, Jan. 1892; Hutchenson, *Report. Sullage* referred to wastewater that contained no human excreta or animal waste, thus distinguishing it from sewage.

25. Hari Ki Pauri is the sacred pond in which people bathe.

26. NAI, Home, Public, 67–72, Jan. 1892; Hutchenson, *Report.*

27. James Lochtefeld, *God's Gateway: Identity and Meaning in a Hindu Pilgrimage Place* (New York: Oxford University Press, 2010).

28. Lochtefeld argues that such conflicts were not confined to Haridwar but were common at Hindu pilgrimage places throughout India. Controlling religious institutions became an important avenue by which to resist colonial power—however symbolic—and to contest the legitimacy of colonial authority, since the government's pledged noninterference in religious affairs made this one of the few areas in which these rights could be confidently asserted.

29. This essay deals largely with the question of drainage/sewerage and pollution. It is pertinent to point out, however, that the question of religious belief also frequently challenged the technological optimism of colonial authorities with respect to the provision of drinking water. See Sandria Freitag, "State and Community: Symbolic Popular Protest in Banaras's Public Arena," in *Culture and Power in Banaras: Community, Performance and Environment, 1800–1980* (Berkeley: University of California Press, 1992), 203–28; Awadhendra Sharan, "From Source to Sink: 'Official' and 'Improved' Water in Delhi, 1868–1956," *Indian Economic and Social History Review* 48, no. 3 (2011): 425–62.

30. David B. Smith, *Report on the Drainage and Conservancy of Calcutta* (Calcutta, 1869). Also see Ishita Pande, *Medicine, Race and Liberalism in British Bengal: Symptoms*

of Empire (New York: Routledge, 2010), esp. ch. 6, "Sensing Modernity: Civility, Class and Citizenship in a Sanitary City."

31. Smith, *Report on the Drainage*, 64.

32. This paragraph largely draws upon Michael S. Dodson, "The Shadows of Modernity in Banaras," in *Banaras: Urban Forms and Cultural Histories* (New Delhi: Routledge, 2012), 143–69.

33. Dodson, "Shadows of Modernity"; also see David Arnold, "The Ecology and Cosmology of Disease in the Banaras Region," in Freitag, *Culture and Power in Banaras*, 246–67.

34. Dodson, "Shadows of Modernity." Dodson prefers to describe these multiple purposes as being ambiguous.

35. Cited in Mridula Ramanna, *Western Medicine and Public Health in Colonial Bombay, 1845–1895* (New Delhi: Orient Longman, 2002), 110.

36. M. K. Gandhi, "Insanitation in Hardvar," *Young India*, 31 Oct. 1929, reproduced in Gandhi, *Social Service, Work and Reform*, 85–87. Also see "Our Insanitation," *Young India*, 19 Nov. 1925; "Sanitation in Ahmedabad," *Young India*, 26 Nov. 1925; "A National Defect," *Young India*, 25 Apr. 1929; "Insanitary Gaya," *Young India*, 3 Feb. 1927.

37. Kelly Alley points out that policies regarding the Ganga had first surfaced in the making of the Ganga and Yamuna canals in the late nineteenth century, and this assumption of authority to divert and channel river water later fed into the controls officials developed over other kinds of flows, such as flows of wastewater running through city sewers and drains. See Alley, *On the Banks of the Gaṅgā*, 136.

38. The cities of Haridwar, Banaras, and Kanpur were all located in this province.

39. NAI, Home, Sanitary, 17–25, Apr. 1893, *Sewage and Drainage of Cawnpore.*

40. NAI, Home, Sanitary, 75–76, June 1890, *Discharge of Sewage into the Ganges at Benares.*

41. NAI, Home, Sanitary, 17–25, Apr. 1893, *Sewage and Drainage of Cawnpore.*

42. NAI, Home, Sanitary, 78–79, Feb. 1890. There would be no harm from the discharge of surface drainage into the Ganges, he wrote, but if the river was to be made into a general sewer for towns on its banks, then the risks could be hardly exaggerated.

43. NAI, Home, Sanitary, 78–79, Feb. 1890.

44. NAI, Home, Sanitary, 44–45, Aug. 1890.

45. NAI, Home, Sanitary, 44–45, Aug. 1890.

46. NAI, Home, Sanitary, B, 2–6, July 1878, *Cawnpore Drainage.*

47. NAI, Home, Sanitary, 75–76, June 1890, *Discharge of Sewage into the Ganges at Benares.*

48. NAI, Home, Sanitary, 75–76, June 1890, *Discharge of Sewage into the Ganges at Benares.*

49. The other means referred to the use of sewage on sewage farms but it was doubtful if even the effluent that remained after use on the farm was free of danger.

50. NAI, Home, Sanitary, 75–76, June 1890, *Discharge of Sewage into the Ganges at Benares.*

51. NAI, Home, Sanitary, 17–25, Apr. 1893, *Sewage and Drainage of Cawnpore.*

52. NAI, Home, Sanitary, 17–25, Apr. 1893, *Sewage and Drainage of Cawnpore.*

53. NAI, Home, Sanitary, B, 2–6, July 1878, *Cawnpore Drainage.*

54. NAI, Home, Municipalities, A, 15–16, June 1901, *Sewerage Scheme for Cawnpore and the Grant of a Loan of Five Lakhs of Rupees to the Cawnpore Municipality.*

55. The extent to which such conditions could be debilitating in the context of sanitary works can be seen in an official noting of the secretary to government, northwest provinces, in the context of proposed works at Agra: "The great and expensive works of sanitary improvement initiated by Sir Auckland Colvin [the previous governor of the province] in the large cities of these provinces have involved the inhabitants in liabilities the liquidation of which must under any circumstances impose a heavy burden upon them, and the provincial resources have now been reduced to such an extent that they can no longer afford such aid in the prosecution of works as was at one time possible." See NAI, Home, Municipalities, 27–31, Oct. 1896.

56. NAI, Home, Municipalities, A, 8–11, May 1898.

57. The difficulty, as he saw it, was the reluctance of house owners to join to the underground system.

58. NAI, Home, Municipalities, A, 15–16, June 1901, *Sewerage Scheme for Cawnpore.*

59. NAI, Home, Municipalities, A, 15–16, June 1901, *Sewerage Scheme for Cawnpore.*

60. NAI, Home, Municipalities, A, 15–16, June 1901, *Sewerage Scheme for Cawnpore.*

61. The point was acknowledged in a joint note of the sanitary commissioner and the sanitary engineer, who observed in the context of the Banaras sewerage works that "in all recent discussions on the principles of sewage disposal in these provinces, it has been recognized that putting sewage into rivers could only be resorted to as a temporary measure, *made compulsory by the pressure of financial necessity*" (emphasis mine). NAI, Home, Sanitary, 17–25, Apr. 1893.

62. This paragraph draws upon Lochtefeld, *God's Gateway.*

63. Alley, *On the Banks of the Gaṅgā*, 136.

64. However, not all within the government were satisfied with the eventual compromise, with some officers reasoning that the first bridges built over the rivers had also been considered sacrilegious, and had the government heeded those objections, India would still have been at the stage of using ferries across the rivers.

65. Cited in Leah Renold, "A Hindu Temple of Learning: The Hybridization of Religion and Architecture," in Dodson, *Banaras*, 170–91, esp. 179.

66. Arnold, "Ecology and Cosmology," 265.

67. Dodson, "Shadows of Modernity."

Chapter 11. Polluted Thames, Declining City

1. Accounts of his wanderings in "Night Walks," in *The Uncommercial Traveller* (London: Chapman and Hall, 1861).

2. Gavin Weightman, *The Making of Modern London, 1815–1914* (London: Sidgwick and Jackson, 1983).

3. Pertinent: Peter Brimblecombe, *The Big Smoke: A History of Air Pollution in London since Medieval Times* (New York: Methuen, 1987); Martin Daunton, ed., *The Cambridge History of Urban Britain*, vol. 3, *1840–1950* (Cambridge: Cambridge University Press, 2001).

4. Christopher Hamlin, "Providence and Putrefaction: Victorian Sanitarians and the Natural Theology of Health and Disease," *Victorian Studies* 28, no. 3 (1985): 382.

5. Charles Dickens, *Our Mutual Friend*, ed. Michael Cotsell (New York: Oxford University Press, 2008); hereafter cited as *OMF*.

6. "Dust; or, Ugliness Redeemed," *Household Words*, 13 July 1850. Ackroyd and Gallagher assign the authorship to R. H. Horne, but without reference. Madeline House, ed., *The Letters of Charles Dickens: The Pilgrim Edition*, vol. 3, *1842–1843*. (Oxford: Clarendon, 1974), 330–31.

7. "An Act for the Amendment and Better Administration of the Laws relating to the Poor in England and Wales," 1834, London Metropolitan Archives.

8. Howard W. Fulweiler, "'A Dismal Swamp': Darwin, Design, and Evolution in *Our Mutual Friend*," *Nineteenth-Century Literature* 49, no. 1 (1994): 50–74.

9. Madeline House and Graham Storey, eds., *The Letters of Charles Dickens: The Pilgrim Edition*, vol. 10, *1862–1864* (Oxford: Clarendon, 1965), 164–65.

10. See House, *Letters*, 3:330–31.

11. "Public Health (London) Act," 1836, CLRO P.D. 51.4, London Metropolitan Archives. Initially the Act excluded London City.

12. Humphrey House, *The Dickens World* (London: Oxford University Press, 1950); Edgar Johnson, *Charles Dickens: His Tragedy and Triumph*, vol. 2 (New York: Simon and Schuster, 1952).

13. Harvey Peter Sucksmith, "The Dust-Heaps in *Our Mutual Friend*," *Essays in Criticism* 23, no. 2 (1973): 206–12; Nancy Aycock Metz, "The Artistic Reclamation of Waste in *Our Mutual Friend*," *Nineteenth-Century Fiction* 34, no. 1 (1979): 59–72.

14. Sucksmith, "Dust-Heaps," 209–10, presents different interpretations of "dust," including historical expression for "money."

15. These focus mainly on homoerotic fantasies ascribed to the intensive involvement with excrements and male rivalry. Ellen Handy, "Dust Piles and Damp Pavements: Excrement, Repression, and the Victorian City in Photography and Literature," in *Victorian Literature and the Victorian Visual Imagination*, ed. Carol T. Christ and John O. Jordan (Berkeley: University of California Press, 1995); Eve Kosofsky Sedgwick, *Between Men: English Literature and Male Homosocial Desire* (New York: Columbia University Press, 1985).

16. Pamela K. Gilbert, *Mapping the Victorian Social Body* (Albany: State University of New York Press, 2004). For a geographer's perspective see Michelle Elizabeth Allen, *Cleansing the City: Sanitary Geographies in Victorian London* (Athens: Ohio University Press, 2008).

17. Hans-Dieter Gelfert interprets *OMF* as Dickens's attempt to address the conflict between his own pursuit for societal advancement and his self-perception as an underdog, disillusioned by the existing societal system. Hans-Dieter Gelfert, *Charles Dickens: Der Unnachahmliche. Biographie* (Munich: Beck, 2011), 276.

18. For the cultural importance of the river for London and England, see Peter Ackroyd, *Thames: Sacred River* (London: Chatto and Windus, 2007).

19. Donald Worster, "Transformations of the Earth: Toward an Agroecological Perspective in History," *Journal of American History* 76, no. 4 (1990): 1092.

20. Donald Worster, *Nature's Economy: A History of Ecological Ideas* (Cambridge: Cambridge University Press, 1994), 370.

21. Marco Armiero, "Seeing like a Protester: Nature, Power, and Environmental Struggles," *Left History* 13, no. 1 (2008): 60.

22. Cited from John Parham, "Green Man Hopkins: Gerald Manley Hopkins and Victorian Ecological Criticism," *Nineteenth-Century Contexts* 25, no. 3 (2003): 258.

23. Gillen D'Arcy Wood, "Introduction: Eco-historicism," *Journal for Early Modern Cultural Studies* 8, no. 2 (2008): 4.

24. Catherine Gallagher, *The Body Economic: Life, Death, and Sensation in Political Economy and the Victorian Novel* (Princeton, NJ: Princeton University Press, 2008), 94.

25. The embankment of the Thames was planned and executed from 1850 onward. Dale H. Porter, *The Thames Embankment: Environment, Technology, and Society in Victorian London* (Akron, OH: University of Akron Press, 1998).

26. Regenia Gagnier, "Twenty-First Century and Victorian Ecosystems: Nature and Culture in the Developmental Niche," *Victorian Review* 36, no. 2 (2010): 19.

27. Barri J. Gold, "Energy, Ecology, and Victorian Fiction," *Literature Compass* 9, no. 2 (2012): 214.

28. Gavin Weightman, *London's Thames: The River That Shaped a City and Its History* (New York: St. Martin's, 2005).

29. Tina Young Choi, "Forms of Closure: The First Law of Thermodynamics and Victorian Narrative," *ELH* 74, no. 2 (2007): 316.

30. "Body snatchers" selling corpses for anatomical research are specifically addressed in *A Tale of Two Cities*, and "Use and Abuse of the Dead" in *Household Words*, 3 Apr. 1858.

31. Philip Hobsbaum, "The Critics and *Our Mutual Friend*," *Essays in Criticism* 13, no. 3 (1963): 233, underlines Lizzy's potential to expressing Dickens's social critique. He dramatizes the "danger of moral pollution," as Michelle Allen phrases it; here referring to the depiction of the Six Jolly Fellowship Porters and its host Abbey Potterson as moral compass for the young woman; *Cleansing the City*, 100.

32. In *Cleansing the City*, 101, Michelle Allen describes in detail how those negative tractions of the "riverside people" stigmatize Lizzy long after declining Miss Peterson's help, e.g., influencing the relationship to her admirers. Here the danger of the city pollution would manifest itself in sexual indecency, driving her to leave town.

33. A profound introduction to the concept of environmental justice in the European research context is offered by Geneviève Massard-Guilbaud and Richard Rodger, *Environmental and Social Justice in the City: Historical Perspectives* (Cambridge: White Horse, 2011).

34. The mechanism of social exclusion is further evidence of Darwinian influence on *OMF*, as "even the characters who are finally redeemed in *Our Mutual Friend* are deeply involved in the self-seeking world of individual advantage." Fulweiler, "Dismal Swamp," 60.

35. John Parham, "Dickens in the City: Science, Technology, Ecology in the Novels of Charles Dickens," *19: Interdisciplinary Studies in the Long Nineteenth Century* 10 (28 Apr. 2010): 8, accessed 7 May 2011, doi:http://doi.org/10.16995/ntn.529.

36. John Ruskin, *Unto This Last* (London: Smith and Elder, 1862), accessed 20 June 2011, http://etext.lib.virginia.edu/toc/modeng/public/RusLast.html.

37. Gallagher, *Body Economic*, 90. Dickens's interest in descriptions of death and death experiences might relate to several friends and family members dying during his work on *Our Mutual Friend*, and his own realization of personal finitude. *OMF*, 937–38.

38. Catherine Gallagher, "The Bio-Economics of *Our Mutual Friend*," in *Fragments for a History of the Human Body*, vol. 3, ed. Michael Feher, Ramona Naddaff, and Nadia Tazi (New York: Zone, 1989).

39. On the influence of love on the laws of natural selection, see Nicola Bown, "What the Alligator Didn't Know: Natural Selection and Love in *Our Mutual Friend*," *19: Interdisciplinary Studies in the Long Nineteenth Century* 10 (28 Apr. 2010), accessed 15 Sep. 2012, doi:http://doi.org/10.16995/ntn.567.

40. Fulweiler, "Dismal Swamp," 51.

41. Edwin Chadwick, *Report on the Sanitary Conditions of the Labouring Population of Great Britain* (London, 1842).

42. Charles Dickens, *The Speeches of Charles Dickens: A Complete Edition*, ed. K. J. Fielding (Atlantic Highlands, NJ: Humanities Press International, 1988).

43. Brian Cheadle, "The Late Novels: *Great Expectations* and *Our Mutual Friend*," in *The Cambridge Companion to Charles Dickens*, ed. John O. Jordan (New York: Cambridge University Press, 2001).

44. Ulrich Beck, *Risikogesellschaft: Auf dem Weg in eine andere Moderne* (Frankfurt am Main: Suhrkamp, 1986).

45. Xueqin Mei, "英国环境史上沉重的一页—泰晤士河三文鱼的消失及其教训" [A heavy page in British environmental history—the disappearance of the salmon from the River Thames and its lessons], *Journal of Nanjing University for Philosophy, Humanities, and Social Sciences* 6 (2013). Also Bill Luckin, *Pollution and Control: A Social History of the Thames in the Nineteenth Century* (Bristol: Adam Hilger, 1986).

46. His impressions are captured in his *Uncommercial Traveller*, evolving at the same time as *OMF*. Dickens here embodies a classic flaneur, but in a setting exotified by his nighttime activity.

47. Franco Moretti, *Atlas of the European Novel, 1800–1900* (London: Verso, 1999), 124–43.

48. Rolf Lindner, *Walks on the Wild Side: Eine Geschichte der Stadtforschung* (Frankfurt: Campus, 2004).

49. Dickens, "Night Walks," 188–89.

50. Peter Ackroyd, *Dickens* (London: Sinclair-Stevenson, 1990). Commonly used quote; however, it is without direct citation.

51. Melanie Kiechle, "Navigating by Nose: Fresh Air, Stench Nuisance, and the Urban Environment, 1840–1880," *Journal of Urban History* 41 (2015): 1–19.

52. Hamlin, "Providence and Putrification," 401.

53. Fulweiler, "Dismal Swamp," 74.

54. William A. Cohen and Ryan Johnson, eds., *Filth: Dirt, Disgust, and Modern Life* (Minneapolis: University of Minnesota Press, 2005), 97.

55. Laura Ball, "Cholera and the Pump on Broad Street: The Life and Legacy of John Snow," *History Teacher* 43, no. 1 (2009): 105–19. On the 1854 event: Howard Brody et al., "Map-Making and Myth-Making in Broad Street: The London Cholera Epidemic, 1854," *Lancet* 356, no. 9223 (2000): 64–68. Relevant: Robert John Morris, *Cholera, 1832: The Social Response to an Epidemic* (London: Holmes and Meier, 1976); Norman Longmate, *King Cholera: The Biography of a Disease* (London: H. Hamilton, 1966). Solid popular science: Steven Johnson, *The Ghost Map: The Story of London's Most Terrifying Epidemic—and How It Changed Science, Cities, and the Modern World* (New York: Riverhead Books, 2006).

56. Baldwin, "How Night Air Became Good Air," 415–16; Christopher Hamlin, *Public Health and Social Justice in the Age of Chadwick: Britain, 1800–1854* (Cambridge: Cambridge University Press, 2009).

57. Parham, "Dickens in the City," 18.

58. "This sense, that a degraded physical environment equates to a hazardous human one—a Victorian 'risk society'—is specifically ecological and it is that point, the extent to which Dickens foreshadowed contemporary ecological thinking." Parham, "Dickens in the City," 3.

59. E.g., Porter, *Thames Embankment*. A detailed contemporary account by leading engineer is Joseph Balgazette, "Balgazette on the Main Drainage of London," 14 Mar. 1865, COL TSD EG 06 01 027, EMD Papers Box 1.27, London Metropolitan Archives.

60. Gandy, "Rethinking Urban Metabolism," 366.

61. I.e., the Wilfers (*OMF*, 316), or later Lizzy (*OMF*, 522). Claire Tomalin, *Charles Dickens: A Life* (London: Viking, 2011), 341, points out the resemblance to Dickens's own recreational patterns.

62. Gilbert, *Mapping the Victorian Social Body*, 140.

63. That the initiated reforms still turned out as inefficient due to the city's growth, becomes apparent with the 1930s new regulations, still featuring waste and excrements in the streets as a major problem: "Public Health (London) Act," 1848, CLRO P.D. 51.4,

London Metropolitan Archives. See also Stephen Halliday, *The Great Stink of London: Sir Joseph Balgazette and the Cleansing of the Victorian Metropolis* (Stroud: Sutton, 1999).

64. Fulweiler, "Dismal Swamp," 65.

65. Fulweiler, "Dismal Swamp," 74.

66. Berbeli Wanning, *Die Fiktionalität der Natur: Studien zum Naturbegriff in Erzähltexten der Romantik und des Realismus* (Berlin: Weidler, 2005).

67. The Kinks, "Waterloo Sunset," *Metrolyrics*, accessed 6 Aug. 2015, http://www.metrolyrics.com/waterloo-sunset-lyrics-kinks.html.

Chapter 12. Living on the River over the Year

This research was supported by a grant from the Russian Foundation for Basic Research 10-04-91005-ANF_a. The authors also express their gratitude to Assoc. Prof. Dr. Martin Schmid (University of Klagenfurt, Vienna) for the significant support during the early stage of this research and to Karen Alexander (University of New Hampshire, USA) for the improvement of English language style.

1. Mikhail V. Lomonosov, "Oda na den' vosshestviia na Vserossiskii prestol ee Velichestva gosudaryni imperatricy Elizavety Petrovny 1747 goda," in *Izbrannye proizvedeniia* (Leningrad: Sovetskii pisatel, 1986), 116.

2. Iogann G. Georgi, *Opisanie stolichnogo goroda Sankt-Peterburga 1790 g.* (Saint Petersburg, 1794), 3–4.

3. Alexei Kraikovski and Julia Lajus, "The Neva as a Metropolitan River of Russia: Environment, Economy and Culture," in *Rivers and Society: From Early Civilizations to Modern Times*, ed. Richard Coopey and Terje Tvedt (London: I. B. Tauris, 2010), 346–53.

4. See, for instance, Paul Keenan, *St Petersburg and the Russian Court, 1703–1761* (Basingstoke: Palgrave Macmillan, 2013); Solomon Volkov, *St. Petersburg: A Cultural History*, trans. Antonina W. Bouis (New York: Simon and Schuster, 2010); W. Bruce Lincoln, *Sunlight at Midnight: St. Petersburg and the Rise of Modern Russia* (New York: Basic Books, 2009); Julie A. Buckler, *Mapping St. Petersburg: Imperial Text and Cityshape* (Princeton, NJ: Princeton University Press, 2005); Anthony Cross, ed., *St. Petersburg, 1703–1825* (Basingstoke: Palgrave Macmillan, 2003); Grigorii Kaganov, *Sankt-Peterburg: obrazy prostranstva* (Saint Petersburg: Izdatelstvo Ivana Limbaha, 2004); Olga Kosheleva, *Liudi Sankt-Peterburgskogo ostrova Petrovskogo vremeni* (Moscow: OGI, 2004); Vladimir Lisovskii, *Sankt-Peterburg: Ocherki arhitekturnoi istorii goroda* (Saint Petersburg: Kolo, 2009).

5. Randall Dills, "The River Neva and the Imperial Façade: Culture and Environment in Nineteenth Century St. Petersburg Russia" (PhD diss., University of Illinois, Urbana–Champaign, 2010), https://www.ideals.illinois.edu/bitstream/handle/2142/18391/Dills_Randall.pdf?sequence=1.

6. Dills, "River Neva," xxi–xxii.

7. Gavrila R. Derzhavin, "Vesna," in *Stihotvoreniia* (Leningrad: Sovetskii Pisatel, 1957), 298.

8. Evgenii V. Anisimov, *Yunyi grad: Peterburg vremen Petra Velikogo* (Saint Petersburg: Dmitrii Bulanin, 2003), 37.

9. Georgi, *Opisanie stolichnogo goroda Sankt-Peterburga 1790 g.*, 15.

10. Aleksandr S. Pushkin, *Mednyi vsadnik* (Leningrad: Nauka, 1978), 11.

11. Elena Keller, *Morskoi Peterburg* (Saint Petersburg: Nevskii rakurs, 2010), 118–19.

12. Valentin Chernenko, *Blagoveshhenskij most-pervyj postojannyj most cherez Nevu: Istorija proektirovanija,stroitel'stva i rekonstrukcii* (Saint Petersburg: Serebrjannyj vek, 2013).

13. Anisimov, *Yunyi grad*, 333.

14. Dmitrii Zasosov and Vladimir Pyzin, *Povsednevnaya zhizn Peterburga na rubezhe XIX–XX vekov* (Moscow: Molodaia Gvardia, 2003), 14–15.

15. Zasosov and Pyzin, *Povsednevnaya zhizn Peterburga*, 15.

16. Keller, *Morskoi Peterburg*, 118–19. For the pictures of this ceremony see Mihail A. Grigoriev, *Peterburg 1910h godov: Progulki v proshloe* (Saint Petersburg: Institut istorii iskusstv, 2005), 88.

17. *Peterburgskaia gazeta*, 22 Mar. 1910.

18. *Russkoe slovo*, 9 Apr. 1905.

19. For example, Grigoriev, *Peterburg 1910h godov*, 92.

20. Andrei I. Bogdanov, *Opisanie Sanktpeterburga, Saint Petersburg* (Saint Petersburg: Filial Arkhiva RAN, 1997), 259.

21. Georgi, *Opisanie stolichnogo goroda Sankt-Peterburga 1790 g.*, 44. For a picture of such a bridge as it was in 1804, see Galina A. Miroliubova et al., *Sankt-Peterburg v akvareliakh, graviurakh i litografiiakh XVIII–XIX vekov* (Saint Petersburg: Arka, 2009), 113.

22. Zasosov and Pyzin, *Povsednevnaya zhizn Peterburga*, 28; *Peterburgskii listok*, 14 Nov. 1911.

23. Grigoriev, *Peterburg 1910h godov*, 94.

24. Nikolai V. Gogol, *Sochinenia* (Moscow: Gosudarstvennoe Izdatelstvo Hudozhestvennoi Literatury, 1956), 238.

25. Alexei Kraikovski , Julia Lajus, and Dmitri Lajus, "Fisheries on the South-Eastern Coast of the Gulf of Finland and the Adjoining River Basins, 15th–18th Centuries," in *Leva vid Östersjöns kust: En antologi om resursutnyttjande på båda sidor av Östersjön ca 800–1800*, ed. Sven Lilja (Stockholm: Södertörn, 2008), 214; Julia Lajus, Alexei Kraikovski, and Dmitry Lajus, "Coastal Fisheries in the Eastern Baltic Sea (Gulf of Finland) and Its Basin from the 15th to the Early 20th Centuries," *PLOS ONE* 8, no. 10 (2013): 7, doi:10.1371/journal.pone.0077059.

26. Anisimov, *Yunyi grad*, 332; Zasosov and Pyzin, *Povsednevnaya zhizn Peterburga*, 13.

27. Gogol, *Sochinenia*, 352.

28. *Peterburgskaia gazeta*, 5 Apr. 1910.

29. Grigoriev, *Peterburg 1910h godov*, 95.

30. *Novosti dnia*, 23 and 24 Mar. 1903.

31. *Rech'*, 6 Mar. 1910.

32. Petr A. Vyazemski, "Pismo Vyazemskoi V. F. 17 aprelia [1828 g.] Peterburg," in *Pushkin, Lermontov, Gogol: Literaturnoe nasledstvo,* vol. 58, ed. Alexandr M. Egolin, (Moscow: Izdatelstvo Akademii Nauk SSSR, 1952), 75.

33. See Keller, *Morskoi Peterburg*, 107.

34. Lars Yu. Erenmalm, "Opisanie goroda Peterburga, vkupe s neskolkimi zamechaniiami," in *Peterburg Petra I v inostrannykh opisaniiakh*, ed. Yuri N. Bespyatykh (Leningrad: Nauka, 1991), 93.

35. Dills, "River Neva," 1–27; Keller, *Morskoi Peterburg*, 103–14; Robert E. Jones, *Bread upon the Waters: The St. Petersburg Grain Trade and the Russian Economy, 1703–1811* (Pittsburgh, PA: University of Pittsburgh Press, 2013).

36. Mstislav V. Dobuzhinski, *Vospominania* (Moscow: Nauka, 1987), 11.

37. Dzhulietta A. Kuchariants, *Ivan Starov* (Leningrad: Lenizdat, 1982), 63–64.

38. Militsa F. Korshunova, *Giacomo Quarenghi* (Leningrad: Lenizdat, 1981), 62–63.

39. For example, Fedor M. Dostoevskiy, "Prestuplenie i nakazanie," in *Sobranie sochinenii v 17 tomah*, vol. 6 (Leningrad: Nauka, 1973), 85.

40. Zasosov and Pyzin, *Povsednevnaya zhizn Peterburga*, 26.

41. *Peterburgskaia gazeta*, 12 July 1910.

42. *Peterburgskaia gazeta*, 12 July 1910.

43. Grigoriev, *Peterburg 1910h godov*, 92.

44. Zasosov and Pyzin, *Povsednevnaya zhizn Peterburga*, 25.

45. Central State Historical Archives of Saint Petersburg—Centralnyi Gosudarstvennyi Istoricheskii Arkhiv (hereafter CGIA SPb), coll. 260, inv. 2, f. 6.

46. CGIA SPb, coll. 260, inv. 2, f. 6.

47. For the general notes on the history of fisheries in Saint Petersburg see Kraikovski and Lajus, "Neva as a Metropolitan River," 359–61.

48. Zasosov and Pyzin, *Povsednevnaya zhizn Peterburga*, 132.

49. Grigoriev, *Peterburg 1910h godov*, 126.

50. See Keller, *Morskoi Peterburg*, 114.

51. For example, *Peterburgskaia gazeta*, 25 July 1908.

52. Galina Miroliubova et al., *Sankt-Peterburg v akvareliakh*, 91; see also Irina Verhovskaya et al., *Iz russkoi zhizni XVIII—nachala XX veka.* (Saint Petersburg: Palace Editions, 2010), 48, 68, 85, 108.

53. A. Miroliubova et al., *Sankt-Peterburg v akvareliakh*, 27.

54. *Sankt-Peterburg: Portret goroda i gorozhan* (Saint Petersburg: Palace Editions, 2003), 78.

55. Keller, *Morskoi Peterburg*, 41.

56. Zasosov and Pyzin, *Povsednevnaya zhizn Peterburga*, 25.

57. Igor V. Zimin, *Povsednevnaia zhizn' Rossiiskogo Imperatorskogo dvora: Vtoraia chetvert' XIX—nachalo XX v. Vzroslyi mir imperatorskih rezidencii* (Moscow: Centrpoligraf, 2010), 166–71.

58. Arkadii M. Gordin, *Pushkinskii Peterburg* (Saint Petersburg: Hudozhnik RSFSR, 1991), no. 107.

59. Zasosov and Pyzin, *Povsednevnaya zhizn Peterburga*, 25.

60. Grigoriev, *Peterburg 1910h godov*, 95.

61. Lajus, Kraikovski, and Lajus, "Coastal Fisheries."

62. Nikolai S. Leskov, "Ostrovitiane," in *Sobranie sochinenii v 11 tomah*, vol. 3 (Moscow: Gosudarstvennoe Izdatelstvo Hudozhestvelloi Literatury, 1957), 14.

63. Ruvim A. Nezhikhovskiy, *Neva i Nevskaia guba* (Leningrad: Gidrometeoizdat, 1981), 34.

64. Nezhikhovskiy, *Neva i Nevskaia guba*, 35–36.

65. See Georgi, *Opisanie stolichnogo goroda Sankt-Peterburga 1790 g.*, 8.

66. For details see Kim Pomeranets, *Tri veka peterburgskih navodnenii* (Saint Petersburg: Iskusstvo-SPb, 2005); Randall Dills, "Cracks in the Granite: Paternal Care, the Imperial Façade, and the Limits of Authority in the 1824 St. Petersburg Flood," *Journal of Urban History* 40, no. 3 (2014): 479–96.

67. Zasosov and Pyzin, *Povsednevnaya zhizn Peterburga*, 21–22.

68. Zasosov and Pyzin, *Povsednevnaya zhizn Peterburga*, 19–20, 22–23.

69. See Lev Lurie, *Pitershiki. Russkii kapitalizm: Pervaia popytka* (Saint Petersburg: BHV-Peterburg, 2011), 205–9.

70. Leskov, "Ostrovitiane," 123–24; *Russkoe slovo*, 7 Nov. 1904.

71. Georgi, *Opisanie stolichnogo goroda Sankt-Peterburga 1790 g.*, 10–11.

72. BV, 23 Oct. 1906.

73. *Novosti dnia*, 29 Oct. 1902.

74. Georgi, *Opisanie stolichnogo goroda Sankt-Peterburga 1790 g.*, 10.

75. Georgi, *Opisanie stolichnogo goroda Sankt-Peterburga 1790 g.*, 45.

76. *Peterburgskaia gazeta*, 12 Oct. 1909.

77. *Peterburgskaia gazeta*, 26 Nov. 1911.

78. "Tochnoe izvestie o kreposti i gorode Sankt-Peterburg i krepotce Kronshlot i ikh okrestnostiah," in *Peterburg Petra I v inostrannykh opisaniiakh*, ed. Yuri N. Bespyatykh (Leningrad: Nauka, 1991), 38.

79. Pushkin, *Mednyi vsadnik*, 10.

80. Georgi, *Opisanie stolichnogo goroda Sankt-Peterburga 1790 g.*, 7.

81. Leskov, "Ostrovitiane," 104.

82. Georgi, *Opisanie stolichnogo goroda Sankt-Peterburga 1790 g.*, 10.

83. *Novosti dnia*, 22 Dec. 1902.

84. For details see Natalia V. Zaitseva, *Neparadnaia zhizn Yusupovskogo dvortsa* (Saint Petersburg: Yusupovskii dvorets, 2006), 23–26.

85. Georgi, *Opisanie stolichnogo goroda Sankt-Peterburga 1790 g.*, 12.

86. For instance, see Zasosov and Pyzin, *Povsednevnaya zhizn Peterburga*, 14; for photo by K. Bulla see Grigoriev, *Peterburg 1910h godov*, 114; see also Dills, "River Neva," 22–26.

87. *Peterburgskaia gazeta*, 17 May 1912.

88. Erenmalm, "Opisanie goroda Peterburga," 93.

89. Zasosov and Pyzin, *Povsednevnaya zhizn Peterburga*, 10.

90. Georgi, *Opisanie stolichnogo goroda Sankt-Peterburga 1790 g.*, 11–12.

91. Friedrich Ch. Weber, "Iz knigi Fridricha Chrisitana Webera 'Preobrazhennaya Rossia,'" in *Peterburg Petra I v inostrannykh opisaniiakh*, ed. Yuri N. Bespyatykh (Leningrad: Nauka, 1991), 121.

92. Georgi, *Opisanie stolichnogo goroda Sankt-Peterburga 1790 g.*, 42; Zasosov and Pyzin, *Povsednevnaya zhizn Peterburga*, 13.

93. See the picture in Oleg N. Zakharov, *Arhitekturnye panoramy nevskih beregov* (Leningrad: Stroyizdat, 1984), 85.

94. Grigoriev, *Peterburg 1910h godov*, 101; for the picture of such a road see Zakharov, *Arhitekturnye panoramy*, 149.

95. See photos in Grigoriev, *Peterburg 1910h godov*, 146–47.

96. Zasosov and Pyzin, *Povsednevnaya zhizn Peterburga*, 12–13; for photo see Grigoriev, *Peterburg 1910h godov*, 113.

97. Yuri N. Bespyatykh, *Archangelsk nakanune i v gody Severnoi voiny 1700–1721* (Saint Petersburg: Blitz, 2010), 71–72.

98. Zasosov and Pyzin, *Povsednevnaya zhizn Peterburga*, 12–13; for photo see Grigoriev, *Peterburg 1910h godov*, 113.

99. *Utro Rossii*, 9 Feb. 1910.

100. See Keller, *Morskoi Peterburg*, 108; Zasosov and Pyzin, *Povsednevnaya zhizn Peterburga*, 12; Grigoriev, *Peterburg 1910h godov*, 101.

101. *Sankt-Peterburg: Portret*, 436–37; Gordin, *Pushkinskii Peterburg*, no. 98.

102. Verhovskaya et al., *Iz russkoi zhizni*, 59.

103. Mihail Gordin, *Vladislav Ozerov* (Leningrad: Iskusstvo, 1991), album 2, without number.

104. Zasosov and Pyzin, *Povsednevnaya zhizn Peterburga*, 10.

105. *Peterburgskaia gazeta*, 19 Nov. 1907 and 4 Feb. 1912.

106. Anisimov, *Yunyi grad*, 359.

107. For instance, see Zasosov and Pyzin, *Povsednevnaya zhizn Peterburga*, 13–14; Grigoriev, *Peterburg 1910h godov*, 102–3.

108. For instance, Igor V. Zimin, *Povsednevnaia zhizn' Rossiiskogo Impertatorskogo dvora: Tsarskaia rabota XIX—nachalo XX v.* (Moscow: Tsentropoligraf, 2011), 119–21.

109. See G. Quarenghi, "1780s," in Korshunova, *Giacomo Quarenghi*, 159; A. Ubigan, "Early XIX c.," in Zimin, *Tsarskaia rabota*, 119; K. Bulla, "1910s," in Grigoriev, *Peterburg 1910h godov*, 117.

110. Zimin, *Tsarskaia rabota*, 120.

111. For details see Vladimir B. Gendrikov, "Traurnye tseremonii v Petropavlovskom sobore," in *Kraevedcheskie zapiski*, vol. 2 (Saint Petersburg: Gosudarstvennyi muzei istorii S. Peterburga, 1994), 306–15.

Chapter 13. "A Ridiculous Failure of Government"

Title quotation: Henry Henderson, as quoted in National Resources Defense Fund, "Railing on the River: EPA Demands Chicago River Cleanup," *National Resources Defense Fund Press Release* (13 May 2011).

1. On the campaign, see *Chicago Tribune*, 10 July 1966; 17 Oct. 1967; 25, 26, and 29 Nov. 1967; 1 Dec. 1967. Rachel Carson, *Silent Spring* (Cambridge, MA: Riverside Press, 1962). Also see Terence Kehoe, "Merchants of Pollution? The Soap and Detergent Industry and the Fight to Restore Great Lakes Water Quality, 1965–1972," *Environmental History Review* 16, no. 3 (1992): 21–46; and Linda J. Lear, *Rachel Carson: Witness for Nature* (New York: Henry Holt, 1997).

2. *Chicago Tribune*, 11 June 1966; 11, 12, and 20 Aug. 1966; 17 Sept. 1966. Also see Craig E. Colten, "Industrial Wastes in Southeast Chicago: Production and Disposal 1870–1970," *Environmental Review* 10, no. 2 (1986): 93–105; Joel Greenberg, *A Natural History of the Chicago Region* (Chicago: University of Chicago Press, 2002); and Libby Hill, *The Chicago River: A Natural and Unnatural History* (Chicago: Lake Claremont, 2000).

3. Lake Michigan covers 22,400 square miles (58,000 square kilometers).

4. *Chicago Tribune*, 8 February, 30 December 1966; 20, 27 June, 8, 9 July, 6, 15, 20 August, 3, 21 September, 26 November 1967.

5. On the local crusade, see note 1, above. On the Clean Water Act, see Terence Kehoe, *Cleaning up the Great Lakes: From Cooperation to Confrontation* (DeKalb: Northern Illinois University Press, 1997); Edmund P. Russell III, "Lost among the Parts per Billion: Ecological Protection at the United States Environmental Protection Agency, 1970–1993," *Environmental History* 2, no. 1 (1997): 29–51; Paul Charles Milazzo, *Unlikely Environmentalists: Congress and Clean Water, 1945–1972* (Lawrence: University Press of Kansas, 2006); and David Stradling and Richard Stradling, "Perceptions of the Burning River: Deindustrialization and Cleveland's Cuyahoga River," *Environmental History* 13, no. 3 (2008): 515–35. Also see Christine Meisner Rosen, "Businessmen against Pollution in Late Nineteenth Century Chicago," *Business History Review* 69, no. 3 (1995): 351–97; Harold L. Platt, "Chicago, the Great Lakes, and the Origins of Federal Urban Environmental Policy," *Journal of the Gilded Age and Progressive Era* 1, no. 2 (2002): 122–53; and Robin Faith Bachin, *Building the South Side: Urban Space and Civic Culture in Chicago, 1890–1919* (Chicago: University of Chicago Press, 2004).

6. Richard J. Daley, as quoted in the *Chicago Tribune*, 13 June 1970. Also see Judith A. Martin and Sam Bass Warner Jr., "Local Initiative and Metropolitan Repetition: Chicago, 1972–1990," in *The American Planning Tradition: Culture and Policy*, ed. Robert Fishman (Washington, DC: Woodrow Wilson Centre, 2000), 263–96. On "envirotechnical systems," see Sara B. Pritchard, *Confluence: The Nature of Technology and the Remaking of the Rhône* (Cambridge, MA: Harvard University Press, 2011). She defines them "as the historically and culturally specific configurations of intertwined 'ecological' and 'tech-

nological' systems, which may be composed of artifacts, practices, people, institutions, and ecologies" (19).

7. Harold L. Platt, *Shock Cities: The Environmental Transformation and Reform of Manchester and Chicago* (Chicago: University of Chicago Press, 2005); and John W. Larson, *Those Army Engineers: A History of the Chicago District U.S. Army Corps of Engineers* (Washington, DC: Government Printing Office, 1980).

8. See Frank E. Dalton, Victor Koelzer, and William J. Bauer, "The Chicago Area Deep Tunnel Project: A Use of the Underground Storage Resource," *Journal of the Water Pollution Control Federation* 41, no. 4 (1969): 515–34; and Frank E. Dalton and Raymond R. Rimkus, "The Chicago Area's Tunnel and Reservoir Plan," *Journal of the Water Pollution Control Federation* 57, no. 12 (1985): 1114–21. Also see Timothy B. Neary, "Chicago-Style Environmental Politics: Origins of the Deep Tunnel Project," *Journal of Illinois History* 4, no. 2 (2001): 83–102. For statistics on the stormwater overflow releases into Lake Michigan, see Metropolitan Water Reclamation District (MWRD), "Reversals to Lake Michigan (1985–Present)," accessed 20 Dec. 2012, https://www.mwrd.org/irj/go/km/docs/documents/MWRD/internet/protecting_the_environment/Combined_Sewer_Overflows/pdfs/Reversals.pdf.

9. *Chicago Tribune*, 14–17 July 1957; 13 Sept. 1957; 14 June 1958; 29 Aug. 1958; 10 Apr. 1963; 25 Jan. 1965; 27 Mar. 1966; 30 Mar. 1967; 4 Oct. 1967. Also see David E. Nye, *American Technological Sublime* (Cambridge, MA: MIT Press, 1994); Maria Kaika, *City of Flows: Modernity, Nature, and the City* (New York: Routledge, 2005); E. Swyngedouw, *Social Power and the Urbanization of Water: Flows of Power* (New York: Oxford University Press, 2004); and Martin V. Melosi, *The Sanitary City: Urban Infrastructure in America from Colonial Times to the Present* (Baltimore: Johns Hopkins University Press, 2000).

10. Samuel P. Hays, *Conservation and the Gospel of Efficiency: The Progressive Conservation Movement 1890–1920* (Cambridge, MA: Harvard University Press, 1959); William H. Wilson, *The City Beautiful Movement* (Baltimore: Johns Hopkins University Press, 1989); Bachin, *Building the South Side*; Carl S. Smith, *The Plan of Chicago: Daniel Burnham and the Remaking of the American City* (Chicago: University of Chicago Press, 2006); and Margaret Garb, "Race, Housing, and Burnham's Plan: Why Is There No Housing in the 1909 Plan of Chicago?," *Journal of Planning History* 10, no. 2 (2011): 99–113. Also see Timothy J. Gilfoyle and Chicago History Museum, *Millennium Park: Creating a Chicago Landmark* (Chicago: University of Chicago Press, 2006).

11. Tom McNamee, as quoted in *Chicago Sun-Times*, 9 May 2005; and Stradling and Stradling, "Perceptions of the Burning River."

12. *Chicago Tribune*, 6 Aug. 1979, for the first quotation; Robert Cassidy, "Our Friendless Chicago River," *Chicago Magazine* (Oct. 1979), accessed 27 Nov. 2012, http://www.chicagomag.com/Chicago-Magazine/August-1979/Our-Friendless-Chicago-River/.

13. *Chicago Tribune*, 19 Mar. 1983.

14. *Chicago Tribune*, 24 May 1985.

15. *Chicago Sun-Times*, 22 Apr. 1990; and MWRD, "Reversals to Lake Michigan."

16. Henry Binford, "Multicentered Chicago," in *The Encyclopedia of Chicago*, ed. James R. Grossman et al. (Chicago: University of Chicago Press, 2004), 548–53; Robert Bruegmann, "Schaumburg, Oak Brook, Rosemont, and the Recentering of the Chicago Metropolitan Area," in *Chicago Architecture and Design, 1923–1993: Reconfiguration of an American Metropolis*, ed. John Zukowsky, Mark Jansen Bouman, and the Art Institute of Chicago (Chicago: Art Institute of Chicago, 1993), 159–78; and Ann Durkin Keating, "Chicagoland: More than the Sum of Its Parts," *Journal of Urban History* 30, no. 2 (2004): 213–30.

17. *Chicago Tribune*, 1–7 Oct. 1986; 6 and 9 Nov. 1986; and MWRD, "Reversals to Lake Michigan."

18. *Chicago Tribune*, 4, 7–8 Oct. 1986; 9 Nov. 1986.

19. *Chicago Tribune*, 9 Nov. 1986; Donald L. Hey, "The Des Plaines River Wetlands Demonstration Project: Restoring an Urban Wetland," in *The Ecological City: Preserving and Restoring Urban Biodiversity*, ed. Rutherford H. Platt, Rowan A. Rowntree, and Pamela C. Muick (Amherst: University of Massachusetts Press, 1994), 83–92; and Martin and Warner Jr., "Local Initiative and Metropolitan Repetition." Also see Ian L. McHarg and American Museum of Natural History, *Design with Nature* (Garden City, NY: Natural History Press, 1969); Reyner Banham, *Los Angeles: The Architecture of Four Ecologies* (1971; repr. Berkeley: University of California Press, 2001); and Richard Joseph Neutra, *Building with Nature* (New York: Universe Books, 1971).

20. Joel Rast, "Creating a Unified Business Elite: The Origins of the Chicago Central Area Committee," *Journal of Urban History* 37, no. 4 (2011): 583–605. Also see *Chicago Tribune*, 28 July 1963; 28 June 1965; 18 July 1977; 1 July 1984; and D. Bradford Hunt and Jon B. DeVries, *Planning Chicago* (Chicago: American Planning Association, 2013).

21. *Chicago Sun-Times*, 13 Mar. 1990; 22 Apr. 1990; 17 Aug. 1990. Also see Oscar Newman, *Defensible Space: Crime Prevention through Urban Design* (New York: Macmillan, 1972).

22. Mayor Richard M. Daley, as quoted in *Chicago Sun-Times*, 13 Mar. 1990 and 2 Sept. 1984. Also see *Chicago Sun-Times*, 31 July 1997; and City of Chicago and Friends of the River, *Chicago River Urban Design Guidelines: Downtown Corridor* (Chicago: Friends of the Chicago River, [1990]).

23. Joanne Alter, as quoted in *Chicago Tribune*, 23 Apr. 1985. Also see *Chicago Sun-Times*, 18 Feb. 1991 and 21 Apr. 1991; and Ruth Glass, *Clichés of Urban Doom and Other Essays* (New York: Blackwell, 1989).

24. *Chicago Tribune*, 25 Apr. 1974; 26 Dec. 1974; 4 Nov. 1983; *Chicago Sun-Times*, 22 Apr. 1990; 18 July 1990; 18 Aug. 1991; 16 Apr. 1992; 15 Oct. 2005. Also see *Chicago Sun-Times*, 18 Feb. 1991 and 21 Apr. 1991.

25. Joyce Peralta, as quoted in *Chicago Sun-Times*, 3 June 1996.

26. *Chicago Tribune*, 15 and 19 June 1984; 23 Jan. 1985; *Chicago Sun-Times*, 1 Apr. 1993. Also see *Chicago Sun-Times*, 28 July 1990.

27. Jill Viehweg, as quoted in *Chicago Sun-Times*, 24 June 1994; and 16 Apr. 1997. Also see *Chicago Sun-Times*, 21 Aug. 2004.

28. *Chicago Sun-Times*, 1 Dec. 1995, for the quotation; 24 June 1994 for the statistics.

29. See *Chicago Sun-Times*, 8 June 1997, for the quotation. Also see Hill, *Chicago River*.

30. See *Chicago Tribune*, 1 June 2010; 12 and 17 May 2011; and *Chicago Sun-Times*, 26 Apr. 2011.

31. Henderson, "Railing on the River."

32. On Emanuel's advocacy in support of saving the Great Lakes while he was a congressman, see his editorial, *Chicago Tribune*, 18 July 2004. On the change of policy at the MWRD, see *Chicago Tribune*, 3–13 June 2010; 31 May 2011. For the denunciation of the MWRD's "junk science," see Ann Alexander, "Chicago River Disinfection Hearings: Science Takes a Hit," 9 Aug. 2010, accessed 22 Dec. 2012, https://www.nrdc.org/experts/ann-alexander/chicago-river-disinfection-hearings-science-takes-hit.

33. US EPA, Letter to the Illinois EPA (3 Nov. 2011), accessed 20 Dec. 2012, http://www.epa.gov/region5/chicagoriver/pdfs/chicagoriver-letter-20111103.pdf. Also see *Chicago Tribune*, 12 and 17 May 2011; 1 June 2011; 6 Nov. 2011; *Chicago Sun-Times*, 8 June 2011; and John T. Slania, "Troubled Waters?," *Time Out Chicago*, no. 396, 27 Sept.–2 Oct. 2012, 12–16.

34. Rahm Emanuel, as quoted in *Chicago Sun-Times*, 20 Sept. 2011. Also see *Chicago Sun-Times*, 7 Nov. 2011; and *Chicago Tribune*, 8 Oct. 2012; *Inside Studio Gang*, exhibit at the Art Institute of Chicago, 2011–2012; and Whet Moser, "Jeanne Gang and Robert Cassidy: Two Generations of Urban Planners on the Chicago River," *Chicago Magazine*, 7 Dec. 2011, accessed 27 Nov. 2012, http://www.chicagomag.com/Chicago-Magazine/The-312/December-2011/Jeanne-Gang-and-Robert-Cassidy-Two-Generations-of-Urban-Planners-on-the-Chicago-River/. The Gang Studio website has information on its Northerly Island and exhibit projects: "Northerly Island," accessed 15 Apr. 2016, http://studiogang.com/project/northerly-island; and "Building: Inside Studio Gang Architects," accessed 15 Jan. 2013, http://www.studiogang.net/work/2012/building.

35. *Chicago Sun-Times*, 26 Apr. 2011.

36. For an interim report, see US Army Corps of Engineers, GLMRIS Natural Resources Team, "Non-Native Species of Concern and Dispersal Risk for the Great Lakes and Mississippi River Interbasin Study," 16 Sept. 2011, accessed 20 Dec. 2012, http://glmris.anl.gov/documents/docs/Non-Native_Species.pdf.

37. US Corps of Engineers, *The Electric Dispersal Barriers*, Chicago, Nov. 2012, accessed 20 Dec. 2012, http://www.lrc.usace.army.mil/Portals/36/docs/projects/ans/docs/ElectricBarrierBrochure.pdf.

Chapter 14. Shared Waters, Shared Conceptions?

1. Der Bombenangriff vom 27. Februar 1945. "Das war das Ende von Mainz," Interview mit einem Zeitzeugen. SWR 26 Feb. 2010: http://www.swr.de/archiv/wissen/bombardierung-von-mainz-februar-1945-2/-/id=6640616/did=6020010/nid=6640616/1g

bikr3/index.html. See also Heinz Leiwig, *Der Tag, als Mainz unterging: 27. Februar 1945* (Gudensberg: Wartberg, 2004).

2. "Bush sieht Mainz von seiner Schokoladenseite"; Wiesbadener Kurier—Rhein-Main-Presse (WK), 9 Feb. 2005.

3. The claim for urban sustainable development was part of the Agenda 21 process, decided upon at the Rio world conference of 1992. Article 28 of the Agenda 21 states: "Because so many of the problems and solutions being addressed by Agenda 21 have their roots in local activities, the participation and cooperation of local authorities will be a determining factor in fulfilling its objectives. . . . As the level of governance closest to the people, they play a vital role in educating, mobilizing and responding to the public to promote sustainable development." United Nations Sustainable Development, *United Nations Conference on Environment & Development, Rio de Janeiro, Brazil, 3 to 14 June 1992, Agenda 21*, accessed 15 Dec. 2014, http://sustainabledevelopment.un.org/content/documents/Agenda21.pdf. See also Stefan Kuhn and International Council for Local Environmental Initiatives, eds., *Handbuch Lokale Agenda 21: Wege zur nachhaltigen Entwicklung in den Kommunen* (Berlin: Umweltbundesamt, 1998), 20–24.

4. Joachim Radkau, *Die Ära der Ökologie: Eine Weltgeschichte* (Munich: Beck, 2011).

5. Paul M. Hohenberg and Lynn Hollen Lees, *The Making of Urban Europe, 1000–1950* (Cambridge, MA: Harvard University Press, 1985); Dieter Schott, *Europäische Urbanisierung (1000–2000): Eine umwelthistorische Einführung* (Cologne: Boehlan, 2014).

6. Dieter Schott, "Stadt und Fluss: Flüsse als städtische Umwelten im 19. und 20. Jahrhundert," in *Beiträge zum Göttinger Umwelthistorischen Kolloquium 2004–2006*, ed. Bernd Herrmann (Göttingen: Universitätsverlag Göttingen, 2007), 145–62.

7. D. Rebentisch, "Auf dem Weg zu einer neuen Stadtkultur am Fluss: Das Beispiel Frankfurt am Main," *Informationen zur Modernen Stadtgeschichte* 2 (2003): 18–20; Steven Hawley, *Recovering a Lost River: Removing Dams, Rewilding Salmon, Revitalizing Communities* (Boston: Beacon, 2011).

8. There are still differences between Wiesbaden underlining the importance of its climatically advantageous location between the river and the Taunus mountain range, and Mainz with its concentration on river and historic city. Thus the main claim of Wiesbaden to the river Rhine lies with the leisure opportunities the river offers, whereas Mainz very strongly stresses its historical and physical closeness to the river. See Stadtarchiv Mainz 100/2000: 56; Stadtarchiv Wiesbaden Pamphlet Box 12/2011.

9. Vom Stapelrecht zur Rheinschifffahrtsakte, Geschichte des Stapelrechts zu Mainz (bis ins 19. Jahrhundert), Staatszeitung 21 Sep. 1958, 38; 28 Sep. 1959, 39; Franz Dumont, Ferdinand Scherf, and Friedrich Schütz, *Mainz: Die Geschichte der Stadt*, 2nd ed. (Mainz: Von Zabern, 1999); Dieter Schott, *Die Vernetzung der Stadt* (Darmstadt: Wissenschaftliche Buchgesellschaft, 1999).

10. The close mental connection of the city to the river Rhine can also be seen in the fact that street plates in Mainz still indicate by their color the direction of the street in

relation to the river; Friedrich Schütz, "Rot und Blau—Die Einführung neuer Straßenschilder und Hausnummern in 1849–1958 in Mainz," *Mainz Zeitschrift* 94/95 (1999–2000): 301–15.

11. C. Bodenbach, "Stadt am Fluss: Beobachtungen zur Stadtentwicklung," *Vivat Wiesbaden* H.7 (2006): 48–50.

12. R. Dörrlamm, "Letzte Chance für AKK-Rückkehr," *Mainz, Vierteljahreshefte für Kultur, Politik, Wirtschaft, Geschichte* H. 2 (1995): 16–20.

13. Sibilla Friedrich-Pauly and Detlev Huber, *Wiesbaden: Eine kurze Stadtgeschichte* (Erfurt: Sutton, 2003).

14. Roman Haimberger, Andreas Hoppe, and Andreas Schäfer, "High-Resolution Seismic Survey on the Rhine River in the Northern Upper Rhine Graben," *International Journal of Earth Sciences* 94, no. 4 (2005): 657–68; Constanze Bückner and Andreas Hoppe, *Kartierte Städte: Mainz und Wiesbaden im Spannungsfeld von Naturraum und Vergesellschaftung* (Frankfurt am Main: Campus, 2012), 19–23, 93–95.

15. For consequences of this so-called "Tulla'sche Rheinkorrektur" see Hessische Landesanstalt für Umwelt, ed., *Verbesserung der Umweltverhältnisse am Rhein, Altrheine, Deiche, Vorländer, Polder, Schöpfwerke*, Studie Teil 1 und Teil 2 (Wiesbaden, 1981), 21–23.

16. German Federal Institute of Hydrology.

17. State Agency for Environment, Water Management, and Trade Control in Rhineland Palatinate.

18. M. Dietz-Lensen, "Strom für Mainz," *Mainz: Vierteljahreshefte für Kultur, Politik, Wirtschaft, Geschichte* 26, H.2 (2007): 40–45.

19. See *DFG-Projektverbund "Nachhaltige Entwicklung von Städten,"* accessed 15 Dec. 2014, http://www.stadtforschung.tu-darmstadt.de/eigenlogik/dfg_projektverbund_nachhaltigkeit/dfg_projekte_nachhaltigkeit.de.jsp.

20. Helmuth Berking and Martina Löw, eds., *Die Eigenlogik der Städte* (Frankfurt am Main: Campus, 2008); Martina Löw, "The Intrinsic Logic of Cities: Towards a New Theory on Urbanism," *Urban Research and Practice* 5, no. 3 (2012): 303–15.

21. Martina Löw, "Eigenlogische Strukturen—Differenzen zwischen Städten als konzeptuelle Herausforderung," in Berking and Löw, *Die Eigenlogik der Städte*, 33–55.

22. Löw, "Eigenlogische Strukturen," 43, my translation.

23. S. Frank, "Eigenlogik der Städte: Forschungsstand und Perspektiven," in *Handbuch Stadtsoziologie*, ed. Frank Eckardt (Wiesbaden: VS Verlag für Sozialwissenschaften, 2011), 14–15.

24. Patrick Le Galès, *European Cities: Social Conflicts and Governance* (Oxford: Oxford University Press, 2002); Dieter Schott and Michael Toyka-Seid, eds., *Die europäische Stadt und ihre Umwelt* (Darmstadt: Wissenschaftliche Buchgesellschaft, 2008).

25. Martin V. Melosi, "The Place of the City in Environmental History," *Environmental History Review* 17, no. 1 (1993): 2.

26. The development of urban environmental history in Europe can be best judged by the continuous row of Roundtables on Urban Environmental History staged between

2000 and 2008. In continuation of this efforts the Technical University of Darmstadt together with the Universiteit Antwerp hosted an ESF Exploratory Workshop on "The Urbanisation of Nature" in March 2014. An overview of the results of this workshop can be found in IMS 2/2014: 98.

27. Bruno Latour, *Science in Action: How to Follow Scientists and Engineers through Society* (Cambridge, MA: Harvard University Press, 1987); Bruno Latour, "On Actor-Network Theory: A Few Clarifications," *Soziale Welt* 47, no. 4 (1996): 369–82; Désirée Schauz, "Historikertag 2010: Wissenschaftsgeschichte," *H-Soz-Kult*, 18 Nov. 2010, accessed 15 Dec. 2014, http://www.hsozkult.de/debate/id/diskussionen-1420.

28. Verena Winiwarter and Martin Schmid, "Umweltgeschichte als Untersuchung sozionaturaler Schauplätze?," in *Umweltverhalten in Geschichte und Gegenwart*, ed. Thomas Knopf (Tübingen: Attempto-Verlag, 2008), 158–73; Martin Schmid, "Die Donau als sozionaturaler Schauplatz: Ein konzeptueller Entwurf für umwelthistorische Studien in der Frühen Neuzeit," in *"Die Natur ist überall bey uns": Mensch und Natur in der Frühen Neuzeit*, ed. Sophie Ruppel and Aline Steinbrecher (Zurich: Chronos, 2009), 59–80; Verena Winiwarter et al., "The Environmental History of the Danube River Basin as an Issue of Long-Term Socio-Ecological Research," in *Long-Term Socio-Ecological Research: Studies in Society-Nature Interactions across Spatial and Temporal Scales*, ed. Simron Jit Singh et al. (Dordrecht: Springer, 2013), 103–22.

29. Winiwarter and Schmid, "Umweltgeschichte," 158, my translations.

30. Theodore R. Schatzki, "Nature and Technology in History," in *History and Theory* 42, no. 4 (2003): 82–93.

31. M. Fischer-Kowalski and H. Weisz, "Society as Hybrid between Material and Symbolic Realms: Toward a Theoretical Framework of Society-Nature-Interaction," *Advances in Human Ecology* 8 (1999): 215–51; see also J. R. McNeill, "Observations on the Nature and Culture of Environmental History," *History and Theory* 42 (Dec. 2003): 5–43.

32. Mainzer Rhein-Zeitung (MRZ), 30 Mar. 1988, 1, 4, 15, 19, 24.

33. H. W. Niemeier, "Wenn der Rhein rein wäre . . . Vom Mainzer Quellwasser zur modernen Wassergewinnungsanlage," *Mainz: Vierteljahreshefte für Kultur, Politik, Wirtschaft, Geschichte* 6 (1987): 21–25; Dumont, Scherf, and Schütz, *Mainz*, 68–69.

34. Dumont, Scherf, and Schütz, *Mainz*, 536–39; Stadtarchiv Mainz 100/2000-80: 702; 100/2000-80: 140.

35. Allgemeine Zeitung (AZ), 5 Dec. 1958.

36. Peter Claus Hartmann, *Kleine Mainzer Stadtgeschichte* (Regenburg: Pustet, 2005), 38–45.

37. AZ 27 Nov. 1969, 9.

38. *Promenade am anderen Ufer: Mainz soll mit AKK und Wiesbaden die rechte Rheinseite aktivieren*, AZ, 1968-77-27: 6.

39. *Der Tiger bleibt nicht allein: Rheinufer bald Freiluftgalerie*, AZ, 14 Nov. 1969, 9; *Neue Bronzeplastik am Rheinufer: "Stürzende" Wilhelm Loth, Das Neue Mainz*, June

1971, 6–7; *Variation mit Röhren: Zu drei neuen Plastiken am Mainzer Rheinufer*, AZ, 3 July 1972, 13.

40. AZ, 27 Nov. 1968; AZ, 3 June 1972.

41. *Erholungsland von Schierstein bis Kostheim*, AZ, 4 Oct. 1974, 9; *Keine durchgehende Promenade am Rheinufer: Wiesbadener CDU fordert Erholung und Gewerbenutzung nebeneinander: "KISA" bleibt*, AZ, 15 July 1977, 13.

42. "Hundert Jahre Mainzer Dreck," *Mainz Magazin* 10 (1974): 3–10, 18–21.

43. Der Rhein—meistbesungene Kloake Europas, AZ, 12 July 1969, 25. See also Mark Cioc, *The Rhine: An Eco-Biography, 1815–2000* (Seattle: University of Washington Press, 2002), 227–35.

44. Vor 50 Jahren: Als die Wasserqualität zum Problem wurde. Die Rheinverschmutzung und die Anfänge des Umweltbewusstseins, Dradio.de 11 June 2008, accessed 5 Feb. 2016, http://www.deutschlandradiokultur.de/vor-50-jahren-als-die-wasserqualitaet-zum-problem-wurde.984.de.html?dram:article_id=153423.

45. Cioc, *Rhine.*

46. *Wiesbadner Tagblatt,* 13 Apr. 1973.

47. *Rheinwasser-Untersuchungsstation Mainz-Wiesbaden*, accessed 5 Feb. 2016, http://www.rheinwasseruntersuchungsstation.de.

48. The pollution of the river Rhine was clearly the most prominent "environmental" question, covered by the local media in the 1970s and 1980s, the number of articles on this topic far outnumbering articles on nuclear energy and air pollution.

49. *Rheinwasser im Qualität des Rhein-Main-Gebiet "unzureichend": Gewässerverschmutzung durch Hoechst-Abwässer*, AZ, 11 Oct. 1980, 4.

50. Jürgen Ruf: Vor 25 Jahren färbte sich der Rhein blutrot, in Stern, 1 Nov. 2011.

51. The Sandoz-Desaster even led to a symbolic "funeral procession" by citizens from both cities. AZ, 15 Dec. 1986.

52. Stadtarchiv Mainz 100/2008-4, 148; 100–20080, 497; Die neue City in der City. Dom-Rathaus-Zentrum am Brand in Mainz eröffnet, sp. ed., AZ, 26 Sep. 1974.

53. Eduard Kreyßig, in *Mainz: Vierteljahreshefte für Kultur, Politik, Wirtschaft, Geschichte* 25 (2005): 8–13.

54. For the rise of the Green Party in local politics see Bückner and Hoppe, *Kartierte Städte*, 46–47.

55. Hessische Landesanstalt für Umwelt, *Verbesserung der Umweltverhältnisse am Rhein*, 6.

56. Stadtarchiv Mainz 100/2004-42: 2009 + 210; Bückner and Hoppe, *Kartierte Städte*, 119–26; G. Jertz, "Das Ried—Naturschutz im Kessel: Unmittelbare Stadtnähe macht es der Laubenheimer Zone nicht leichter," *Mainz: Vierteljahreshefte für Kultur, Politik, Wirtschaft, Geschichte* 7 (1987): 10–12.

57. MRZ, special Easter edition 1988.

58. Bürgerwunsch wird erfüllt: Ufergrün und Hafenmündung gutgeheißen, AZ, 12 July 1993, 13.

59. "Wohnen am Rhein," *Mainz: Vierteljahreshefte für Kultur, Politik, Wirtschaft, Geschichte* 31, no. 2 (2011): 34.

60. Bückner and Hoppe, *Kartierte Städte*, 95–98.

61. *Architektursommer Rhein-Main 2011*, 50, http://www.wiesbaden.de/leben-in-wiesbaden/planen/architektur/resuemee.php

62. Stadtplanungsamt Mainz, ed., *Rheinufer: Geschichte, Bestand, Analyse* (Stadt Mainz: Stadtplanungsamt, 1997); Stadtplanungsamt, *Rhein-Ufer-Forum Mainz Schlussbericht* (Mainz: Dezernat für Planung, Bauen und Sport, 2000).

63. *Mainz*, 31 Jan. 2001, 76.

Chapter 15. Revitalization of a Tamed River.

1. "So lang die grüne Isar durch d' Münchnerstadt noch geht . . . so lang stirbt die Gemütlichkeit in München gar net aus," quoted from: Claudia Preis, "Münchner Volksleben in Lied und Wort: Volkssänger-Unterhaltung in München," in *München-Sound: urbane Volkskultur und populäre Musik*, ed. Johannes Moser and Eva Becher (Munich: Herbert Utz, 2012), 72.

2. This concept of an "ecosociotechnical system" follows up on the extension of the model of the "sociotechnical system" by Hans Lenk (*Global TechnoScience and Responsibility: Schemes Applied to Human Values, Technology, Creativity and Globalisation* [Berlin, Münster: LIT, 2007], 176). The concept of the sociotechnical system was originally created to describe the reciprocal interrelationship between human beings and machines in the context of labor studies by the Tavistock Institute in London. The German philosopher of technology Günter Ropohl expanded the idea of the sociotechnical system to describe a "system of action, in which personal and social functional units have been aggregated with technical systems" (Günter Ropohl, *Eine Systemtheorie der Technik: zur Grundlegung der Allgemeinen Technologie* [Munich: Hanser, 1979], 180). Referring to Ropohl, Lenk has gone even further in the last few years by involving environmental matters and embedding sociotechnical systems into their surrounding ecosystems: "I would even like to add to the theory of socio-technical systems . . . also the *ecological* embedding and the respective perspectives by talking of ecological *and* socio-technological systems or *eco-socio-technical systems*" (Lenk, *Global TechnoScience*, 176). Compared to the concept of "envirotechnical systems" (Sara B. Pritchard, *Confluence: The Nature of Technology and the Remaking of the Rhône* [Cambridge, MA: Harvard University Press, 2011], 19), the focus is more on the social dimension of the systems.

3. Klement Trockner, Achim Paetzold, Ute Karaus, Cécile Claret, and Jürg Zettel, "Ecology of Braided Rivers," In *Braided Rivers: Process, Deposits, Ecology and Management*, Special publication Number 36 of the International Association of Sedimentologists, ed. Gregory H. Sambrook et al., 339–60 (Malden, MA: Blackwell Publishing, 2006).

4. Michael Schattenhofer, "Aus der Geschichte der Isarflößerei," in *Die Isar—Ein Lebenslauf*, ed. Maria-Louise Plessen (Munich: Hugendubel, 1983), 64–78, 64.

5. Schattenhofer, "Aus der Geschichte der Isarflößerei," 66.

6. Plessen, *Die Isar*, 196.

7. Schattenhofer, "Aus der Geschichte der Isarflößerei," 64–78.

8. Helga Lauterbach, *Von Flossmeistern und Flösserbräuchen: Geschichte und religiöses Brauchtum der Isar- und Loisachflößer* (Munich: Erich Wewel Verlag, 1992), 31–32.

9. Schattenhofer, "Aus der Geschichte der Isarflößerei," 77–78.

10. Christine Rädlinger, Stadtarchiv München, ed., *Geschichte der Münchner Stadtbäche* (Munich: Franz Schiermeier, 2004); Christine Rädlinger, *Neues Leben für die Isar* (Munich: Franz Schiermeier, 2011), 54–55 and 82.

11. Rädlinger, *Neues Leben für die Isar*, 60.

12. Bayerisches Landesamt für Statistik und Datenverarbeitung, ed., *Festschrift 200 Jahre amtliche Statistik in Bayern 1808–2008* (Munich: Bayerisches Landesamt für Statistik, 2008), 112.

13. Thomas P. Hughes, "The Evolution of Large Technological Systems," in *The Social Construction of Technological Systems: New Directions in the Sociology and History of Technology*, ed. Wiebe E. Bijker, Thomas P. Hughes, and Trevor J. Pinch (Cambridge, MA: MIT Press, 1987), 77.

14. Thomas P. Hughes, *Networks of Power: Electrification in Western Society, 1880–1930* (Baltimore: Johns Hopkins University Press, 1993), 335.

15. Hughes, *Networks of Power*, 336.

16. Carl Zell, *Geschichte der Elektrizitätsversorgung Münchens: Gewidmet den Städtischen Elektrizitätswerken München zu deren 50-jähriges Jubiläum* (Munich: Theodor Dietz, 1949), 11–12.

17. Hughes, *Networks of Power*, 337.

18. On the occasion of a devastating flood catastrophe in the fifties (houses in Bad Tölz, Munich, and Landshut were swept away), it was decided to build the Sylvenstein Reservoir with two functions: innovative flood protection by allowing rainwater to fill up the reservoir but also guaranteeing a minimum current of water (ten cubic meters per second) in the river, for the purpose of diluting and removing drained-off sewage disposals.

19. Hughes, *Networks of Power*, 334.

20. Hughes, *Networks of Power*, 465.

21. Hughes, *Networks of Power*.

22. Hughes, *Networks of Power*, 337.

23. Quoted in Walther von Miller, *Oskar von Miller: Pioneer of Power Economy; Founder of the "Deutsches Museum" and "Ingenieurbüro OvM"* (Munich: Bruckmann, 1982), 63.

24. Helga Lauterbach, *Von Floßmeistern und Flößerbräuchen: Geschichte und religiöses Brauchtum der Isar- und Loisachflößer* (Munich: Erich Wewel, 1992), 12–13.

25. Lauterbach, *Von Floßmeistern*, 22.

26. "Schön ist die grüne Isar, schön ist die Flößerei; Wenn kommt das Walchensee-Projekt, dann ist's mit unser'm G'werb vorbei!" (Lauterbach, *Von Floßmeistern*, 23).

27. Plessen, *Die Isar*, 262–63.

28. Rädlinger, *Neues Leben für die Isar*, 82.

29. Johann Karl, "Die Isar—ein Gebirgsfluß im Wandel der Zeiten," *Jahrbuch des Vereins zum Schutz der Bergwelt* (1998): 63.

30. Rädlinger, *Neues Leben für die Isar*, 84.

31. Rädlinger, *Neues Leben für die Isar*, 82.

32. Rädlinger, *Neues Leben für die Isar*, 83.

33. Trockner et al., "Ecology of Braided Rivers," 345–47.

34. An example of a temporarily extinct fish in the upper Isar is the Streber (*Zingel streber*), which came back after elimination of the interruptions (see Rädlinger, *Neues Leben für die Isar*).

35. The stone curlew (*Burhinus oedicnemus*) was widespread until 1850; in 1896 only three were shot in the north of Munich near Ismaning (Walter Wüst, *Avifauna Bavariae: Die Vogelwelt Bayerns im Wandel der Zeit*, 3rd ed. [Munich: Ornithologischen Gesellschaft in Bayern, 1986], 514ff). Gull-billed tern (*Gelochelidon nilotica*) were bred in masses on the gravel islands until 1900; e.g., in 1830 in the north of Munich near Ismaning 116 individuals were shot during one day.

36. Wüst, *Avifauna Bavariae*, 540.

37. Johannes Schnell, Landesfischereiverband Munich, oral information (2013).

38. Animals like arthropods (Blauflügelige Ödlandschrecke [*Oedipoda caerulescens*], Blauflügelige Sandschrecke [*Sphingonotus caerulans*]) also existed on the gravel banks of Alpine rivers. They still exist in secondary habitats on the gravel banks near Munich railway station (in Nikolaus Döring et al., *Von Schotterbänken an der Isar und der Kieswüste am Hauptbahnhof: Begleitheft zum Videofilm "Lebensraum Kieswüste,"* [Munich: Isar-Allianz, 2000]).

39. The characteristic gravel bank plant *Myricaria germanica* (Deutsche Tamariske) was once commonly found along the whole Isar. Now it still exists in a few small colonies near the Alps. Birgit Reif et al., "Wildflussleitarten an der Oberen Isar," *Landesbund für Vogelschutz* (Hilpoltstein: Landesbund für Vogelschutz, 2006): 20.

40. Alfred Ringler, *Obere Isar und ihre Zuläufe* (Lenggries: Notgemeinschaft Rettet die Isar, 2001).

41. Nikolaus Döring, "Anforderungen an Mindestwasserstrecken," in *Die Isar—Problemfluss oder Lösungsmodell*, ed. Theodor Strobl et al. (Laufen an der Salzach: Bayerische Akademie für Naturschutz und Landschaftspflege, 1997), 96.

42. Ralf Sartori, ed. *Die neue Isar* (Munich: Buch and Media, 2010), 65.

43. The communication between the opposing parties was a quite delicate subject to handle and was deliberately and prudently managed by Nikolaus (Nico) Döring.

44. Julia Düchs, *Wann wird's an der Isar wieder schön? Die Renaturierung der Isar in München* (Munich: Herbert UTZ, 2014), 43–44.

45. Nico Döring, *Studie zum Gewässerpflegeplan Mühltal: Darstellung der umgesetzten Teilmaßnahmen, Vorschläge für eine ökologisch und flussmorphologisch wirkungsvol-*

le Vervollständigung (Auftraggeber: Gregor Louisoder Umweltstiftung, 2012), http://www.glus.org/fileadmin/downloads/bayernwild/studie-text-030213-kl.pdf.

46. Nico Döring, "Gründung der Isar-Allianz und das Ringen um Flussrenaturierungen; Mühltal / Eine Chance auf vielen Ebenen," in Sartori, *Die neue Isar*, 68ff.

47. After serious discussions with the electric power lobby, no consensual long-term sustainable agreement was found. Following a political administrative decision in 1994, nearly 20 percent of the water had to remain in the Isar River. The river beside the canal had to be reshaped according to the common guiding principles at the expense of the operators of the hydropower plants. (Source: Vollzug der Wasser- und Naturschutzgesetze. Bewilligung der Gewässerbenutzung für das Wasserkraftwerk Mühltal an der Isar. Landratsamt Bad Tölz-Wolfratshausen, Held 1995).

48. Rädlinger, *Neues Leben für die Isar*, 86.

49. Rädlinger, *Neues Leben für die Isar*, 86–87.

50. Rädlinger, *Neues Leben für die Isar*, 5.

51. Christiane Lutz, "Hochwasser in München: Die Isar bleibt zahm," *Süddeutsche Zeitung*, 3 June 2013.

52. Bayerisches Landesamt für Umwelt, *Abtötung bzw. Inaktivierung ausgewählter Krankheitserreger in Abwasser durch UV-Strahlung*, 2014.

53. Regierung von Oberbayern, ed., *Restwasseruntersuchung Mühltal/Isar* (Munich: Regierung von Oberbayern, 1993).

54. Rädlinger, *Neues Leben für die Isar*, 70.

55. Paul J. Crutzen and Eugene F. Stoermer, "The 'Anthropocene,'" *Global Change Newsletter* 41 (2000): 17–18.

56. Ulrich Beck, *Die Erfindung des Politischen: Zu einer Theorie reflexiver Modernisierung* (Frankfurt am Main: Suhrkamp, 1993).

57. Dr. Nico Döring, now at the Environmental Academy, founded the Isar Alliance. The first coordinator and spokesperson (1993–2012) of the Isar Alliance, he was the motor for the New IsarPlan, and is convinced that by "being dedicated to go for utopias beyond mainstream thinking, it is possible to move mountains through uniting multidisciplinary competencies with creative strategies, and as a result, recreate more rivers" (cited in Sartori, *Die neue Isar*, 65).

Chapter 16. Union Is a Raging River, or Remembering Fez as the River Remembers

Epigraph: Jonathan Star, ed., *Rumi: In the Arms of the Beloved* (New York: Penguin Group, 2008), 26.

1. Richard Weller, "An Art of Instrumentality: Thinking through Landscape Urbanism," in *The Landscape Urbanism Reader*, ed. Charles Waldheim (New York: Princeton Architectural Press, 2006), 71.

2. Weller, "Art of Instrumentality," 71.

3. Leonard Lawlor, "Memory Becomes Electra," review of *Archive Fever: A Freudian Impression*, by Jacques Derrida, *Review of Politics* 60, no. 4 (1998): 796.

4. Andreas Huyssen, *Twilight Memories: Marking Time in a Culture of Amnesia*, (New York: Routledge, 1995), 3–5.

5. Huyssen, *Twilight Memories*, 7.

6. Bureau EAST, "Resuscitating the Fez River," *Analysis and Planning: 2010 ASLA Professional Awards*, accessed Jan. 2014, https://www.asla.org/2010awards/492.html.

7. Aziza Chaouni and Takako Tajima, "Landscape INFRAtecture: A New Approach for the Revitalization of Medinas," in *Culture & Space in the Built Environment Network*, ed. Hulya Turgut Yildiz (Istanbul: IAPS, 2009), 1–11.

8. Jamil M. Abun-Nasr, *A History of the Maghrib in the Islamic Period* (Cambridge: Cambridge University Press, 1987).

9. Simon O'Meara, *Space and Muslim Urban Life: At the Limits of the Labyrinth of Fez* (New York: Routledge, 2007), 6.

10. Stefano Bianca, *Urban Form in the Arab World: Past and Present* (London: Thames and Hudson, 2000), 284.

11. UNESCO, "Committee Decisions," World Heritage Committee, *World Heritage Convention, United Nations Education, Scientific and Cultural Organization*, January 2014.

12. *INDEX: Award 2009 Exhibition Catalogue*, ed. INDEX: Design to Improve Life (Copenhagen: INDEX: Award, Aug. 2009), 2, accessed 13 May 2016, http://designto improvelife.dk/resuscitating-the-fez-river/.

13. Chaouni and Tajima, "The River Returns: River Remediation and Urban Development Scheme Fez, Morocco," *Holcim Global Awards 2009* 34 (2009): 14–21.

14. In-Keun Lee, "Cheong Gye Cheon Restoration Project: A Revolution in Seoul, ICLEI Local Governments for Sustainability," International Council for Local Environmental Initiatives, Seoul Metropolitan Government, 2006, 1–63.

15. Caroline Bâcle, *Lost Rivers* (Montreal: Catbird Films, 2012), DVD.

16. Chaouni and Tajima, "Landscape INFRAtecture," 1–11.

17. Aziza Chaouni and Liat Margolis, "Water Infrastructure and Innovative Technologies in Arid Climates," in *Sustainable Architecture and Urban Development* (Amman, Jordan: CSAAR Press, 2010).

18. Chaouni and Margolis, "Water Infrastructure."

19. Henri Bergson, *Matter and Memory*, trans. Nancy Margaret Paul and William Scott Palmer (Zone Books, 1988), 61–76.

20. Maryam Keramat Ardakani and Seyyed Saeed Ahmadi Oloonabadi, "Collective Memory as an Efficient Agent in Sustainable Urban Conservation," *Procedia Engineering* 21, no. 1 (2011): 985–88.

21. Aziza Chaouni and Takako Tajima, "Cultured Infrastructures," in *Infrastructure as Architecture: Designing Composite Networks*, ed. Katrina Stoll, Scott Lloyd, and Stan Allen (Berlin: Jovis, 2010), 86–93.

22. Aziza Chaouni and Takako Tajima, "Fez River Rehabilitation Project," *Topos: The International Review of Landscape Architecture and Urban Design* 68 (2009): 33–38.

23. United Nations Environment Program International Environmental Technology Centre, *Phytoremediation: An Environmentally Sound Technology for Pollution Prevention, Control and Remediation: An Introductory Guide to Decision-Makers* (Osaka, Japan: UNEP, Division of Technology, Industry and Economics, International Environmental Technology Centre, 2002).

24. James E. Landmeyer, *Introduction to Phytoremediation of Contaminated Groundwater: Historical Foundation, Hydrologic Control, and Contaminant Remediation* (New York: Springer, 2011), 362–64.

25. Aziza Chaouni and Takako Tajima, "Reconstructing the Fez River: Stitching a Twenty-First Century City from Medieval Fabric," in "Commercial Streetscape," special issue, *International New Landscape* 38 (Apr. 2012): 92–103.

26. Chaouni and Tajima, "Reconstructing the Fez River," 92–103.

27. Judith Butler, *Excitable Speech: A Politics of the Performative* (New York: Routledge, 1997), 8.

28. B. Y. Zhang, J. S. Zheng, and R. G. Sharp, "Phytoremediation in Engineered Wetlands: Mechanisms and Applications," *Procedia Environmental Sciences* 2 (2010): 1315–25.

29. Hannah Arendt, *The Human Condition* (Chicago: University of Chicago Press, 1958), 198–99.

30. Butler, *Excitable Speech*, 8.

31. Tony Judt, "A la recherche du temps perdu," *New York Review of Books*, 3 Dec. 1998.

32. Aleida Assmann, "Re-framing Memory: Between Individual and Collective Forms of Constructing the Past," in *Performing the Past: Memory, History, and Identity in Modern Europe*, ed. Karin Tilmans, Frank van Vree, and Jay Winter (Amsterdam: Amsterdam University Press, 2010), 35–36.

33. James E. Young, "The Counter-Monument: Memory against Itself in Germany Today," *Critical Inquiry* 18, no. 2 (1992): 267–96.

34. "Viking River Cruises ®—Book a River Cruise at 2 for 1 Prices!," accessed 10 Jan. 2014, http://www.vikingrivercruises.com/.

CONTRIBUTORS

Sabine Barles is a professor of urban planning at Université Paris 1 Panthéon-Sorbonne. Her main research topics concern environmental history and history of urban technology (eighteenth through twentieth centuries), urban metabolism, territorial ecology, and socioecological trajectories. Paris and more generally cities have long been her main center of interest; she has recently developed research on regional and rural issues.

Christoph Bernhardt is the head of the Department for Historical Research/Scientific Collections at the Leibniz Institute for Research on Society and Space (IRS) in Erkner near Berlin. He teaches at the Technical Universities of Darmstadt and Berlin. He recently published the book *Im Spiegel des Wassers* (Böhlau, 2016) on the environmental history of the Upper Rhine, 1800–2000. Bernhardt is the managing editor of the book series Beiträge zur Stadtgeschichte und Urbanisierungsforschung. For more details see www.irs-net.de.

Stéphane Castonguay is a history professor at the Université du Québec à Trois-Rivières. He coedited *Metropolitan Natures* (University of Pittsburgh Press, 2011) with Michèle Dagenais, and *Urban Rivers* (University of Pittsburgh Press, 2012) with Matthew Evenden, both in the History of the Urban Environment series. His most recent book, *Le gouvernement des ressources naturelles: Sciences et territorialités de l'État québécois, 1867–1939* (Presses de l'Université Laval, 2016), deals with the historical geography of state formation and natural resource exploitation.

Igor Iwo Chabrowski is a postdoctoral fellow in the Centre for China Studies at the Chinese University of Hong Kong. He is the author of a monograph on the society and culture of the boat workers in Sichuan, China (*Singing on the River: Sichuan Boatmen and Their Work Songs, 1880s–1930s* [Brill, 2015]). Currently, he researches twentieth-century Chinese opera as a social and political institution.

Nico Döring is a board member of the Environmental Academy Munich and responsible for biodiversity education. He is the main initiator and cofounder of the Isar-Allianz (Isar Alliance) and was highly active in the realization of two of its early projects: the restoration of the Isar Mühltal and IsarPlan Munich. Both plans combined utilization and protection of the river, flood control, leisure activities, and biodiversity. Nico is one of the main authors of *Die Neue Isar*, I–IV (Buch and Media, 2010–2012), in which all promoters for a new Isar River are represented.

Shelley Hornstein is a professor of architectural history and cultural heritage at York University, Toronto, Canada. Her work explores the relationship of memory and place to architectural and urban sites. She is currently writing a book on architectural tourism and heritage sites and another on architectural permanence and cultural memory. Her most recent book, *Losing Site: Architecture, Memory and Place*, was published by Ashgate in 2011.

Georg Jochum is a senior researcher at the Technical University of Munich, Department of Sociology. He is the author of articles in the areas of sociology of labor, environmental sociology, and cultural history. Most recently he has started research on national variants of the discourse on sustainable development. He is coeditor of an anthology on sustainable work (*Nachhaltige Arbeit: Soziologische Beiträge zur Neubestimmung der gesellschaftlichen Naturverhältnisse*, 2016).

Agnes Kneitz is an assistant professor for world history at the Renmin University of China. She graduated from Ludwig-Maximilians-University, Munich in 2013 with a dissertation on perceptions of environmental justice in nineteenth-century literature. She is a member of the Center for Ecological History at RUC, and an affiliate of the Rachel Carson Center for Environment and Society in Munich. Her research focuses on environmental and social questions in the context of greater Europe in the long nineteenth century, and on interdisciplinarity in the environmental humanities.

Martin Knoll is a professor for the history of European regions at the University of Salzburg, Austria. He is the author of a monograph on the perception of nature and society in early modern topographical literature (*Die Natur der menschlichen Welt* [transcript, 2013]). With Reinhold Reith he coedited *An Environmental History of the Early Modern Period: Experiments and Perspectives* (LIT, 2014). He has conducted research in city-hinterland relations and most recently in the field of regional change triggered by tourism.

Alexei Kraikovski is an associate professor at the European University at Saint Petersburg, Russia. His research interests are mostly connected with the interrelations between the Russian nation and the water environment. He is completing a monograph on the history of modernization of the Russian marine harvesting in the eighteenth century. Some parts of this research have already been published both in Russian and in English. Recently he has started research on the history of the historic manor Schloss Fall (Estonia).

Julia Lajus is an associate professor in the Department of History at the National Research University Higher School of Economics at Saint Petersburg. Her research interests include environmental history and history of technology, water history, and history of field sciences. She has published four coauthored and edited books in Russian and twenty-five papers in international journals and chapters in edited collections in English. In 2011–2015 she served as vice-president of the European Society of Environmental History.

Uwe Lübken is a professor of American history at Ludwig-Maximilians-University, Munich. His most recent publications include a book on the history of flooding of the Ohio River (2014), and two special issues of the journal *Global Environment* (on environmental change and migration, 2012, and on small islands and natural hazards, 2015, coedited with Rebecca Hofmann). Together with Greg Bankoff and Jordan Sand he coedited the volume *Flammable Cities* (University of Wisconsin Press, 2012). His current work explores the intersections of mobilities and the environment.

Geneviève Massard-Guilbaud is a professor of environmental, economic and social history at the École des Hautes Études en Sciences Sociales and a researcher at the International Research Centre for Environment and Development, Paris. She is the author of monographs on Algerian immigration to interwar France and on nineteenth-century industrial pollution. As an editor and author she has broadly published mainly in the fields of environmental and urban history, and is currently completing a book on environmental management and democracy in Nantes and the Loire estuary (1750–1950).

Harold L. Platt is a professor of history emeritus at Loyola University Chicago. He has written several books on the interactions of the built and the natural environments. They include *City Building in the New South* (Temple University Press, 1983); *The Electric City* (University of Chicago Press, 1991); *Shock Cities* (University of Chicago Press, 2005); and most recently, *Building the Urban Environment: Visions of the Organic City in the United States, Europe, and Latin America* (Temple University Press, 2015). He is currently completing another book with the tentative title "Sinking Chicago: Climate Change and the Remaking of a Flood-Prone Environment."

Vladimir Sánchez-Calderón is a part-time teacher at Universidad de los Andes (Colombia). He is a member of the Environmental History Research Group at Universidad Nacional de Colombia and of the Rivers and History group at the Latin American and Caribbean Society for Environmental History. He has written about the role assigned to nature in the contemporary environmental historiography. His interests deal with the intermingling between urban disasters and socioenvironmental inequalities.

Dieter Schott is a professor of modern history with a focus on urban and environmental history at the Technical University of Darmstadt, Germany. He recently published a textbook on the environmental history of European urbanization (*Die europäische Urbanisierung, 1000–2000* [Böhlau, 2014]). He has edited and coedited several collected volumes on urban and environmental history, among them *Resources of the City* (Ashgate, 2005), with Bill Luckin and Geneviève Massard-Guilbaud. His special field of research is the relationship of cities and rivers post-1800.

Dirk Schubert is a professor in the Department of Urban Planning, Comparative Planning History, Housing and Urban Renewal at the HafenCity University Hamburg. His research focuses on urban history, planning history, history of housing, and urban renewal, as well as studies on the revitalization of harbor and waterfront areas in seaport regions and city/port interface areas. His latest two books deal with the urban thinker and writer Jane Jacobs.

Awadhendra Sharan is an associate professor at the Centre for the Study of Developing Societies, Delhi. His research interests are in the fields of urban and environmental studies. Sharan is the author of *In the City, Out of Place: Nuisance, Pollution and Dwelling in Delhi, c.1850–2000* (New Delhi: Oxford University Press, 2014). His current research is focused on cultures of waste and pollution in colonial India, in particular the history of urban air pollution in the nineteenth and twentieth centuries and on the interrelated histories of rivers and cities in colonial north India.

David Stradling is a professor of history at the University of Cincinnati. He is the author of *Smokestacks and Progressives: Environmentalists, Engineers, and Air Quality in America, 1881–1851* (Johns Hopkins University Press, 1999), *Making Mountains: New York City and the Catskills* (University of Washington Press, 2007), and *The Nature of New York: An Environmental History of the Empire State* (Cornell University Press, 2010). His most recent book is *Where the River Burned: Carl Stokes and the Struggle to Save Cleveland* (Cornell University Press, 2015), which he cowrote with his brother, Richard Stradling.

Vanessa Taylor is a lecturer in environmental history and research fellow at the University of Greenwich. Her recent research has focused on the environmental governance of the Thames River, alongside Sarah Palmer. Earlier postdoctoral research with Frank Trentmann at Birkbeck College on "liquid politics" explored the historic formation of the water consumer. She is currently researching energy transitions and everyday life in the twentieth century as part of the Material Cultures of Energy team.

Michael Toyka-Seid has been a teacher and researcher in urban and environmental history for many years. At the center of his interests are sanitary history and urban noise in cities of the nineteenth and twentieth centuries and conservation and rehabilitation in urban planning. His most recent project dealt with urban politics and sustainable development in two German towns since World War II. Current research interests include the search for physical and mental borders between town and hinterland in urban ecology.

Index